Automating
SOLIDWORKS 2023
Using Macros

Mike Spens

SDC Publications
P.O. Box 1334
Mission, KS 66222
913-262-2664
www.SDCpublications.com
Publisher: Stephen Schroff

ISBN-13: 978-1-63057-571-7
ISBN-10: 1-63057-571-2

Printed and bound in the United States of America.

Introduction

I had an engineering professor once tell me that a good engineer is a lazy engineer. He was not trying to tell me to sit around all day in front of a computer, adjusting line colors and thicknesses on a CAD drawing (or whatever would burn the daylight hours), and never accomplish anything. His point was that a lazy engineer will be motivated to do more with less time and less effort.

It makes my skin crawl to do tedious, repetitive work. Computers are made for that, not me. One of the most exciting aspects of SOLIDWORKS is its robust programming interface or API. The software was built from the ground up to automate.

This book has been a work in progress for many years, even before the original Visual Basic for Applications interface was implemented. I hope the information provided within can help you cut time out of your tedious days so you can do the things you enjoy in engineering and design. Or golf, ski, bike, take naps in the boss's office, stroll between the bathroom and the water cooler, take longer lunches, go home early ...

Continued – Tools or Handcuffs?

I have intentionally left the original introduction to this book essentially as it was written when the first version was published. I still feel the same way about tedious tasks. And, yes, I still get excited over macros that do pretty much anything. As an engineer, it's often the small things in life that float your boat. However, my feelings about mountains of custom software have changed. It takes time for things to come full circle in life. It might only take a few short months in the software industry.

When we create a macro for SOLIDWORKS, it is an exciting new adventure. We go at it full force. But we fail to take a moment to think of the long term consequences. For example, have you ever heard of Fergiware? I expect not. You might know it as Bob 1.0, DaveWare or MikeSoft. Fergiware was a custom tool that LaVerl Ferguson wrote for a company where I did some consulting. I don't really know if LaVerl was his first name, but he left his name dangling over his employer like the sword of Damocles. Someone at the company fell in love with LaVerl's macro, took it seriously, and decided it was a necessary piece of their business rather than a temporary, time-saving tool. LaVerl proceeded to exit the beloved company for greener pastures. No one else knew how to read or modify his code. Add a few major releases of their CAD software into the mix and stir briskly. It took several years for the Director of Engineering to realize they were relying on software that had no maintenance plan, technical support, upgrade path or future reliability. Without much effort, he found another software tool to replace Fergiware. The new tool did what the business needed and was backed by a strong company with a bright future (SOLIDWORKS to be specific) rather than a bright employee with a strong future in a different company.

I will be the first to admit that I am not the best salesperson. The story I shared might make you reconsider the macro idea that inspired you to buy this book. It might be better for my wallet to sing praises to custom code like the sage consultant who looks for problems to solve rather than solving problems. But I would rather encourage the reader to look down the road a little further. Broaden your perspective and think of the results and consequences. Will your macro become the Fergiware of your company? What is its life expectancy? Are you ready to maintain it year after year? Are you willing to pull the plug if SOLIDWORKS or someone else comes up with a better way to accomplish the same objective? Or will your company become as reliant on your macro as we are on oil in the United States? Do not be like my favorite spoof on motivational posters showing a beautiful sunset over a silhouetted shipwreck that reads, "Mistakes – It could be that the purpose of your life is to serve as a warning to others." Create tools, not handcuffs.

Version Notes

Over the last several years there have been numerous changes to SOLIDWORKS as well as the SOLIDWORKS API. For example, APIs have been added for the SOLIDWORKS DocumentManager, SOLIDWORKS Simulation, eDrawings, SOLIDWORKS PDM Professional, FeatureWorks, Toolbox, and others. The .NET languages of Visual Basic.NET and C# have become mainstream. SOLIDWORKS supports Visual Studio for .NET macros. Each of these changes can have an effect on how you write code to automate SOLIDWORKS. Since there are so many advantages to the .NET platform for building macros as well as migrating them to full applications, this book will focus on Visual Basic.NET macro projects while also including several C# examples. I hope you find it a helpful reference and springboard for your own macros to help improve the way you work with SOLIDWORKS.

* To download the examples and files referenced in this book, please visit...

http://www.sdcpublications.com

Table of Contents

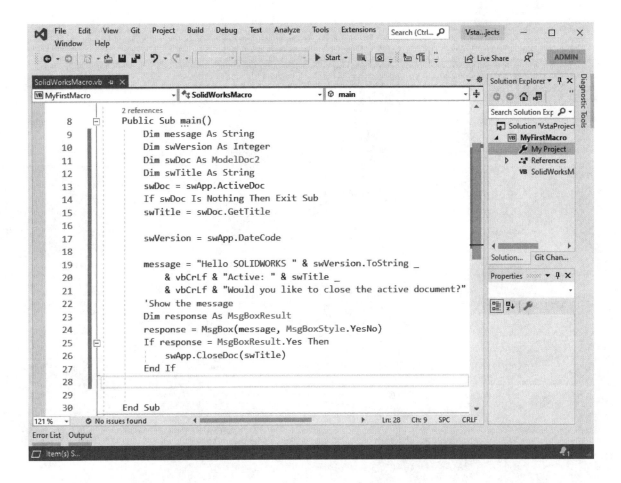

- **Creating a New Macro**

- **Data Types and Variables**

- **Connecting to SOLIDWORKS**

- **Connecting to Documents**

Introduction

This chapter is dedicated to the fundamentals of Visual Basic.NET including code structure, language use and features that you will use throughout this book. If you are already comfortable with VB.NET or C#, move ahead to the other chapters that focus on the use of the SOLIDWORKS APIs. Otherwise, welcome to the world of code development!

The examples in this chapter serve no purpose other than helping you understand Visual Basic.NET fundamentals. I am not assuming to write a comprehensive language manual, but this chapter should give you a good start with Visual Basic. It includes an assortment of personal recommendations as well. If you are new to writing code, I would encourage using a companion reference for Visual Basic.NET as you learn. I have found my copy of "Programming Microsoft Visual Basic.NET" by Francesco Balena to be an invaluable reference guide. The 1400-page behemoth does not read like James Patterson, but it sure gets the job done. There are scores of Visual Basic.NET books out there. Take your pick. Online resources are seemingly endless with sample code to do nearly anything you can imagine. That is the first place I go for tough challenges and creative ideas.

Visual Studio Tools for Applications (VSTA)

Visual Studio Tools for Applications, or VSTA, is the primary development environment that will be used throughout this book. SOLIDWORKS introduced VSTA 1.0 many years ago to support the updated Visual Basic.NET and C# languages. It conveniently included its own development environment. However, VSTA 1.0 was discontinued as of SOLIDWORKS 2021. With SOLIDWORKS 2018 and later, VSTA 3.0 is available using the Visual Studio 2015 or 2019 development environment. Use of VSTA 3.0 requires a separate installation of Visual Studio 2015 or 2019. Even though there are newer versions of Visual Studio available, there is only one that is supported per release. For example, Visual Studio 2019 is functionally supported at the publication of this book for SOLIDWORKS 2023. Visual Studio 2015 is supported for earlier versions of SOLIDWORKS.

SOLIDWORKS still includes VBA, or Visual Basic for Applications, development environment. It is simple, forgiving, and is based on the older Visual Basic language structure. It is still a useful tool for quick, portable macros and we will explore its use in conjunction with Microsoft Excel.

If you already have a supported version of Visual Studio installed, you can skip the installation step here.

Installing Visual Studio 2019

1. If you are using SOLIDWORKS 2023, download and install Visual Studio 2019. If you do not have a paid version available, you can install the community edition for education purposes from the Microsoft Visual Studio site.

   https://visualstudio.microsoft.com/vs/older-downloads/

2. Scroll down to the list of older versions, expand 2019 and select Download. You will be prompted to sign in or create an account.
3. After logging in, search for "Visual Studio Community 2019" if you are looking for the free version. Tap download.
4. When prompted for the development environment, select ".NET desktop development" and select Install.

This is the free community edition. You will need a Microsoft account to log in and use Visual Studio Community Edition beyond the 30 day evaluation period. Review the license agreement to ensure you are in compliance. If you are not, or if you have it available already, install any paid version of Microsoft Visual Studio.

5. If you installed Visual Studio after installing SOLIDWORKS, you will also need to download and install Microsoft Visual Studio Tools for Applications 2015 (VSTA) from the Microsoft Download Center. Search for VSTA 2015 or use the address below.

    https://www.microsoft.com/en-US/download/details.aspx?id=49031

Installing VSTA – SOLIDWORKS 2017 and Earlier

If you are still using SOLIDWORKS 2017 or older, you must install VSTA through the SOLIDWORKS Installation Manager rather than a full Visual Studio installation. Verify the installation of VSTA by modifying your SOLIDWORKS installation from Windows Settings, Apps and Features. When presented with the list of components to install, expand the SOLIDWORKS items at the top of the list and scroll down. Make sure Visual Studio Tools for Applications (VSTA) is checked and finish the installation.

Note that all screen images in this book are based on Visual Studio 2019 rather than VSTA 1.0. Most of the differences are only cosmetic. Code syntax and structure are the same.

Create a New Macro

Now it is time to create your first macro in SOLIDWORKS.

1. Start SOLIDWORKS and create a new macro by selecting Tools, Macro, New or by clicking 📄 on the Macro toolbar.

2. Save the macro with the name *MyFirstMacro* after choosing the Save as type to be "SW VSTA VB Macro (*.vbproj)" to the directory of your choice.

Note: SOLIDWORKS macros can also be saved in VBA and C# format. If you prefer C#, you will need to be familiar with C# syntax and usage. Some samples are provided in this book, but descriptions and suggestions are based on VB.NET. VBA will be introduced in a later chapter.

Visual Studio Interface

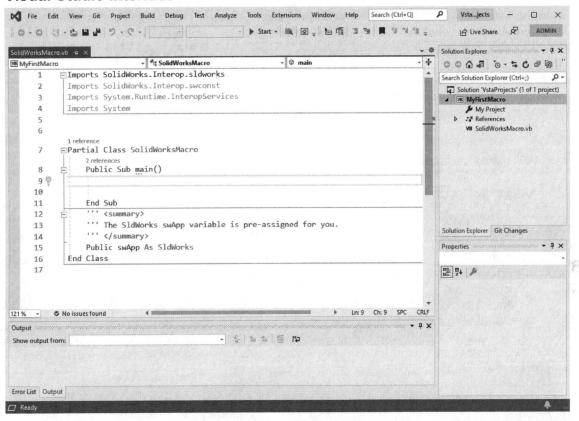

Once saved, Visual Studio opens with an initial project and one code file named *SolidWorksMacro.vb*. The Solution Explorer panel on the right is a quick place to navigate your project code files and general project settings in My Project. The new macro only has one code file, but macro projects can have many files of different types. They are all used to build the resulting macro as a Dynamic Link Library or dll. More about those later.

The Properties pane on the right gives quick access to the properties of anything you select. Select the project name MyFirstMacro at the top of the Solution Explorer and the Properties pane will display the project file name *MyFirstMacro.vbproj* and its path. Properties are used extensively when creating forms with buttons, text boxes and other user controls.

The open code files will display in tabs across the center of the screen. The three dropdown boxes at the top of the code window are quick navigation tools. These are especially helpful as your code gets more complicated. For example, from the center dropdown called Class Name, select SolidWorksMacro. From the right dropdown called Method Name, select main. The cursor will jump inside the main procedure block within the class named SolidWorksMacro. More about classes, methods and procedures in a moment.

Across the top of the interface you will find the typical menus and buttons for common operations. We will explore many of these as they are used with the exception of the common tools like save, copy and paste.

Code Description

What is all this stuff? Every macro will have a similar initial structure. From top to bottom, the *SolidWorksMacro.vb* code file contains Imports statements, a Class, a main procedure and a variable for the SOLIDWORKS application itself.

```vb
Imports SolidWorks.Interop.sldworks
Imports SolidWorks.Interop.swconst
Imports System.Runtime.InteropServices
Imports System

Partial Class SolidWorksMacro

    Public Sub main()



    End Sub

    '''<summary>
    '''The SldWorks swApp variable is pre-assigned for you.
    '''</summary>
    Public swApp As SldWorks

End Class
```

Imports

The first few lines define code libraries or namespaces that might be used in the macro. Namespace is a term for an existing library of features and capabilities you can use in your macro or application. Two are SOLIDWORKS namespaces and two are general System namespaces. The dot structure, like SolidWorks.Interop.sldworks, is the foundation of Object-oriented programming.

Here's an analogy. The structure is like an address. If I wanted to visit the Eiffel Tower using Object-oriented programming, the namespace might be Earth. To get to the tower itself, you might get there through Earth.Europe.France.Paris.

Importing namespaces helps simplify your code. Using the same analogy, I could directly access Paris in my code if the code file included the statement shown below.

```vb
Imports SolidWorks.Interop.sldworks
Imports SolidWorks.Interop.swconst
Imports System.Runtime.InteropServices
Imports System
Imports Earth.Europe.France
```

There is one important step before using the Imports statements in a code module. You must first add them as a reference in the project itself.

3. Verify references to the SOLIDWORKS libraries by selecting Project, MyFirstMacro Properties from the Visual Studio menus. Select the References tab on the left. Verify references to SolidWorks.Interop.SldWorks and SolidWorks.Interop.swconst.

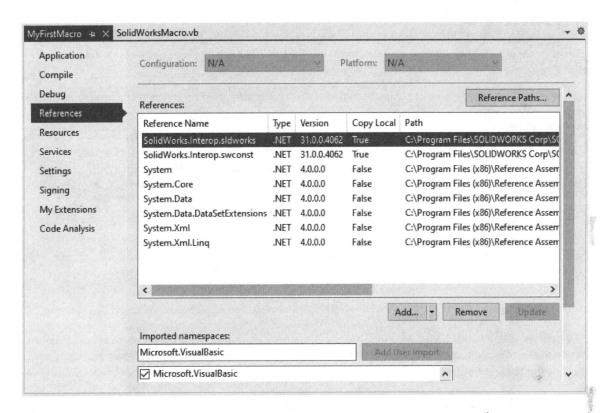

The version number indicates the SOLIDWORKS version. SOLIDWORKS 2021 is the 29th version. Service packs and specific build numbers follow.

Additional references can be added by clicking the Add button below the list of references.

Hint: if you would like to Import a reference project wide, simply check the reference in the list of Imported namespaces. By importing them at the project level you do not need to use the Imports statement in your code module. Just be aware of the possibility of conflicting definitions from two different namespaces. After all, there is also Paris, Idaho, USA.

.NET Structure

You can loosely compare the structure of a .NET macro to a SOLIDWORKS assembly. The macro project itself is like a top assembly. It contains all of the pieces needed to make the macro work. If you view the Solution Explorer on the right side of the interface, MyFirstMacro is the project name at the top of the tree. Assembly is also the term used in the .NET language to refer to a complete project.

The first sub assembly in a .NET assembly is a Namespace. All of the pieces of the assembly are contained in a Namespace. This part of the structure, just like the project itself, is defaulted to the name of your saved macro.

Classes

Classes in Visual Basic.NET are the fundamental container for code. Every recorded macro has a class automatically created for you. If your macro does not have any user interface forms you can get away without creating any additional classes. However, if you add forms to your macro, additional classes will automatically be created.

The first real structure to a macro comes at the declaration of a class named SolidWorksMacro. A declaration is a code statement that defines a variable or procedure.

```
Partial Class SolidWorksMacro
```

The fact that it is declared as a Partial Class is not important to the functionality of the macro itself. But if you are curious, class definitions can be broken into pieces in a project. There is a hidden code module in a macro project that does some of the hard work for you. This hidden code module contains the other piece of the class named SolidWorksMacro. It is not something you should ever need to edit.

Procedures

The next section is the entry point or procedure in the macro. It is common programming practice to name the primary procedure main. The name main should make sense. A Sub procedure can take variables and modify them with a series of loops and commands. The double parentheses are where variables could be passed into this procedure as arguments. Main procedures typically do not have any arguments passed to them since they are the first to run.

The main procedure is declared with the Public statement. This defines the procedure's scope. Public scope means this procedure will be available to any code that connects to this class instance. More on connecting to class instances later.

The close of the procedure is marked with the statement End Sub.

```
Public Sub main()

End Sub
```

Variable Declarations

A new macro has only one variable declared in the SolidWorksMacro class. This is the variable that is automatically connected to the running session of SOLIDWORKS.

```
'''<summary>
'''The SldWorks swApp variable is pre-assigned for you.
'''</summary>
Public swApp As SldWorks
```

Like the main procedure, the variable named swApp is declared with the Public statement, making it available to all procedures in the SolidWorksMacro class as well as to code that connects to this class instance.

Variables must be declared with a data type. The As statement precedes the declared type. In this case, the SldWorks type is a member of the SolidWorks.Interop.sldWorks namespace and can contain a reference to the SOLIDWORKS application.

Think of variables as a custom size box. A variable's data type defines the size of the box and what can fit in the box. In reality, the type defines the variable's storage capacity in memory. If you try to fill the variable (box) with a mismatched data type, you will get errors when you run your macro.

Hello World

It's time for the ubiquitous "Hello World" code test.

4. Add the following code inside the main procedure block in the SolidWorksMacro class.

```
Public Sub main()
    Dim message As String
    message = "Hello World"
    'Show the message
    MsgBox(message)

End Sub
```

First, a variable named message is declared as a String data type. The next line initializes the variable with a value. Enclosing text in quotes defines everything inside the quotes as a string data type (so it fits in the box). The Dim statement is the most common declaration statement. It is for variables that will only be used inside the class or procedure.

Notice the green color of the line "'Show the message." An apostrophe in Visual Basic is the comment character. Anything following an apostrophe is not compiled and won't be run. Use comments frequently to help you remember the purpose of the code and to help others who might have to edit your macros.

The last line calls the MsgBox (or message box) function from the general Visual Basic namespace. The message variable is passed to the function as an argument and results in the string being displayed to the user. Arguments are always enclosed in parentheses in Visual Basic.NET.

5. Run the macro by selecting Debug, Start or click ▶. The code will run through the main procedure and you will see the following. Clicking OK will close the message box.

6. If needed, stop the debugger by selecting Debug, Stop Debugging or click ■.

Data Types

Familiarize yourself with some of the standard Visual Basic data types and what they store.

Data Type	Definition
Boolean	True or False (sometimes also indicated by 1 for True and 0 for False)
Byte	Single 8-bit numbers from 0 to 255
Currency	64-bit numbers ranging from 922,337,203,685,477.5808 to 922,337,203,685,477.5807
Date	64-bit numbers with dates ranging from 1 January 100 to 31 December 9999 and times from 0:00:00 to 23:59:59
Double	64-bit floating point numbers ranging from -1.7977E+308 to -4.9407E-324 for negative values and from 4.9407E-324 to 1.7977E+308 for positive values
Integer	16-bit numbers ranging from -32,768 to 32,767 (no decimals)
Long	32-bit numbers ranging from -2,147,483,648 to 2,147,483,647
Object	32-bit reference to an object (any class or data type)
Single	32-bit numbers ranging from 3.402823E38 to 1.401298E-45 for negative values and from 1.401298E-45 to 3.402823E+38 for positive values
String	Can contain approximately two billion characters

Object Data Type

Object is the catch-all data type. It is the universal box that fits everything. It also uses the most storage. You could be lazy and declare every variable as an object type but you would end up with a bloated macro.

Classes are a special type of object. Classes like our SolidWorksMacro class can contain properties and methods. Properties are values like name or color, commonly Public variables, while methods do something with or without inputs (arguments). Methods can be either procedures or functions. Procedures take action, like depositing money in the bank. Functions take action and return something, like making a withdrawl. In our first example, MsgBox is a function. It can return the results of how the user interacts with the dialog. The user might have clicked OK, or might have closed the message box by clicking X in the top-right corner.

Interface is the term commonly used for a class from an application or the application reference itself. The variable swApp is declared as the data type SldWorks. Through the rest of this book, SldWorks will be refered to as an interface, though it would be valid to also call it a class or even an object.

SOLIDWORKS Data Types

Familiarize yourself with a few of the SOLIDWORKS data types, or interfaces, along with some examples of their use. There are nearly 400 so this is a condensed list. The full list is available in the Help documentation.

Data Type	Definition
ISldWorks	A reference to the SOLIDWORKS application interface providing access to create new files and access to open documents
IModelDoc2	The general container interface for parts, assemblies and drawings with access to document settings, the FeatureManager and selections
IPartDoc	A special IModelDoc2 interface with methods and properties specific to parts
IAssemblyDoc	A special IModelDoc2 interface with methods and properties specific to assemblies
IDrawingDoc	A special IModelDoc2 interface with methods and properties specific to drawings
ISketch	The interface to a sketch and its properties and methods
ISheet	The interface to each drawing sheet including properties and methods for name, size and scale
IView	The interface to each view on a drawing as well as the drawing sheet

SOLIDWORKS Application

Now it is time to try more data types as well as properties and methods of the SOLIDWORKS application interface.

7. Modify your main procedure as shown to declare a new Integer variable type and show the SOLIDWORKS DateCode (build version). Modify or add everything in bold font.

```
Public Sub main()
    Dim message As String
    Dim swVersion As Integer

    swVersion = swApp.DateCode
```

```
    message = "Hello SOLIDWORKS " & swVersion.ToString
      'Show the message
    MsgBox(message)

End Sub
```

A new variable `swVersion` is declared as an integer type. The new variable is then set to the version of the running SOLIDWORKS application by calling the `DateCode` method from the `swApp` variable which has been automatically connected to the `SldWorks` interface.

ToString

The `message` variable has been expanded to include the SOLIDWORKS version converted to a string data type. Remember that the `message` variable can only hold string data types. Since `swVersion` is an integer, it has to be converted to a string before it will fit into `message`. Most Visual Basic data types can be converted to a string by calling their `ToString` function.

Combining Strings

Multiple strings can be combined, or concatenated, in several ways. The first, and simplest is to use the & character. The only requirement is that the combined elements must be strings. It is similar to how you might concatenate in an Excel formula.

8. Start the macro again to see the new message box, including the SOLIDWORKS DateCode property.

SOLIDWORKS Documents

Everything has a path or address in object-oriented programming. Whether building SOLIDWORKS macros or connecting to Microsoft Excel, finding the path to the needed object, class or interface is the first step. After you have connected that interface to a variable, you can interact with its methods and properties. SOLIDWORKS documents are a great place to start.

9. Open an existing SOLIDWORKS part (*.sldprt). Any part will work including a newly saved part.
10. Modify the main procedure as shown to connect to the active document and return its title in the message box.

```
Public Sub main()
  Dim message As String
  Dim swVersion As Integer
  Dim swDoc As ModelDoc2
```

```
Dim swTitle As String
swDoc = swApp.ActiveDoc
swTitle = swDoc.GetTitle

swVersion = swApp.DateCode

message = "Hello SOLIDWORKS " & swVersion.ToString _
& vbCrLf & "Document: " & swTitle
    'Show the message
MsgBox(message)
```

End Sub

11. Start the updated macro to see the new message.

12. Stop Debugging after clicking OK on the message box.

Visual Basic Constants

Before exploring the SOLIDWORKS methods, notice the term vbCrLf in the newly combined message variable. It has not been declared as a variable, so where does it come from and what does it do? vbCrLf is a Visual Basic constant representing a carriage-return and linefeed character. It gives your string the format of having hit the Enter key while typing and separates the two lines. A constant differs from a variable in that they don't change. Their values remain constant. Use the following table as a reference for some of the more common Visual Basic constants for manipulating strings.

VB Constant	Definition
vbCrLf	A combined carriage-return and linefeed character – like using the Enter key while typing
vbTab	Adds a tab character

Line Continuation

A brief comment about Visual Basic line formatting. Other code languages use an explicit line termination character. Java and C# use the semi-colon. Visual Basic compiles each line up to the return character. So instead of adding a termination character at the end of each line, Visual Basic uses a line continuation character _ (underscore) when you would like to wrap a line. The code that defines the value of the message variable gets too long for me to show across the page, so it is continued by adding an underscore at the end of the line that continues on the next. You can use as many line continuation characters as needed to make your code easier to read.

ISldWorks.ActiveDoc

After declaring two new variables, the first line that does any work attaches the variable named swDoc to current active document by calling swApp.ActiveDoc.

```
swDoc = swApp.ActiveDoc
```

IModelDoc2

The ActiveDoc property of the SOLIDWORKS interface will return an IAssemblyDoc, IPartDoc or IDrawingDoc. These are interfaces. That is the reason for the I in front of the name, though you won't use that in VB.NET code. Each of these three types are part of the more general IModelDoc2 interface as described in the table of SOLIDWORKS objects and is used here because ActiveDoc could be any of the three. So how do you know which to use? Now is a good time to look at the SOLIDWORKS API Help.

API Help

The SOLIDWORKS API Help is an invaluable companion to this book. If you are serious about automating SOLIDWORKS, you must learn to effectively navigate and read the API Help. This book covers many SOLIDWORKS API calls, but there are thousands. I could not hope to document them all here if I tried.

There are two formats for the API Help. By default, the help system is set to Use Web Help. The advantage of web help is that you are always accessing the latest up-to-date build of the help documentation. However, the web help format lacks some functionality referenced by this book. Even though you will not have the latest content, I recommend turning off the Use Web Help option for API help. Toggle it on or off depending on your preference and needs. All references to the API help here will reference local help.

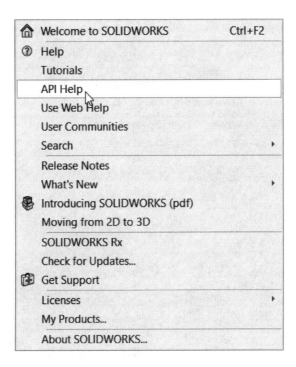

Look up the ISldWorks interface and explore its properties and methods.

13. Open the API Help from the Help menu in SOLIDWORKS.

This is a standard help file with all of the detailed calls to the SOLIDWORKS API and its components as well as many of the other SOLIDWORKS product APIs. It includes language syntax references for Visual Basic.NET, VBA (Visual Basic 7), C# and C++.

14. Select the Index tab and type "ISldWorks." Then double-click on ISldWorks Interface from the list.

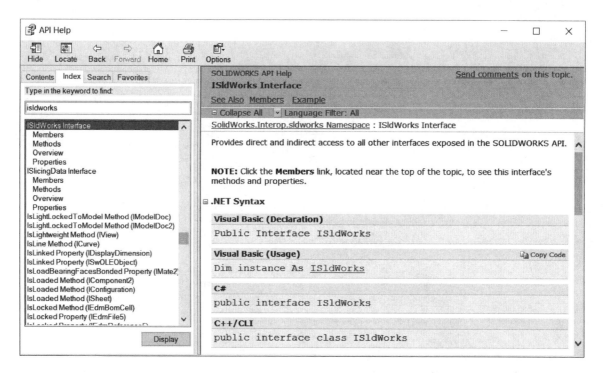

The reason each interface is documented here with the preceding 'I', like ISldWorks, instead of the SldWorks you see in your code, is based on the way it is documented in the API Help. Each interface has quick links to its members (properties and methods) and an overview. The overview is the best place to start. Look over the description.

Towards the top of the overview page you will notice that ISldWorks comes from the namespace SolidWorks.Interop.sldworks described earlier in the chapter.

Use the .NET Syntax section to understand the Visual Basic usage. This will mean more later.

Use the Example section for example code that shows how you might use the interface, method or property. Make sure you choose the VB.NET examples for Visual Basic.NET macros.

The Remarks section will explain details about usage and best practices. Don't skip reading the remarks.

The Access Diagram gives you a graphical PDF diagram of the surrounding parent and child interfaces. This can be a great way to understand how to get to the right interface.

See Also allows you to link to other interfaces, methods and properties related to the current API help topic.

Find the ActiveDoc method in the API Help.

15. From the ISldWorks Interface help page, select the Members link at the top of the page. This page displays all of the properties and methods of the ISldWorks interface.
16. Select the first Public Property link to ActiveDoc.

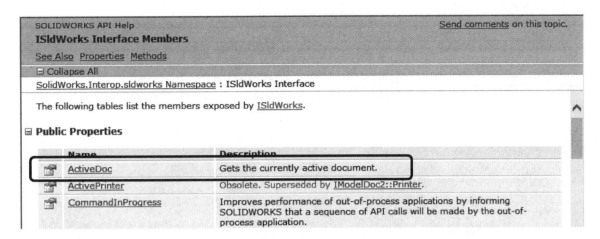

As expected, the ActiveDoc Property help page describes the property as getting the currently active document.

From the Visual Basic (Declaration) section, ActiveDoc is defined As `System.Object`. This should look similar to declaring a variable. Object is the generic data type for holding anything including classes and interfaces.

The API help makes it easy to use the property in the Visual Basic (Usage) section. Simply click the Copy Code link and paste it into your macro. Notice the variable names need to be updated in your code with a meaningful name instead of instance and value.

Move down on the page to the Property Value section. It explains the returned value. ActiveDoc returns a "model document or Nothing or null if the operation fails."

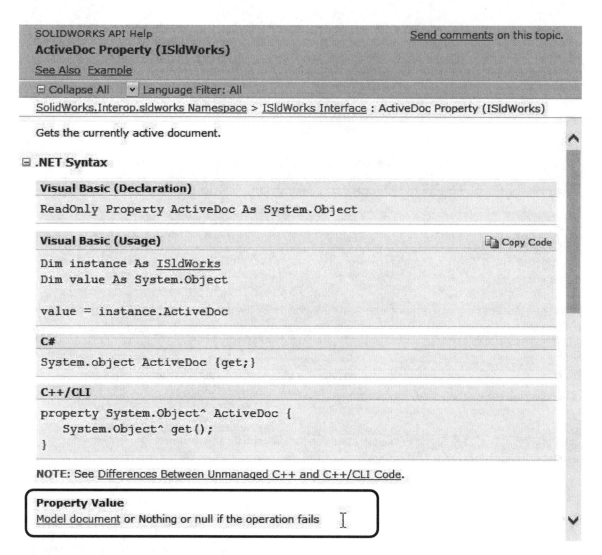

SOLIDWORKS API Help   Send comments on this topic.
ActiveDoc Property (ISldWorks)
See Also Example

⊟ Collapse All ▾| Language Filter: All

SolidWorks.Interop.sldworks Namespace > ISldWorks Interface : ActiveDoc Property (ISldWorks)

Gets the currently active document.

⊟ **.NET Syntax**

Visual Basic (Declaration)

```
ReadOnly Property ActiveDoc As System.Object
```

Visual Basic (Usage)   ▣ Copy Code

```
Dim instance As ISldWorks
Dim value As System.Object

value = instance.ActiveDoc
```

C#

```
System.object ActiveDoc {get;}
```

C++/CLI

```
property System.Object^ ActiveDoc {
    System.Object^ get();
}
```

NOTE: See Differences Between Unmanaged C++ and C++/CLI Code.

Property Value
Model document or Nothing or null if the operation fails

17. Click the Model document hyperlink to navigate to the interface returned by ActiveDoc. You are taken to the IModelDoc2 Interface page of the API help.

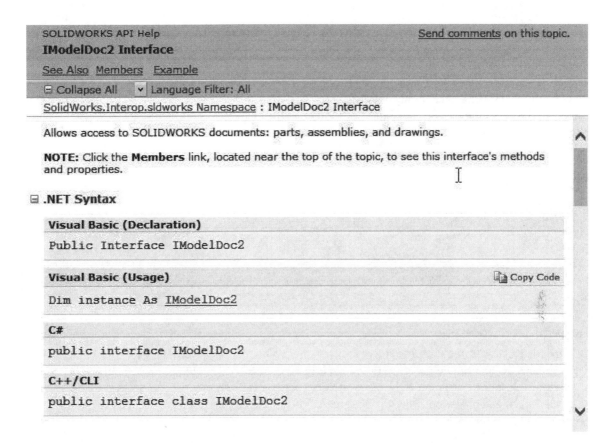

This is how we know to use a variable declared as the ModelDoc2 data type for the active document. Take a moment to read the description and Remarks section of IModelDoc2. You will be using this interface extensively.

ActiveDoc is not the only way to get to the IModelDoc2 interface. The Accessors section lists all of the different paths to get to IModelDoc2. Look through the Accessors section whenever you need to find a path to an interface.

IModelDoc2.GetTitle

Navigate deeper into the API help to find the GetTitle method of IModelDoc2.

18. From the top of the IModelDoc2 help page, select the Members link. Scroll down into the Public Methods until you find GetTitle.

Read its description to get a quick idea of what it does. This method gets the title of the document that appears in the active window's title bar.

19. Click the GetTitle link to open its help page.

The variable used to collect the document's title was declared as a String data type. Look at the Visual Basic (Declaration) section and notice that GetTitle is defined As System.String. It matches. Now you know how to declare the right type of variable for the interface, method or property. By the way, we don't have to explicitly declare the `swTitle` variable with its namespace as `System.String` since our code file has the statement `Imports System` at the top. We already have that part of the address defined for the entire code file.

Be warned that GetTitle may or may not display the open document extension. If your message box displayed the file name without the extension, you have the Windows Folder view setting to show "File name extensions" off.

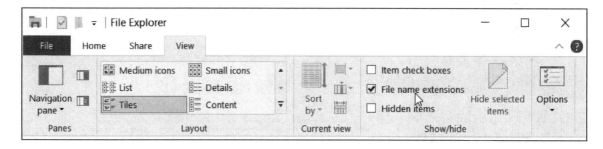

Logical Statements

To introduce logical statements into the macro, let's get some feedback from the user through the message box. As mentioned earlier, MsgBox is a function. It can return the user's response. The macro should then close the document if the user answers Yes. Logical statements compare two objects like 45 > 90, red = green or swApp IsNot swDoc. Each of these statements gives a result of True or False and is commonly used for decision making in a macro.

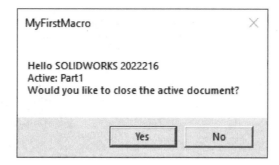

If Then Statements

If Then statements are perfect for testing a logic statement and taking action when the logic is True. The syntax is shown below.

```
If [logic statement] Then
  [do something here]
End If
```

You can also shortcut an If Then statement as a single line if there is only one action to take. It might look like this.

```
If [logic statement] Then [do one thing]
```

If Then statements can also be expanded to include additional checks with an ElseIf and a final operation if the previous checks are all False with an Else. You can use as many ElseIf lines as needed but they must follow the first If statement and are always before an Else statement as follows.

```
If [logic statement] Then
    [do something here]
ElseIf [different logic statement] Then
    [do something different here]
Else
    [if all else fails, do something here]
End If
```

The message box needs to be modified with Yes and No buttons to test the user's response. The result of the message box function needs to be collected in a variable so it can act as our logical statement in an If Then block.

 20. Modify the macro code as shown below in bold.

```
Public Sub main()
    Dim message As String
    Dim swVersion As Integer
    Dim swDoc As ModelDoc2
    Dim swTitle As String
    swDoc = swApp.ActiveDoc
    swTitle = swDoc.GetTitle

    swVersion = swApp.DateCode

    message = "Hello SOLIDWORKS " & swVersion.ToString _
    & vbCrLf & "Active: " & swTitle _
    & vbCrLf & "Would you like to close the active document?"
    'Show the message
    Dim response As MsgBoxResult
    response = MsgBox(message, MsgBoxStyle.YesNo)
    If response = MsgBoxResult.Yes Then
        swApp.CloseDoc(swTitle)
    End If

End Sub
```

Another line in the message variable asks the user if they would like to close the active document. A new variable `response` has been declared as `MsgBoxResult`. This is the data type returned by a message box. Then rather than just calling the message box blindly, its result is now captured. Whenever a function is called and a value is returned, the right side of the statement is run before its result is sent back to the variable on the left.

An additional argument is passed to `MsgBox` to create the Yes and No buttons instead of the default OK. `MsgBoxStyle.YesNo` changes the look of the message box. There are several other styles you can use to get just the right feedback from the user.

Now look over the If Then block. The logical statement is comparing response to MessageBoxResult.Yes. Yes is another Visual Basic constant within the MessageBoxResult data type. If the logical statement is True, the inside of the If Then block runs and the active document is closed. You should begin to recognize that CloseDoc is a method of the SOLIDWORKS application, the variable swApp in this macro. It takes the name of the document to close as the only argument. You can pass just the file name from SldWorks.GetTitle, or you can pass the full file path from IModelDoc2.GetPathName. More on full file paths later.

String.Format

The System String class has a built-in alternative for concatenating multiple strings. If you prefer an old-school style to concatenation, use this method.

```
someString = String.Format("Hello SOLIDWORKS {0}", swVersion.ToString)
```

The first argument is a string containing at least one digit in curly brackets {0}. Each unique digit in curly brackets requires a corresponding string value in the following arguments. The digit represents the index of the argument. Replacing the entire message would look like the following. Notice the placement of the newline in multiple places {1}.

```
message = String.Format( _
    "Hello SOLIDWORKS {0}{1}Active:{2}{1}Would you like to close the active
document?", _
    swVersion.ToString, vbCrLf, swTitle)
```

IntelliSense

As you typed the code for the message box style you should have noticed some interactive feedback from VSTA. Hovering your mouse over a method or function like MsgBox will display its arguments and returns (if any) in a tool tip. When entering an argument or typing the method or property name after the dot following the object, Microsoft IntelliSense will display a filtered list of possible options.

```
        Dim swDoc As ModelDoc2
        Dim swTitle As String
        swDoc = swApp.ActiveDoc
        swTitle = swDoc.GetTitle


        swVersion = swApp.DateCode


        message = "Hello SOLIDWORKS " & swVersion.ToString _
    & vbCrLf & "Active: " & swTitle _
    & vbCrLf & "Would you like to close the active document?"
        'Show the message
        Dim response As MsgBoxResult
        response = MsgBox(message, yes)

        If response = MsgBoxResult.    MsgBoxStyle.YesNo
            swApp.CloseDoc(swTitle)    MsgBoxStyle.YesNoCancel
        End If                         vbYes
                                       vbYesNo
                                       vbYesNoCancel
```

My Project
▷ References
SolidWorksMacro.vb

Solution Explorer | Git Changes

Properties

MsgBoxStyle.YesNo = 4
Yes and No buttons. This

Make use of IntelliSense and let it help you! In the image above, I only typed "yes" and IntelliSense brought me to the exact syntax I needed. Hitting Tab (or nearly any other key on the keyboard) will complete the code with the right syntax.

When you type a class or interface variable name and then the following dot, IntelliSense will display all methods and properties from the object for easy access.

You can manually trigger IntelliSense as well. Start by typing one or more characters of a variable or common Visual Basic function and then hit Ctrl+Space. The IntelliSense pop-up will display everything available that matches the characters you typed. Use the arrow keys on the keyboard to navigate the list if needed and hit Tab to complete the entry.

Save typing, code faster and more accurately by using IntelliSense.

21. Run a final test of your new macro. Try hitting both No and Yes while a document is active to test your logical statement.

Common Errors

Unless you mistyped, there is a good chance the macro works as expected. However, for more reasons than are imaginable, errors happen. Users do things you don't expect. What would happen if no files were open in SOLIDWORKS and you ran the macro? Let's find out.

22. Close all open documents in SOLIDWORKS and then run your macro. You should get the following message.

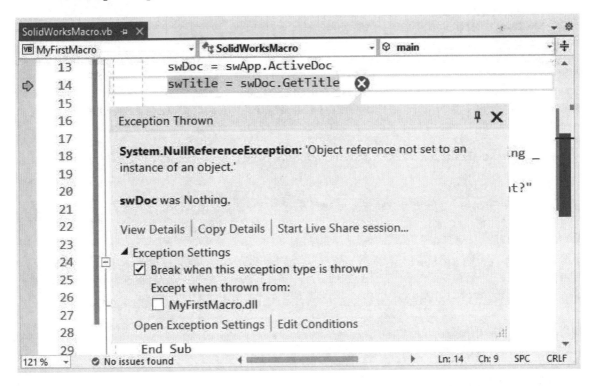

A NullReferenceException error is one of the most common when writing macros. Notice the first statement below the error title bar.

System.NullReferenceException: 'Object reference not set to an instance of an object.'

The problem line is highlighted in yellow. The Visual Studio debug engine has run to that line and paused because of the problem. So what happened? In the error, an object reference is our variable – swDoc in this case. It hasn't been set to an instance of an object. The previous line should return the active document from the SOLIDWORKS interface, but there is no active document. So the ActiveDoc method returns Nothing. In our box analogy, Nothing is an empty box. In fact, you can empty an interface variable by setting it equal to Nothing.

Whenever you get this error, look above the highlighted line for the problem. Find the line that was supposed to set the variable to something. That is probably where it failed.

23. Stop the debugger by clicking ■.
24. Add an If block as shown to exit the macro if nothing is open.

```
Public Sub main()
  Dim message As String
  Dim swVersion As Integer
  Dim swDoc As ModelDoc2
  Dim swTitle As String
  swDoc = swApp.ActiveDoc
  If swDoc Is Nothing Then Exit Sub
  swTitle = swDoc.GetTitle
  ...
```

The If Then statement is conveniently consolidated to one line since the processing code is also a single statement. The Exit Sub statement can be used anywhere in your code to exit a Sub procedure without continuing.

25. Do another debug run to make sure the error has been solved. If nothing is open, nothing happens.

26. Save and close your macro by clicking Save All 🖫 and then closing the Visual Studio interface.

Conclusion

This may not be the macro you have always been searching for, but you are beginning to understand the basics. Use this chapter as a reference as you build your skill through the following chapters. They will dive deeper into the SOLIDWORKS API as well as VB.NET structures and techniques.

C# Example Code

```
public void Main()
{
    string message;
    int swVersion;
```

```csharp
ModelDoc2 swDoc;
string swTitle;

swDoc = (ModelDoc2)swApp.ActiveDoc;
if(swDoc == null)
{
    return;
}

swTitle = swDoc.GetTitle();
swVersion = swApp.DateCode();

message = "Hello SOLIDWORKS " + swVersion.ToString()
    + System.Environment.NewLine + "Document: " + swTitle
    + System.Environment.NewLine
    + "Would you like to close the active document?";
//message = string.Format("Hello SOLIDWORKS {0} \nActive:{1}\nWould you
  like to close the active document?", swVersion.ToString(), swTitle);
//show the message
DialogResult diaRes;
diaRes = MessageBox.Show(message, "Close document",
    MessageBoxButtons.YesNo);
if(diaRes == DialogResult.Yes)
{
    swApp.CloseDoc(swTitle);
}

return;
}
```

One Button PDF Publishing

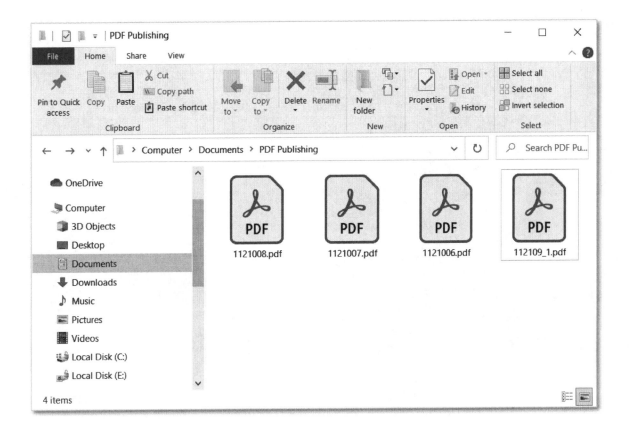

- **Recording a Macro**

- **Save As Different Formats**

- **System.IO Namespace**

Introduction

A fellow SOLIDWORKS user asked if there was an easy way to create PDFs. He wanted PDFs published to the same folder as the drawing, using the same name – a perfect exercise to introduce saving SOLIDWORKS files in different formats. It also makes use of some string manipulation and the file path of the active document. You could expand this macro to save any other format such as IGES, DXF or eDrawings.

Record the Save As Action

The best way to learn the SOLIDWORKS API is through the macro recorder. When you record a macro in SOLIDWORKS, each user interaction step is documented. The resulting macro can be saved in Visual Basic, Visual Basic.NET or C#.NET language format. There are some limitations when recording, but most basic functionality is available.

1. Open any drawing in SOLIDWORKS and start recording a macro by selecting Tools, Macro, Record or click **II●** on the Macros toolbar.

2. Create a PDF of the open drawing by selecting File, Save As. From the file type list, select Adobe Portable Document Format (*.pdf). Use the current directory and default file name and click Save.

3. Stop the macro recording by selecting Tools, Macro, Stop or click **■** and save the macro as type SW VSTA VB Macro (*.vbproj) and name it *SavePDF.vbproj*.

4. Edit the new macro by selecting Tools, Macro, Edit or click 🖼 and browse to the newly created macro project file named *SavePDF.vbproj*.

Hint: Turn on the SOLIDWORKS option to "Automatically edit macro after recording" from Tools, Options, System Options, General so you can skip the editing step when recording new macros.

Your code should look something like the following.

```
Imports SolidWorks.Interop.sldworks
Imports SolidWorks.Interop.swconst
Imports System.Runtime.InteropServices
Imports System

Partial Class SolidWorksMacro

  Public Sub main()

    Dim swDoc As ModelDoc2 = Nothing
    Dim swPart As PartDoc = Nothing
    Dim swDrawing As DrawingDoc = Nothing
    Dim swAssembly As AssemblyDoc = Nothing
    Dim boolstatus As Boolean = False
    Dim longstatus As Integer = 0
    Dim longwarnings As Integer = 0
```

```
    swDoc = CType(swApp.ActiveDoc,ModelDoc2)
    Dim myModelView As ModelView = Nothing
    myModelView = CType(swDoc.ActiveView,ModelView)
    myModelView.FrameState = CType(swWindowState_e.swWindowMaximized,Integer)
    swDoc.ClearSelection2(true)
    '
    'Save As
    longstatus = swDoc.SaveAs3("C:\Automating SOLIDWORKS\Drawing1.pdf", 0, 2)
  End Sub

  ''' <summary>
  ''' The SldWorks swApp variable is pre-assigned.
  ''' </summary>
  Public swApp As SldWorks

End Class
```

The first line doing work in the main procedure attaches the variable named swDoc to current active document using swApp.ActiveDoc.

```
swDoc = CType(swApp.ActiveDoc, ModelDoc2)
```

Most macros will have this same line of code. Those of you who are already familiar with Visual Basic.NET may find the format of the line a little unusual. The CType Visual Basic function is used in the recorded code to verify that the data type being returned is in fact a IModelDoc2 interface.

If you are not already familiar with the CType function, the same line of code can be simplified as follows. This format might make a little more sense if you are new to Visual Basic. The object oriented format of Visual Basic uses the Class.Method format. In this example, swApp is the variable that references the SOLIDWORKS Application class and ActiveDoc is a method of the class that returns a reference or pointer to the active document's interface.

```
swDoc = swApp.ActiveDoc
```

Extra Code

Every macro recording adds code that you may not need to use. For example, there are several lines of code related to the ModelView interface. The ModelView relates to screen display functionality. This macro will not need any reference to screen display so the extra lines of code can be removed for simplicity. It also has variables declared for parts, drawings and assemblies that are not used in the code.

5. Delete the following lines of unnecessary code.

```
Dim swPart As PartDoc = Nothing
Dim swDrawing As DrawingDoc = Nothing
Dim swAssembly As AssemblyDoc = Nothing
...
Dim myModelView As ModelView
myModelView = CType(swDoc.ActiveView, ModelView)
```

```
myModelView.FrameState = CType(swWindowState_e...
```

Finally, the macro saves the active document by calling swDoc.SaveAs3. As a reminder, swApp and swDoc are simply variable names that represent the SOLIDWORKS application (ISldWorks) interface and the IModelDoc2 interface respectively.

IModelDoc2.SaveAs3

The SaveAs3 method of IModelDoc2 is actually an obsolete method. It is a good practice to use the latest version of an API call in your macros, but not always necessary. ModelDocExtension.SaveAs3 is the current method for saving a copy of a file. However, its argument structure is more complicated. For simplicity, we will use the recorded SaveAs3 method since we have not yet introduced the ModelDocExtension interface.

The SaveAs3 method is perfect for common cases. It can save a drawing, part or assembly with a new name or it can save the file into another format. Not only can it be used to save a PDF from a drawing, but it can be used to save a DWG. From a model it can save to Parasolid, IGES or STEP.

Look over the structure and arguments of SaveAs3 here. Reference the API help for a more detailed description and additional remarks.

value = IModelDoc2.SaveAs3(NewName, SaveAsVersion, Options)

- **NewName** is simply the new name of the file to save as a string. This should include the full path to the new file name, not just the name itself. This name could have a SOLIDWORKS file extension, or another extension if you are trying to save the file as a different format.

- **SaveAsVersion** is really only used if you are trying to save the file as a Pro/Engineer part or assembly. Otherwise it is passed 0 as seen in the recorded code.

- **Options** are from swSaveAsOptions_e enumeration and include the following list. These options can be added together if needed to get the combined option. A value of 2 as recorded in the code saves a copy.

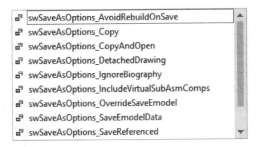

Changing Filename and Paths

If you run the recorded macro with a different drawing open, it will simply overwrite the original PDF with the output from the active document. The name and path of the PDF will be the same as the initial recording. The path and file name are hard-coded into the macro. This might be OK if you intend to copy and rename the new PDF every time you publish. But that doesn't pass the test of being even somewhat useful.

Visual Basic includes easy file-based string manipulation that can be used to change the extension of the file name string to PDF.

6. Edit the main procedure code to match the following, adding new variables and the final MsgBox function. Several of the unnecessary lines of code have been removed below.

```
Public Sub main()
    Dim swDoc As ModelDoc2 = Nothing
    Dim longstatus As Integer = 0
    swDoc = CType(swApp.ActiveDoc, ModelDoc2)
    Dim FilePath As String = ""
    Dim NewFilePath As String = ""
    FilePath = swDoc.GetPathName
    NewFilePath = IO.Path.ChangeExtension(FilePath, ".PDF")
    '
    'Save As
    longstatus = swDoc.SaveAs3(NewFilePath, 0, 0)
    MsgBox("Saved " & NewFilePath, MsgBoxStyle.Information)
End Sub
```

The first modification removes additional unnecessary variables at the top of the procedure and removes all ModelView modifications.

The example uses two variables to manipulate the file path in order to replace the default SOLIDWORKS extension ".SLDDRW" with the ".PDF" extension.

IModelDoc2.GetPathName

The GetPathName function of the IModelDoc2 interface will get the full path and file name of the document. Since our goal is to have the PDF published to the same directory, we simply need to modify the extension and we are ready for the SaveAs3 operation.

System.IO.Path Class

The System.IO namespace contains an invaluable set of tools for dealing with files. It includes methods for getting directories and all files in the directory. Use this .NET namespace or library in any Visual Basic.NET or C# macro. Notice the Imports System line towards the top of the macro code. The Path class of System.IO has several methods and properties for manipulating file name and path strings. Since the System namespace was already imported, you do not have to type System.IO.Path.ChangeExtension to use the method. If you were to add Imports System.IO at the top of the code, you could shorten the call to Path.ChangeExtension. Recall that the Imports statements are essentially addresses to libraries of methods and procedures.

Path.ChangeExtension

The ChangeExtension method of the Path interface is a quick and easy way to change the file extension. You simply pass the file name to be modified as the first argument followed by the new extension including the period as the second. This method returns the new file name with the new extension.

One last line of code has been added to inform the user that the PDF has been created. It is a simple Visual Basic Message Box. While not necessary, the message box helps the user know that the macro finished successfully.

Debug

The macro is ready to test. If you want to make your macro more robust, add the error checking for no open documents described in the previous chapter.

7. Open a different drawing in SOLIDWORKS and run your macro by clicking Start Debugging ▶ in VSTA or by hitting F5 on the keyboard. Verify that the new PDF was created.

8. Stop the macro if necessary. Save your changes to the modified macro and close Visual Studio.

Running Macros

There are several ways to run macros without running Visual Studio. You can run macros in SOLIDWORKS by clicking ▶ from the Macro toolbar or going to Tools, Macro, Run. You are presented with a browse dialog to go find the compiled macro. Where is it?

Macro Folder Structure

When you saved the macro, a folder was created with the name you specified. When you go back to edit the macro, the project file *SavePDF.vbproj* is located in the *SavePDF* folder. When you debug or build the macro, the macro dll file that is actually run, *SavePDF.dll*, is located in the *bin* folder along with its required supporting files.

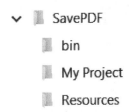

To share this macro with other users you only need to copy the contents of the *bin* folder. None of the other files need to be shared to run the compiled macro.

Custom Macro Buttons

Adding a custom button to a SOLIDWORKS toolbar is a great way to run commonly used macros.

9. Open a document in SOLIDWORKS and select Tools, Customize. Select the Commands tab. Select Macros from the categories list.

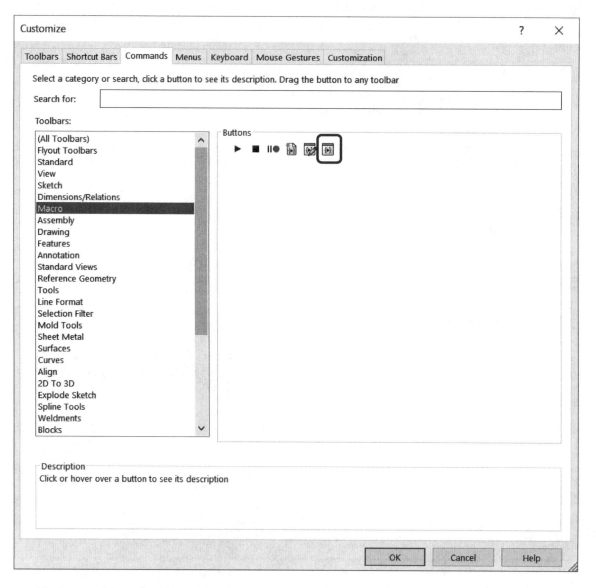

10. Drag and drop the New Macro Button onto any toolbar to launch the Customize Macro Button tool.

11. Fill out the Customize Macro Button dialog. Browse to the compiled macro dll named *SavePDF.dll*. As mentioned earlier, it is stored under the project folder name under *bin*. In this example it would be found in …*SavePDF\bin*.

12. Set the start method to main since that is the only procedure in the macro.

13. Add your own bitmap for a custom icon and fill out any desired tooltip and prompt.

14. Click OK to finish the custom macro button and add it to the toolbar.

Hint: you can further customize a macro button by right-clicking on it.

Conclusion

Now every time you want to publish a PDF, simply click your new macro button.

To convert to IGES, Parasolid and other formats, simply change the extension of the file name using the SaveAs3 method. Record the steps first for a quick and easy export macro.

C# Example Code

```csharp
public void Main()
{

    ModelDoc2 swDoc = null;
    int longstatus = 0;
    swDoc = ((ModelDoc2)(swApp.ActiveDoc));
    string FilePath = "";
    string NewFilePath = "";
    FilePath = swDoc.GetPathName();
    NewFilePath = System.IO.Path.ChangeExtension(FilePath, ".PDF");
    //
    // Save As
    longstatus = swDoc.SaveAs3(NewFilePath, 0, 0);
    MessageBox.Show("Saved " + NewFilePath, "Save message",
        MessageBoxButtons.OK, MessageBoxIcon.Information);

    return;
}
```

Model Dimensions Using Excel VBA

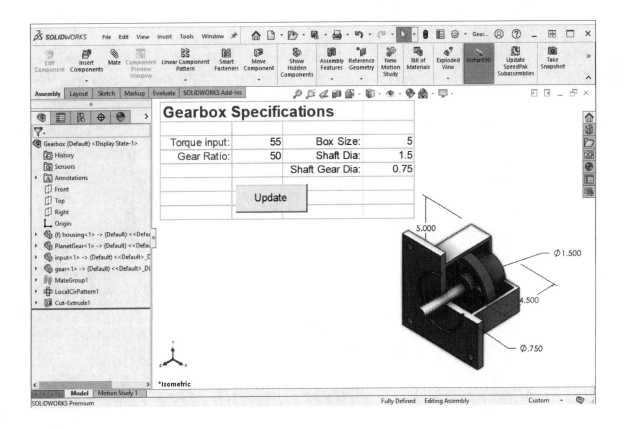

- **SOLIDWORKS VBA Macros**

- **Dimension Parameters**

- **VBA in Microsoft Excel**

- **Selection Methods**

Introduction

One of the most fundamentally accessible features in SOLIDWORKS is a dimension. Design tables are a great way to use Microsoft Excel's spreadsheet functionality to drive them. This exercise will cover another method of controlling dimensions through Excel's Visual Basic for Applications, or VBA, interface and API. VBA is based on the older Visual Basic language structure. Even though it is very similar to Visual Basic.NET, it is not the same. However, to make the code compatible with VBA in Excel, this chapter will use the older VBA style macro recording methods.

Through this chapter, you will learn to create an assembly containing an embedded Excel spreadsheet that controls part level dimensions and hides a part.

Changing Dimensions

Dimensions are easy to modify through the SOLIDWORKS API. A simple macro recording will show you how. This exercise changes dimensions from the assembly level, but controlling dimensions of parts is just as simple. Download the example files referenced at the beginning of the book to begin this exercise or use your own assembly.

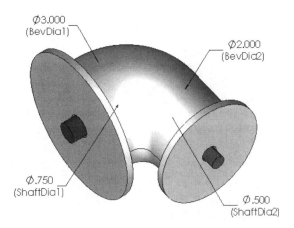

Recording the Macro

1. Open the assembly *BEVELGEARBOX.sldasm* from the downloaded example files. For better dimension name visibility, select View, Hide / Show, Dimension Names. (*Hint: If you are not using the example files from the download, you can still complete the exercise. Simply use any assembly and keep track of which dimensions you change by name.*)

2. Select Tools, Macro, Record or click ❚❙● to start recording.

3. Increase the value of *ShaftDia1*, *BevDia1*, *ShaftDia2* and *BevDia2* by 0.5 inches by double-clicking on each dimension.

4. Select Edit, Rebuild or click 🔋 .

5. Stop the macro by selecting Tools, Macro, Stop or clicking ■ .

6. Save the macro with the name *dimensions.swp* as an SW VBA Macro rather than a VSTA macro.

7. Edit the macro *dimensions.swp* by selecting Tools, Macro, Edit or clicking 🖉.

SOLIDWORKS VBA Macros

Your main procedure should be similar to the following code. The code below has been modified with underscores (line continuation character) to fit the width of the page.

```
Dim swApp As Object

Dim Part As Object
Dim boolstatus As Boolean
Dim longstatus As Long, longwarnings As Long

Sub main()

Set swApp = Application.SldWorks

Set Part = swApp.ActiveDoc
Dim myModelView As Object
Set myModelView = Part.ActiveView
myModelView.FrameState = swWindowState_e.swWindowMaximized
boolstatus = Part.Extension.SelectByID2 _
    ("ShaftDia1@Sketch1@bevelgear-1@BevelGearbox", _
    "DIMENSION", -0.05444898332424, 0.01124738657035, _
    -0.04436129312303, False, 0, Nothing, 0)
Dim myDimension As Object
Set myDimension = _
    Part.Parameter ("ShaftDia1@Sketch1@bevelgear.Part")
myDimension.SystemValue = 0.0254
boolstatus = Part.Extension.SelectByID2 _
    ("BevDia1@Sketch1@bevelgear-1@BevelGearbox", _
    "DIMENSION", 0.0371129077578, 0.07372578805571, _
    0.04736675418872, False, 0, Nothing, 0)
Set myDimension = _
    Part.Parameter ("BevDia1@Sketch1@bevelgear.Part")
myDimension.SystemValue = 0.0889
boolstatus = Part.Extension.SelectByID2 _
    ("ShaftDia2@Sketch1@bevelgear2-1@BevelGearbox", _
    DIMENSION", 0.1098860309409, -0.1378177510392, _
    -0.07059706833881, False, 0, Nothing, 0)
Set myDimension = _
    Part.Parameter ("ShaftDia2@Sketch1@bevelgear2.Part")
myDimension.SystemValue = 0.015875
boolstatus = Part.Extension.SelectByID2 _
    ("BevDia2@Sketch1@bevelgear2-1@BevelGearbox", _
    "DIMENSION", 0.1606982503071, -0.04076881955329, _
    0.02588814510427, False, 0, Nothing, 0)
Set myDimension = _
    Part.Parameter ("BevDia2@Sketch1@bevelgear2.Part")
myDimension.SystemValue = 0.0635
boolstatus = Part.EditRebuild3 ()
Part.ClearSelection2 True
```

End Sub

VBA Code Structure

The structure of the macro code for a VBA project is very similar to the Visual Basic.NET format we saw in the first chapter. Its file structure is a little simpler. First, you will notice that there are no class definitions. VBA requires a separate code window for classes rather than being able to declare them in your code module. Also, there are no Imports statements. References to outside libraries can only be set by going to Tools, References in the VBA interface menu.

8. Verify references to the two major SOLIDWORKS libraries by selecting Tools, References. Make sure that *SldWorks 2023 Type Library* is selected as well as *SOLIDWORKS 2023 Constants Type Library*, or whichever major version you are using.

These two libraries, along with a few others, are selected by default when you record a VBA macro in much the same way as the two Imports statements reference similar libraries in a VSTA macro.

One negative aspect of a recorded VBA macro is that the declarations are not explicit with data types. For example, notice the first declaration of swApp as Object. In VSTA macros it is declared as SldWorks. Recall that Object is a generic data type that could hold anything. So the code compiler has to figure out what kind of object swApp is going to be based on how it is first used. This is referred to as late binding. VSTA macros are recorded with early binding, a better programming practice that uses exact type matching.

Another syntax difference is the Set statement. Whenever you are setting a variable to an interface or class in Visual Basic (specifically VBA), you must use the Set statement. This is not the case in Visual Basic.NET in VSTA macros.

Optimizing the Code

As you have seen, SOLIDWORKS will frequently record more code than a macro would require. Stepping through the code body can help determine which variables and lines can be removed.

The first is the use of the model view, declared with the variable myModelView. Until we explore optimizing the interface for speed, we won't be making use of the model view.

Each time you modify a dimension in SOLIDWORKS, the macro records the action of selecting the dimension, getting the dimension as well as changing its value. The selection of dimensions is not necessary. Keeping the selection code will not prevent the macro from running correctly, but it will cause unnecessary processing.

9. Optimize your macro code by deleting or commenting out each line of code that uses the Part.Extension.SelectByID2 method. You will also only need the variables swApp, Part and boolstatus declared in the general declarations section above Sub main().

10. Add the Option Explicit statement at the top of your code.

After deleting the extra code, your main procedure should look something like this.

```
Option Explicit
```

```
Dim swApp As Object
Dim Part As Object
Dim boolstatus As Boolean
Sub main()

Set swApp = Application.SldWorks

Set Part = swApp.ActiveDoc
Dim myDimension As Object
Set myDimension = Part.Parameter ("ShaftDia1@Sketch1@bevelgear.Part")
myDimension.SystemValue = 0.0254
Set myDimension = Part.Parameter ("BevDia1@Sketch1@bevelgear.Part")
myDimension.SystemValue = 0.0889
Set myDimension = Part.Parameter ("ShaftDia2@Sketch1@bevelgear2.Part")
myDimension.SystemValue = 0.015875
Set myDimension = Part.Parameter ("BevDia2@Sketch1@bevelgear2.Part")
myDimension.SystemValue = 0.0635
boolstatus = Part.EditRebuild3 ()
Part.ClearSelection2 True
End Sub
```

Code Description

Look over the steps of the main procedure. VBA macros do not have the hidden code that connects to the SOLIDWORKS application, so that is the first line. The active document is set to a variable using ActiveDoc. After attaching to SOLIDWORKS and the active document, an additional variable is declared named myDimension. It will be set to a specific dimension object prior to changing its value. The process of getting each dimension and then changing its value is repeated for all four dimensions you modified as you recorded the macro. First, the myDimension variable is set to a specific Parameter in the active document (the variable named Part in the recorded code).

Parameter

value = IModelDoc2.Parameter (stringIn)

Parameter is a method of the IModelDoc2 interface that returns an IDimension interface. In the recorded code, IModelDoc2 is represented in the macro by a variable named Part.

- **value** is a pointer to the dimension object specified by stringIn.

- **stringIn** is a string argument representing the full dimension name. Since the dimension was accessed at the assembly level, the part name must also be part of the dimension name as in "ShaftDia1@Sketch1@bevelgear.Part".

SystemValue

SystemValue is a property of the IDimension interface. It simply sets the parameter's value. The value must be passed as a double data type and is always in units of meters. If you check the API Help, you will notice that the SystemValue property is now obsolete. SystemValue does not take configurations into account. The SetSystemValue3 method allows you to make changes to a dimension at the configuration level. For simplicity, this macro will continue to use the obsolete SystemValue property. Sometimes a simpler, older API call can perform the desired operation without unnecessary detail.

Following the dimension changes are two additional lines of code that complete the macro prior to End Sub.

```
boolstatus = Part.EditRebuild3 ()
Part.ClearSelection2 True
```

EditRebuild3

EditRebuild3 is yet another method of IModelDoc2. You can probably guess what it does by its name. It rebuilds the model. It returns True if the rebuild was successful or False if it was not. There are no arguments to this method, so it is followed by empty parentheses.

ClearSelection2

You can again guess what this method of IModelDoc2 does. It clears out any current graphical selections. Since you have removed all references to selections using SelectByID2, this line is also unnecessary.

Using VBA in Excel

For sake of example, if you wanted to have the two shafts related to each other, it could be done with SOLIDWORKS equations. What if you wanted to determine if the first shaft is greater than a given size, then use one equation? If the first shaft is less than or equal to a given size, use a different equation. Not many users know how to do that with SOLIDWORKS equations (but it is possible). Microsoft Excel may be a more familiar place to create logic statements. Excel can also reference data from different sources to help define geometry and size. You can make Microsoft Excel and SOLIDWORKS talk to each other using their common VBA macro capabilities.

Excel Command Buttons

11. Open the macro enabled Microsoft Excel spreadsheet named *BevelGearbox.xlsm* from the file downloads referenced at the front of this book. You can create your own spreadsheet if you are working with your own assembly model.

The spreadsheet already has cells containing the values to be controlled. It will essentially be the user interface for the macro. A button can be added to your spreadsheet that will activate a macro inside Excel.

Microsoft Excel does not have the macro tools visible through the user interface by default and must be turned on to make them accessible.

12. In Excel, enable the Developer tab by clicking File, Options. Select the Customize Ribbon category on the left and turn on the Developer checkbox in the Main Tabs listing. Click OK. You will also need to enable macros by selecting Macro Security ⚠ from the Developer Tab. Change the macro settings to Disable all macros with notification. In the security warning message, select ⬛ Enable Content ⬛ to enable macro code in the spreadsheet.

(Hint: as a general rule, you should not enable all macros in Excel. The macro settings block potentially malicious macros from acting as viruses.)

13. From the DEVELOPER tab, select Insert, select the Button tool, then click in the worksheet window to insert a button.

14. Select New on the Assign Macro dialog. This will display the VBA interface for Excel with a new procedure for the new button's click event. Any code put into the new procedure will be run any time a user clicks on the button in the spreadsheet.

You can now copy the SOLIDWORKS macro you recorded into the VBA procedure for this button.

15. Switch back to the SOLIDWORKS VBA interface and copy all the code between, but not including, Sub main() and End Sub.

16. Switch back to the Excel VBA interface.

17. Paste your copied code above the End Sub line in the Excel macro.

This procedure already has a name defined by the name of the button that was created. When the button is clicked, the code in the procedure will be executed.

```
Private Sub Button1_Click()
Set swApp = Application.SldWorks
Set Part = swApp.ActiveDoc
Dim myDimension As Object
Set myDimension = Part.Parameter ("ShaftDia1@Sketch1@bevelgear.Part")
myDimension.SystemValue = 0.0254
Set myDimension = Part.Parameter ("BevDia1@Sketch1@bevelgear.Part")
myDimension.SystemValue = 0.0889
Set myDimension = Part.Parameter ("ShaftDia2@Sketch1@bevelgear2.Part")
myDimension.SystemValue = 0.015875
Set myDimension = Part.Parameter ("BevDia2@Sketch1@bevelgear2.Part")
myDimension.SystemValue = 0.0635
boolstatus = Part.EditRebuild3 ()
Part.ClearSelection2 True
End Sub
```

Excel.Range Method

The Excel API gives access to one or many cells or the active worksheet by using its Range method. It expects a string argument of the cell range. For a single cell, use the cell name like "A1". For several cells, the string should be the bounds like "A1:B10". By default, the Range method of a single cell returns its text value.

18. Modify your code as shown below to access the Excel values.

```
Private Sub Button1_Click()
```

```
Set swApp = Application.SldWorks
Set Part = swApp.ActiveDoc
Dim myDimension As Object
Set myDimension = Part.Parameter ("ShaftDia1@Sketch1@bevelgear.Part")
myDimension.SystemValue = Excel.Range("B1") * 0.0254
Set myDimension = Part.Parameter ("BevDia1@Sketch1@bevelgear.Part")
myDimension.SystemValue = Excel.Range("B2") * 0.0254
Set myDimension = Part.Parameter ("ShaftDia2@Sketch1@bevelgear2.Part")
myDimension.SystemValue = Excel.Range("B3") * 0.0254
Set myDimension = Part.Parameter ("BevDia2@Sketch1@bevelgear2.Part")
myDimension.SystemValue = Excel.Range("B4") * 0.0254
boolstatus = Part.EditRebuild3 ()
End Sub
```

The code retrieves specific cell values by using the Excel.Range method. The value is used as the input for dimension parameters. Adding a multiplier of 0.0254 will convert the inch value shown on the spreadsheet to meters. Divide by 1000 if entering values in millimeters on the worksheet. Remember that SOLIDWORKS API calls require an input value in meters for linear units.

GetObject

Another line of code must be changed before the macro will work using Excel's VBA environment. The line that connects to SOLIDWORKS using Application.SldWorks only works while in the SOLIDWORKS VBA environment. Since the code now runs in Excel's VBA interface, you must change this line to use the Visual Basic GetObject method to connect to another running application. The code below will only connect to SOLIDWORKS 2023(version 31). If you wanted to connect to a different version of SOLIDWORKS, change the application string. For example, "SldWorks.Application.28" would connect to SOLIDWORKS 2020 and so on.

19. Modify the code shown in bold type to declare swApp and Part and to connect to SOLIDWORKS 2021 using GetObject. Make sure to delete or comment out the line that references Set swApp = SldWorks.Application.

```
Private Sub Button1_Click()
Dim swApp As Object
Dim Part As Object
Set swApp = GetObject ( ,"SldWorks.Application.31")
'Set swApp = SldWorks.Application

Set Part = swApp.ActiveDoc
Dim myDimension As Object
Set myDimension = Part.Parameter ("ShaftDia1@Sketch1@bevelgear.Part")
myDimension.SystemValue = Excel.Range("B1") * 0.0254
Set myDimension = Part.Parameter ("BevDia1@Sketch1@bevelgear.Part")
myDimension.SystemValue = Excel.Range("B2") * 0.0254
Set myDimension = Part.Parameter ("ShaftDia2@Sketch1@bevelgear2.Part")
myDimension.SystemValue = Excel.Range("B3") * 0.0254
Set myDimension = Part.Parameter ("BevDia2@Sketch1@bevelgear2.Part")
myDimension.SystemValue = Excel.Range("B4") * 0.0254
```

```
boolstatus = Part.EditRebuild3 ()
End Sub
```

20. Select File, Close and Return to Microsoft Excel. While you are in Design Mode, you can edit the text and size of the button as needed. Right-click on the button to modify. Turn off Design Mode when finished by clicking the Design Mode button.

The Microsoft Excel spreadsheet is ready to control the SOLIDWORKS assembly. Try changing the values in the cells and then click the button. Because the macro reads the cell values, you can use Excel formulas and functions or additional worksheets to drive them. The options are practically limitless!

OLE Objects in SOLIDWORKS

Rather than using an external file to control the model, you can use OLE capability to copy and paste this Excel spreadsheet into our assembly.

21. Select all the required cells (A1 to C7 if your spreadsheet looks like the image below) and copy them. Switch to your SOLIDWORKS assembly and paste. You may need to zoom to fit to see the newly added spreadsheet.

Now your spreadsheet is embedded in your assembly, along with all of the macro code. The original spreadsheet is no longer needed.

22. To activate the embedded spreadsheet, double-click on it. Select Enable Macros in the Security Notice in Excel.

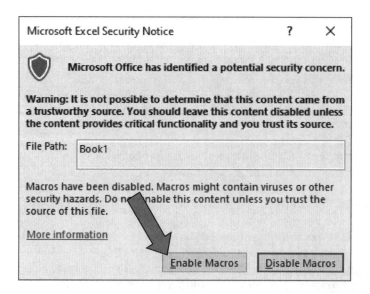

Change the cell values and click your command button to watch the assembly update.

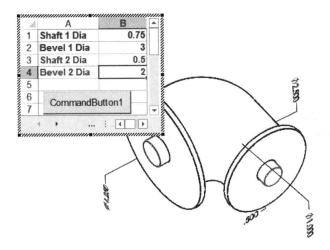

Selection Methods

Many SOLIDWORKS API methods require a selection before an operation, much in the same way a user would work. Once selections are made, a variety of methods can act on them. Selections can be made based on the object name, by its physical location in the graphics area, or after connecting a variable to an object's interface.

An easy place to start is to hide and show components. Record the steps in a macro to generate some preliminary code. The focus will be selecting components. This same method can be employed to select nearly all SOLIDWORKS objects.

23. Using the same SOLIDWORKS assembly, start recording another macro.

24. Right-click on the component named *bevel housing* from the FeatureManager and hide it by selecting Hide components ⌧.

25. Right-click on the *bevelgear* component from the graphics area and hide it as well.

26. Stop the macro and save it as *HidePart.swp* (SW VBA Macro).

27. Edit the macro that was just saved. The Sub main() procedure should look like the following.

```
Sub main()

Set swApp = Application.SldWorks

Set Part = swApp.ActiveDoc
boolstatus = Part.Extension.SelectByID2("bevel housing-1@BevelGearbox", _
   "COMPONENT", 0, 0, 0, False, 0, Nothing, 0)
Part.HideComponent2
Part.ClearSelection2 True
boolstatus = Part.Extension.SelectByRay(5.71499999999787E-02, _
   -1.86285395596997E-03, 3.36769657588434E-02, 0.376397825030699, _
   0.546609266460829, -0.748026060462678, 1.95371669784035E-03, _
```

```
  2, False, 0, 0)
Part.HideComponent2
Part.ClearSelection2 True
End Sub
```

SelectByID2 Method

The first component selection, done in the FeatureManager, uses the SelectByID2 method of IModelDocExtension. SelectByID2 is a common selection method and allows you to select objects by name or by x, y, z coordinate location.

It does have a limitation, however. If selecting by coordinate location, if another component is in front based on rotation, it can result in an unexpected selection. The following is a summary of its structure with more detail available in the API help.

value = IModelDocExtension.SelectByID2 (Name, Type, X, Y, Z, Append, Mark, Callout, SelectOption)

- **Name**, quite simply, is the name of the object as a string. Since you name most objects in SOLIDWORKS, this one should make perfect sense. However, there are some objects that SOLIDWORKS names for you. If an empty string is passed here (""), SOLIDWORKS assumes you are going to pass an x, y, z location to select.

- **Type** is similar to SOLIDWORKS selection filters. In the example, you selected a component. So the string passed is "COMPONENT." What if you wanted to select a feature by name? The string passed would be "BODYFEATURE." SOLIDWORKS refers to these as selection types. You can view a complete list of these selection types by searching the API help index for swSelectType_e.

- **X, Y,** and **Z** values are only necessary if you are selecting an object by its coordinate location. If the name of the object is passed a non-empty string, the three coordinate values are ignored.

- **Append** is a Boolean value. A value of True means the selection will be added to the existing selection as if you held down the Control key. A value of False will build a new selection and all other existing selections will be cleared. See the API Help for more details.

- **Mark, Callout and SelectOption** see the API Help for more details. These are not commonly modified in a recorded macro. The default value is often adequate. However, for selections that will be used for feature creation, you will often need a Mark value. This will define what the selection will be used for in the feature.

- **value** is a returned Boolean value. This can be helpful if you need to check if the user selected something prior to using a control. If the method returns False you can inform the user that something needs to be selected before they can continue.

IModelDocExtension

It is worth mentioning that IModelDocExtension is a result of a fundamental limitation of a class or interface. Each interface can only have a certain number of methods and properties before it is essentially full. The IModelDoc2 class or interface was full, but there were still additional methods and properties needed. So IModelDocExtension was created as a sub class. As a result, you will find methods and properties that reference the IModelDocExtension interface that relate back to operations you would typically perform on the model itself. There is no particular reason why a method or property relates to the extension. It is simply a matter of when the API call was implemented – before or after IModelDoc2 reached its capacity.

SelectByRay

The component selected in the graphics area used a different method, SelectByRay, also from IModelDocExtension. Ray selections are made by projecting a vector, starting at the tip of the mouse cursor, in a direction perpendicular (or normal) to the screen. The vector is given a very small radius and the first entity the ray contacts is selected. Imagine a cylinder being projected towards the face like the image below, but much smaller.

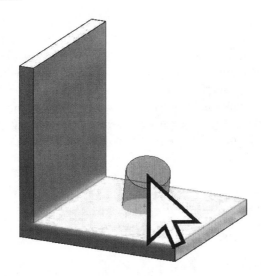

value = IModelDocExtension.SelectByRay(WorldX, WorldY, WorldZ, RayVecX, RayVecY, RayVecZ, RayRadius, TypeWanted, Append, Mark, Option)

- **WorldX**, **WorldY** and **WorldZ** represent the x, y and z coordinates of the end of a vector measured in the model's coordinate system. In a recording, it is the location of the tip of the mouse cursor projected onto the first visible face, edge or vertex.

- **RayVecX**, **RayVecY** and **RayVecZ** create the unit vector's direction, also in the model's coordinate system.

- **RayRadius** can be considered a small target zone at the tip of the mouse cursor. In the image above, it would be the radius of the cylinder. Notice the small recorded values that help avoid selection mistakes on playback.

- **TypeWanted** filters the selection type. These come from swSelectOption_e. Use 2 for faces, 1 for edges and 3 for vertices.

- **Append** is passed False to result in a single selection, clearing all previous selections. Pass True to add the selected entity to the existing selections (like holding down Ctrl).

- **Mark** is an integer used when an operation needs multiple selections. The API help outlines the required marks for each.

- **Option** comes from swSelectOption_e and imitates using the Shift key during selection.

SelectByID2 can also be used for selections from the graphics window, but it is based on view orientation. If the view changed from one run to another, SelectByID2 could fail to select anything or even select the wrong entity. SelectByRay provides a more robust selection for graphics objects.

HideComponent

The line following the SelectByID2 method simply hides the currently selected component(s).

```
Part.HideComponent2
```

The order of operations is critical. To hide a component in an assembly, you first select the component(s) to be hidden. You then use the HideComponent2 method of the assembly to perform the action. There is an alternative method to using HideComponent2 if you would rather not use a selection to drive the operation. You could get the IComponent2 interface first. The IComponent2 interface has a property named Visible that can be used to get or set the specific component's visibility. The question then becomes how to get IComponent2 for a specific part or sub assembly. This will be covered in a later chapter. A little searching through the API help may give you a solution right away if you are not patient enough to wait.

28. Remove the unnecessary code lines that reference the variable myModelView. There are three. Remember, there is no problem leaving them, but they are not required.

29. Copy the code between Sub main() and End Sub.

30. Switch back to SOLIDWORKS and activate the Excel spreadsheet again by double-clicking on it. (*Hint: your SOLIDWORKS toolbars will be overtaken by the Excel toolbars and interface temporarily.*)

31. Using the Excel commands in SOLIDWORKS, from the Developer tab, select Insert, Button.

32. Add another command button to your spreadsheet.

33. Select New from the Assign Macro dialog.

34. Once the Excel VBA interface is open and the new sub procedure is visible, paste in the copied code after the procedure call line as shown below. Make sure you are using your recorded code for the SelectByRay method and not the numeric values below.

```
Private Sub Button2_Click()

  Set swApp = Application.SldWorks
  Set Part = swApp.ActiveDoc
  boolstatus = Part.Extension.SelectByID2("bevel housing-1@BevelGearbox", _
    "COMPONENT", 0, 0, 0, False, 0, Nothing, 0)
  Part.HideComponent2
```

```
  Part.ClearSelection2 True
  boolstatus = Part.Extension.SelectByRay(5.71499E-02, _
    -1.862853E-03, 3.3676965E-02, 0.376397, _
    0.5466092, -0.7480260, 1.953716E-03, _
    2, False, 0, 0)
  Part.HideComponent2
  Part.ClearSelection2 True
End Sub
```

Just like the previously copied macro, the SOLIDWORKS application interface has to be connected using the general GetObject. Variables should be declared that have not been copied.

35. Add the declarations for swApp, Part and boolstatus and change the Application.SldWorks method to use GetObject.

```
Private Sub Button2_Click()
Dim swApp As Object
Dim Part As Object
Dim boolstatus As Boolean
Set swApp = GetObject( ,"SldWorks.Application.31")

Set Part = swApp.ActiveDoc
boolstatus = Part.Extension.SelectByID2("bevel housing-1@BevelGearbox", _
  "COMPONENT", 0, 0, 0, False, 0, Nothing, 0)
Part.HideComponent2
Part.ClearSelection2 True
boolstatus = Part.Extension.SelectByRay(5.714999999E-02, _
  -1.8628539E-03, 3.3676965E-02, 0.3763978, _
  0.54660926, -0.7480260, 1.95371669E-03, _
  2, False, 0, 0)
Part.HideComponent2
Part.ClearSelection2 True
End Sub
```

36. The new button will now hide the part named *bevel housing* using the SelectByID2 method and the part named *gearbox* using SelectByRay.

If you are looking for a way to hide any selected component, simply remove the lines of code that use SelectByID2 or SelectByRay. If the user pre-selects any components prior to activating the macro, they will be hidden using the same button.

Conclusion

You can tie almost any application together with Visual Basic as long as it has an API. It becomes a simple task to copy and paste code from one application to another after recording the desired operation. Understanding some of the basic API calls for each application will also help you make the connections. You can use this same strategy in any Microsoft Office application or tool that supports VBA or VSTA. You can also reverse the process and read and write to Excel from a SOLIDWORKS macro. You would need to add a reference to the Microsoft Excel Object library in the SOLIDWORKS macro and use GetObject to connect to a running instance of Excel. Use "Excel.Application" as the class name. There are many Visual Basic examples of communicating with Excel online.

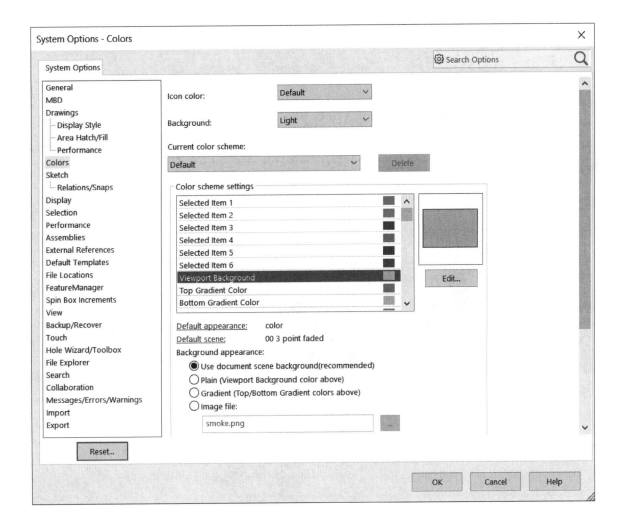

- **Getting and Setting Options**

- **SOLIDWORKS Interop Libraries**

Introduction

As you begin to build your own macros you may want to programmatically change system option and document properties as part of your code. For example, while automatically creating a part or drawing you might want to set the units to something other than the default setting of the template used. You may choose to alter the way the interface behaves for performance reasons. Or you may simply want to make it easier to toggle between background appearance, display of dimension names or grid settings. Through this chapter we will explore the different techniques involved in getting and setting these option settings using macros. As exercises you will build macros that toggle the display of the gradient background for parts and assemblies, increase and decrease the decimal precision display and toggle transparency of selected parts in an assembly.

Getting and Setting Options

This section will focus on getting and setting options from a new macro. When you build your own macros you may wish to record the desired settings first to make their modification easier. Recording can also make it easier to find the desired setting. We will look over a few of the option settings and how to access them. They can be broken into six basic categories.

> **Get/SetUserPreferenceToggle**
> **Get/SetUserPreferenceIntegerValue**
> **Get/SetUserPreferenceDoubleValue**
> **Get/SetUserPreferenceStringValue**
> **Get/SetUserPreferenceStringListValue**
> **Get/SetUserPreferenceTextFormat**

The method used depends on what type of setting you are trying to control and whether you are trying to get the value or set it. The following code will check the system options for the background appearance setting. It will then switch between Use document scene background and Plain (Viewport Background color above). (*Hint: Create a keyboard shortcut for this one so you can turn off the scene background while you are sketching or to capture a quick screen image with a nice clean white background.*)

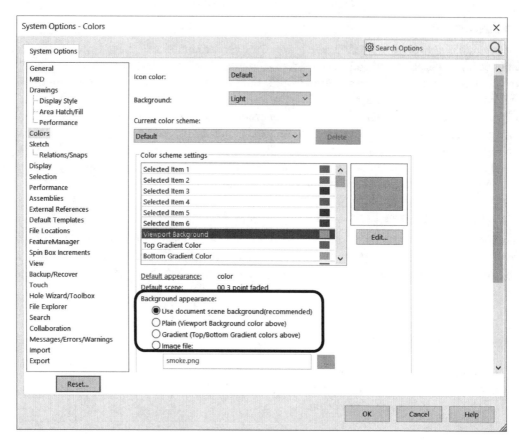

Background Appearance Macro

1. Create a new macro by selecting Tools, Macro, New or clicking ⬚.

2. Save the macro as a SW VSTA VB macro with the name *background.vbproj*.

3. Add the following code to the new macro.

```
Public Sub main()
  Dim result As Boolean
  Dim swModel As ModelDoc2 = Nothing
  swModel = swApp.ActiveDoc

  If swApp.GetUserPreferenceIntegerValue( _
  swUserPreferenceIntegerValue_e.swColorsBackgroundAppearance) = _
  swColorsBackgroundAppearance_e.swColorsBackgroundAppearance_DocumentScene _
  Then

    result = swApp.SetUserPreferenceIntegerValue( _
    swUserPreferenceIntegerValue_e.swColorsBackgroundAppearance, _
    swColorsBackgroundAppearance_e.swColorsBackgroundAppearance_Plain)

  Else
```

```
   result = swApp.SetUserPreferenceIntegerValue _
      (swUserPreferenceIntegerValue_e.swColorsBackgroundAppearance, _
      swColorsBackgroundAppearance_e. _
      swColorsBackgroundAppearance_DocumentScene)

   End If

   swModel.GrahicsRedraw2
End Sub
```

Early Binding

Notice how the variable swModel was declared as ModelDoc2, rather than declaring it as a general Object as was done in the VBA example. IModelDoc2 is a specific interface from the SolidWorks.Interop.SldWorks library. If you look lower in the class code, you will also notice that swApp is declared as SldWorks (or the ISldWorks interface). This is called Early Binding. Because the declaration is specific, the compiler does not have to guess its type.

In Visual Studio, Early Binding has a few direct benefits. First, your macros will run faster. Your application will know exactly what interface is being used and can quickly determine if your code is correct for that interface.

Second, the Visual Studio environment makes use of the Microsoft IntelliSense technology. As you type the period after swApp you will notice a list of possible methods and procedures for that interface become available. As you get to the call you want, simply hit the Space or Tab key to finish the word as shown in the image below.

As you continue to build the syntax of the call with a space or parenthesis, you will also see the structure of the call below the line in a tip box. As you become more comfortable with this technique, it can save you a lot of time typing and referencing the API help.

```
1 reference
Partial Class SolidWorksMacro
    2 references
    Public Sub main()
        Dim result As Boolean
        Dim swModel As ModelDoc2 = Nothing
        swModel = swApp.ActiveDoc

        If swApp.GetUserPreferenceIntegerValue()
            ISldWorks.GetUserPreferenceIntegerValue(UserPreferenceValue As Integer) As Integer
```

swApiHoleWizardItemImportStatus_e
swApiToolboxItemExportStatus_e
swApiToolboxItemImportStatus_e
swApp
swAppCallBackCmd_e
swAppearanceTargetType_e
swApplicationType_e
swAppNotify_e
swArcEndCondition_e

```
    End Sub
    ''' <summary>
    ''' The SldWorks swApp variable is pre-assi
    ''' </summary>
    Public swApp As SldWorks
End Class
```

If ... Then...Else Statements

After capturing the active document, a little Boolean logic is needed and an If Then statement works well. Rather than collecting the return value from GetUserPreferenceIntegerValue, storing it in a variable, and then using the variable for the logical test, the method is used directly in the logical test. You can build your code either way. If you need the value more than once, it is best practice to store it in a variable for re-use. Also consider the complexity of the logical test. Using a method inside the logic statement makes for code that is more challenging to read for you and other users.

Get/SetUserPreferenceIntegerValue

The two SOLIDWORKS API calls used in this example are GetUserPreferenceIntegerValue and SetUserPreferenceIntegerValue. The macro calls for an integer value because the setting in the options dialog is a series of radio button options. The structures of the two methods are quite simple.

value = ISldWorks.GetUserPreferenceIntegerValue (UserPrefValue)

- **UserPrefValue** is an integer value and represents which option setting to get.

- GetUserPreferenceIntegerValue has a return **value** of an integer. This integer represents which specific option out of a group of options is selected. It could also represent a numeric value for an option as long as the numeric value is an integer.

In this example the constant swColorsBackgroundAppearance is used in the argument for UserPrefValue. This constant is a member of the swUserPreferenceIntegerValue_e enumeration. This constant comes from the SolidWorks.Interop.swconst library. The library is referenced automatically whenever you create a new macro and is also available to IntelliSense. There will be more discussion about these constants later in the chapter.

return = ISldWorks.SetUserPreferenceIntegerValue (UserPreferenceValue, Value)

The structure for SetUserPreferenceIntegerValue is almost identical to the Get version with a couple meaningful differences. The first is that the second argument Value is used to tell SOLIDWORKS which option to set. The second is that the return value is simply True or False if the method was successful or not.

- **UserPreferenceValue** again requires a long value of which toggle is being set.

- **Value** requires an integer that represents which option to choose or what numeric value to set.

- **return** is a Boolean True or False if the method was successful or not.

SetUserPreferenceToggle

Here is another example of changing option settings. This time it will change the setting of a checkbox. Checkboxes in options are associated with GetUserPreferenceToggle and SetUserPreferenceToggle. A Boolean value is returned from the Get method indicating whether the checkbox is checked. The Set method requires you to pass a Boolean value as the second argument. This code may not seem extremely valuable right now, but it must be used any time you want to add dimensions in a macro. If you leave this common option setting on while dimensioning a sketch, the Modify dialog will pop up every time your macro adds a dimension. Not exactly efficient if you are trying to automate a process and your macro stops every time a dimension is placed.

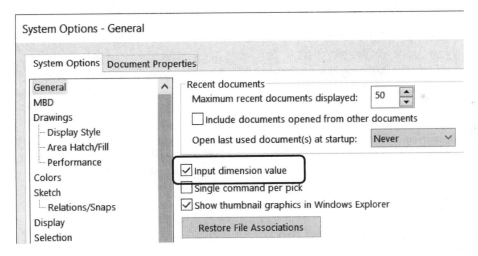

Turn Off Input Dimension Value

```
Public Sub main()

    Dim OldSetting As Boolean = False
    OldSetting = swApp.GetUserPreferenceToggle( _
        swUserPreferenceToggle_e.swInputDimValOnCreate)
    swApp.SetUserPreferenceToggle ( _
        swUserPreferenceToggle_e.swInputDimValOnCreate, _
        False)

End Sub
```

Setting System Options vs. Document Properties

There is a simple rule to follow when trying to distinguish between system options and document properties for SOLIDWORKS. System options are a method of the application interface – swApp in the example. Document properties are a method of the IModelDocExtension interface with a slight alteration to the structure. For example, if you wanted to get or set settings from the currently open document you could use the following code inside your main procedure in a VSTA macro.

Increase Decimal Places by One

```
Public Sub main()

Dim swDoc As ModelDoc2 = Nothing
Dim swDocExtension As ModelDocExtension = Nothing
swDoc = swApp.ActiveDoc
swDocExtension = swDoc.Extension
Dim CurrentSetting As Integer = 0
Dim NewSetting As Integer = 0
CurrentSetting = swDocExtension.GetUserPreferenceInteger _
  (swUserPreferenceIntegerValue_e.swUnitsLinearDecimalPlaces, _
  swUserPreferenceOption_e.swDetailingDimension)

NewSetting = CurrentSetting + 1

swDocExtension.SetUserPreferenceInteger _
  (swUserPreferenceIntegerValue_e.swUnitsLinearDecimalPlaces, _
  swUserPreferenceOption_e.swDetailingDimension, NewSetting)

End Sub
```

Running this macro will increase the dimension display precision of the current document by one place. It is a handy macro to assign to your "+" key to effortlessly increase the number of decimal places shown on your dimensions. It would be smart to have the "-" key decrease the decimal places. The sister to this macro would be the following.

Decrease Decimal Places by One

```
Sub main()

Dim swDoc As ModelDoc2 = Nothing
Dim swDocExtension As ModelDocExtension = Nothing
swDoc = swApp.ActiveDoc
swDocExtension = swDoc.Extension
Dim CurrentSetting As Integer = 0
Dim NewSetting As Integer = 0
CurrentSetting = swDocExtension.GetUserPreferenceInteger _
  (swUserPreferenceIntegerValue_e.swUnitsLinearDecimalPlaces, _
  swUserPreferenceOption_e.swDetailingDimension)
```

```
NewSetting = CurrentSetting - 1

swDocExtension.SetUserPreferenceInteger _
  (swUserPreferenceIntegerValue_e.swUnitsLinearDecimalPlaces, _
  swUserPreferenceOption_e.swDetailingDimension, NewSetting)

End Sub
```

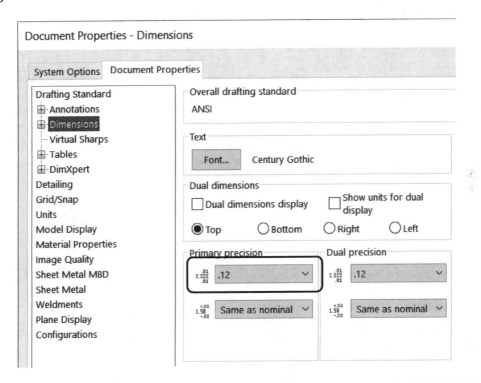

In both of these macros you will notice the setting is an integer setting again. In SOLIDWORKS, click select Tools, Options, and then select the Document Properties tab. Select the Dimensions category. Notice that the setting for Primary precision is a list. The integer setting is used for lists as well.

GetUserPreferenceInteger

There is a slight structural difference for document settings. Since there are now individual settings for each dimension type, the decimal precision can be set generally, or for each dimension separately. The previous code uses the `swDetailingDimension` option to set the general setting. If you wanted to set the linear dimension setting only, you would use `swDetailingLinearDimension` instead. *(Hint: Use IntelliSense to view all of the different options and settings that are available.)*

SOLIDWORKS Constants

SOLIDWORKS constant enumerations (categorized lists of constants represented by integers) are used in many API calls beyond system options and document properties. We briefly discussed the use of constants earlier in the chapter. They are critical to getting or setting the right option.

Object Browser

Once your project has a reference to the SolidWorks.Interop.swconst library, you can browse or search the library using the Object Browser.

4. To access the Object Browser select View, Object Browser or click 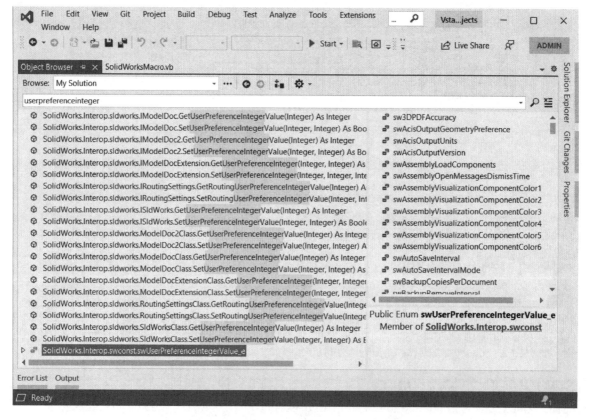 in Visual Studio.

5. Type "userpreferenceinteger" in the search text and click search 🔍 or Enter.

6. Select the enumeration, shown in the search results with the following symbol ⬤.

You will be presented with a list of all of the constants (⬤) to be used with GetUserPreferenceInteger and SetUserPreferenceInteger. To reference a specific constant, first enter the name of the enumeration and then type a period to activate IntelliSense. The names are often descriptive enough to choose from. Be forewarned, it may take some experimentation to determine the exact constant to drive the desired option setting. There are several that have very similar names and are sometimes difficult to differentiate without trying them. They are generally named so that they are grouped by what they change. They are named starting with "sw" followed by the object or setting they change.

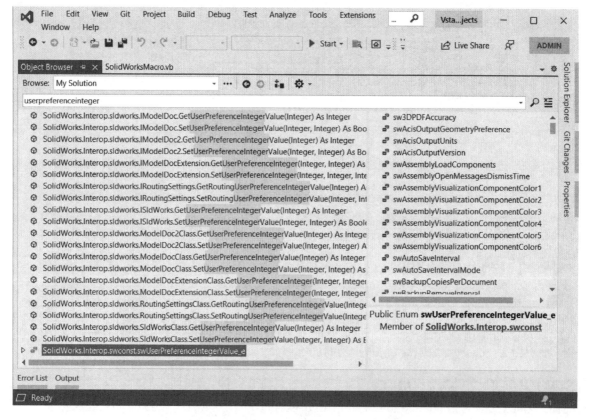

Conclusion

Effectively changing options through code will enhance any automated process. Use them frequently as core development tools to build applications. Do not forget to change them back to where they were when your program started. That means always getting the current setting and storing it to a variable first. There is nothing more frustrating than having a macro or program that changes your system settings and then does not set them back.

C# Example (background)

```csharp
public void toggleBackgroundAppearance()
{
    Boolean result;
    ModelDoc2 swModel;
    swModel = ((ModelDoc2)(swApp.ActiveDoc));
    int bgd = swApp.GetUserPreferenceIntegerValue((int)
      swUserPreferenceIntegerValue_e.swColorsBackgroundAppearance);

    if(bgd == (int)swColorsBackgroundAppearance_e
       .swColorsBackgroundAppearance_DocumentScene)
    {
        result = swApp.SetUserPreferenceIntegerValue((int)
           swUserPreferenceIntegerValue_e.swColorsBackgroundAppearance,
           (int)swColorsBackgroundAppearance_e
           .swColorsBackgroundAppearance_DocumentScene);
    }
    swModel.GraphicsRedraw2();
    return;
}

public void toggleInputDimVal()
{
    Boolean OldSetting = false;
    OldSetting =
swApp.GetUserPreferenceToggle((int)swUserPreferenceToggle_e.swInputDimValOnCr
eate);

swApp.SetUserPreferenceToggle((int)swUserPreferenceToggle_e.swInputDimValOnCr
eate, !OldSetting);
    return;
}

public void increaseDecimalPlacesByOne()
{
    ModelDoc2 swDoc = null;
    ModelDocExtension swDocExtension = null;
    swDoc = (ModelDoc2)swApp.ActiveDoc;
    swDocExtension = (ModelDocExtension)swDoc.Extension;
    int CurrentSetting = 0;
    int NewSetting = 0;
    CurrentSetting = (int)swDocExtension.GetUserPreferenceInteger
        ((int)swUserPreferenceIntegerValue_e.swUnitsLinearDecimalPlaces,
        (int)swUserPreferenceOption_e.swDetailingDimension);
    NewSetting = CurrentSetting + 1;
```

```
    swDocExtension.SetUserPreferenceInteger
        ((int)swUserPreferenceIntegerValue_e.swUnitsLinearDecimalPlaces,
        (int)swUserPreferenceOption_e.swDetailingDimension, NewSetting);

    return;
}

public void decreaseDecimalPlacesByOne()
{
    ModelDoc2 swDoc = null;
    ModelDocExtension swDocExtension = null;
    swDoc = (ModelDoc2)swApp.ActiveDoc;
    swDocExtension = (ModelDocExtension)swDoc.Extension;
    int CurrentSetting = 0;
    int NewSetting = 0;
    CurrentSetting = (int)swDocExtension.GetUserPreferenceInteger
        ((int)swUserPreferenceIntegerValue_e.swUnitsLinearDecimalPlaces,
        (int)swUserPreferenceOption_e.swDetailingDimension);
    NewSetting = CurrentSetting - 1;

    swDocExtension.SetUserPreferenceInteger
        ((int)swUserPreferenceIntegerValue_e.swUnitsLinearDecimalPlaces,
        (int)swUserPreferenceOption_e.swDetailingDimension, NewSetting);

    return;
}
```

Material Properties

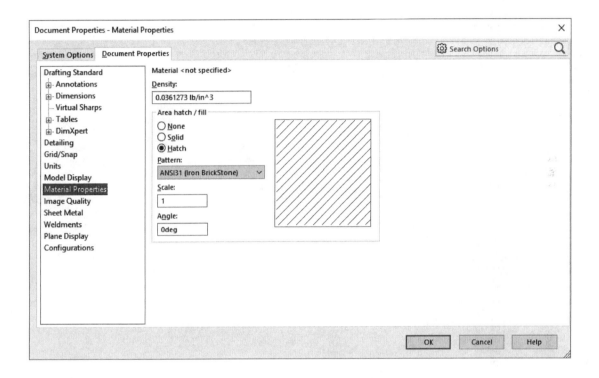

- **Basic Material Properties**

- **Adding Forms**

- **Arrays**

- **Working with Assemblies**

- **Selection Manager**

- **Verification and Error Handling**

Introduction

This exercise is designed to introduce model materials settings along with several Visual Basic.NET language tips. It will also examine a method for parsing through an assembly tree to change some or all of the parts in the assembly.

As an additional preface to this chapter, remember that SOLIDWORKS sometimes does things better than your macros might. For example, you can select multiple parts at the assembly level and set their materials at once. This chapter was originally written before you could do so. Use this chapter to better understand the tools available through Visual Basic.NET and the SOLIDWORKS API rather than as a functional tool. No matter how clever you get with your macros, at some point someone else might come up with the same idea. In the perfect world, you can and should obsolete your macros as SOLIDWORKS adds the functionality to the core software.

Part 1: Basic Material Properties

The first part will walk through creation of a tool that allows the user to select a material from a pull-down list or combo box. When they click an OK button, the macro should apply the selected material to the active part. The user should be able to cancel the operation as well.

We could take the approach of recording the initial macro, but the code for changing materials is simple enough that we will build it from scratch in this example.

User Forms

Most applications and macros need user interaction at some point. It might be a question of what to do next – like the message box control in the first chapter asking a yes or no question. More commonly, developers organize user input in a custom dialog box or form. These allow users to input text, click buttons to activate procedures, and select options. This example builds a form that looks like the one shown using a drop down list or ComboBox and two buttons.

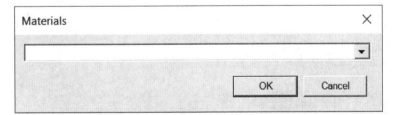

1. Start a new SW VSTA VB macro and save it as *materials.vbproj*.

2. Add a form to your macro by selecting Project, Add Form (Windows Forms).

3. Choose the Dialog template from Common Items and click Add.

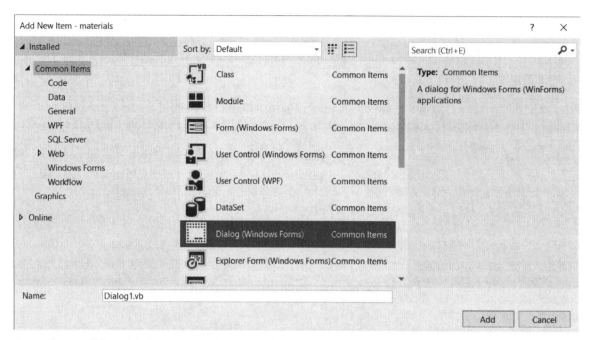

A new form will be added to your project named *Dialog1.vb* and will be opened for editing. The Dialog template has two pre-defined Button controls for OK and Cancel. To add additional controls to the form you will need to access the controls Toolbox.

4. Enable the Toolbox by selecting View, Toolbox (Ctrl+Alt+X). It will display on the left pane in Visual Studio.

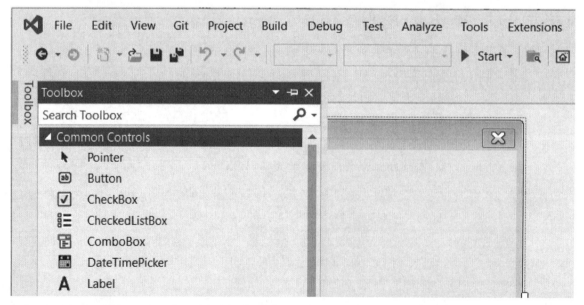

A ComboBox control needs to be added to the form.

5. Click on the Toolbox and optionally select the pushpin to keep it visible as you build your form.

6. Expand the Common Controls group and drag and drop the ComboBox control 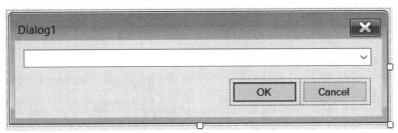 from the Toolbox onto your form.

After adding the ComboBox and resizing, your form should look something like the following. An effective form is one that is compact enough to not be intrusive while still being easy to read and use.

Object Properties

Each form control has properties you can change to affect its visual display as well as its behavior. The properties panel is visible on the right side of VSTA under the Solution Explorer.

7. Select the OK button on the form. The Properties window on the right side of the screen will list the control's properties.

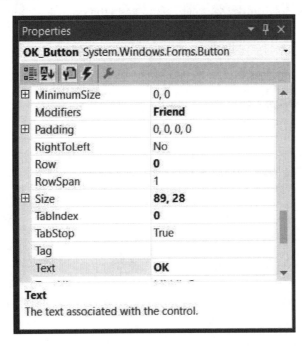

8. Review the following properties that were pre-defined by the use of the Dialog template.

- Text = OK. This is the text that is visible to the user. Use an ampersand (&) before a character to assign the Alt-key shortcut for the control. Entering &OK defines Alt-O as the shortcut. OK

- (Name) = OK_Button. This name is what your code must reference to respond to the button or to change its properties while your macro is running.

9. Select the Cancel button and review its Text and Name properties as well.

10. Click anywhere inside the form, not on a control, and change its Text property to "Materials." This changes the title shown on the form.

11. Review the following additional properties of the form.

- AcceptButton = OK_Button

- CancelButton = Cancel_Button

⊟ **Misc**	
AcceptButton	**OK_Button**
CancelButton	**Cancel_Button**
KeyPreview	False

12. Select the ComboBox and set its Name to "MaterialsCombo."

Show the Dialog

The macro should display the form at the right time. If you run the macro right now, it will not do anything since the main procedure is empty.

13. Switch back to the SolidWorksMacro.vb tab and add the following code in your main procedure.

```
Sub main()

    'Initialize the dialog
    Dim MyMaterials As New Dialog1
    Dim MyCombo As Windows.Forms.ComboBox
    MyCombo = MyMaterials.MaterialsCombo

    'Set up materials list
    Dim MyProps(2) As String

    MyProps(0) = "Alloy Steel"
    MyProps(1) = "6061 Alloy"
    MyProps(2) = "ABS PC"

    MyCombo.Items.AddRange(MyProps)

    MyMaterials.ShowDialog()
End Sub
```

14. Start the macro and test.

You should see your new dialog box. This is a result of MyMaterials.ShowDialog() at the bottom of the procedure. Every user form has a ShowDialog method that makes the form visible to the user and returns the user's action. ShowDialog also pauses code execution of the current procedure and passes control to the form until it is closed. So any code written after ShowDialog will not run until the form closes.

Materials ☒

> Alloy Steel
> 6061 Alloy
> ABS PC

If you click on the combo box, you should see Alloy Steel, 6061 Alloy and ABS PC listed.

15. Close the running macro and return to Visual Studio. If needed, click Stop Debugging ■.

There are a few steps required to make a form or dialog visible. The first is to declare a variable named MyDialog as a new instance of the Dialog1 class. Even though you have created a dialog in the project, it is not created or used at run time until you reference it. It's worth mentioning that the name of the class does not always have to match the name of the file as it does in this example.

16. Review the code behind Dialog1.vb by right-clicking on it in the Solution Explorer and selecting View Code.

```
Imports System.Windows.Forms

Public Class Dialog1

    Private Sub OK_Button_Click(ByVal sender …
        [Additional code here]

    End Sub
    ...
End Class
```

Notice that the code in the form itself is declared as a public class named `Dialog1`. You can change the name of the class without changing the name of the vb code file. In fact, a single code file can contain many classes, though it makes it more difficult to manage and reuse.

17. Switch back to the *SolidWorksMacro.vb* tab to return to the main procedure.

The first time you use any class, whether it be code or a dialog, you must use the New keyword before you can reference it. This is distinctly different than Visual Basic (VBA). VBA creates new instances of forms and dialogs as soon as they are referenced. It may seem that the .NET method of referencing forms is a little more complicated, but it has real benefits.

Windows.Forms Namespace

A variable named MyCombo was declared as Windows.Forms.ComboBox and was set to the MaterialsCombo control from the instance of the form named MyMaterials. The declaration defines the namespace or library a control comes from. The ComboBox class is a child of the Forms namespace which is a child of the Windows namespace. To add another level of complexity, the Windows namespace is a member of the System namespace which has already been referenced by the Imports statement at the top of the code window. Think of namespaces as libraries of pre-build elements.

Since the macro will reference several components from the Windows.Forms namespace, it will make the code less wordy to import that namespace.

18. Add the following Imports statement to the top of the code window to reference the namespace. Notice that this same imports statement was automatically added to the *Dialog1.vb* code.

```
Imports SOLIDWORKS.Interop.sldworks
Imports SOLIDWORKS.Interop.swconst
Imports System.Runtime.InteropServices
Imports System
Imports System.Windows.Forms
```

Now the declaration of MyCombo can be simplified as follows.

```
Dim MyCombo As ComboBox
```

Now that there is a reference to the ComboBox control, it is populated with an array of values.

Arrays

An array is simply an ordered list of values. The general syntax to declare an array is Dim variablename(x) As type.

MyProps(2) was declared as a string data type. In other words, you made room for three text elements in that one variable. "Wait! I thought you declared two?" Arrays count from a zero element, so a size of 2 gives room for 3 values.

Arrays can also be declared and populated with values in one line. The original code should help you understand an array by element. But you could replace this code...

```
Dim MyProps(2) As String

MyProps(0) = "Alloy Steel"
MyProps(1) = "6061 Alloy"
MyProps(2) = "ABS PC"
```

With this single line...

```
Dim MyProps() As String = {"Alloy Steel", "6061 Alloy", "ABS PC"}
```

The difference is that the array size cannot be explicitly declared since it will be defined by the array enclosed in brackets { }.

ComboBox.Items.AddRange Method

To populate the combo box with the array, you must tell the macro where to put them. Typing MyCombo.Items.AddRange(MyProps) tells the procedure that you want to populate the items (or list) of the MyCombo control with the values in the MyProps array by using the AddRange method. The

ComboBox control automatically creates a row for each element in the array. If you wanted to add items one-at-a-time rather than en masse, you could use the Add method of the Items property.

DialogResult

After a user has selected the desired material, it should be applied to the active part when clicking OK. If the user clicks Cancel, we would expect the macro to close without doing anything. At this point, either button simply continues running the remaining code in the main procedure – which is nothing.

19. Modify the main procedure to add processing of the DialogResult.

```
...
MyProps(0) = "Alloy Steel"
MyProps(1) = "6061 Alloy"
MyProps(2) = "ABS PC"
MyCombo.Items.AddRange (MyProps)

Dim Result As DialogResult
Result = MyMaterials.ShowDialog ()

If Result = DialogResult.OK Then
    'Assign the material to the part

End If
End Sub
```

The ShowDialog method of a form will return a value from the System.Windows.Forms.DialogResult enumeration once the form is closed or dismissed. Any code after ShowDialog will then be run. Since we have added the Imports System.Windows.Forms statement in this code window, the code can be simplified by declaring Result as DialogResult. You probably noticed that when you typed If Result = , IntelliSense immediately gave you the logical choices for all typical dialog results.

As a result of the If statement, when the user chooses Cancel, the main procedure will find the If statement False, skip the inner code, and run to the end.

Setting Part Materials

It is time to modify SOLIDWORKS materials. The next step will be to set the material based on the material name chosen in the drop down.

20. Add the code inside the If statement to set material properties as follows.

```
If Result = DialogResult.OK Then
    'Assign the material to the part
    Dim Part As PartDoc = Nothing
    Part = swApp.ActiveDoc
    Part.SetMaterialPropertyName2 ("Default",
    "SOLIDWORKS Materials.sldmat", MyCombo.Text)
End If
```

IPartDoc Interface

Notice the new declaration of the Part variable as IPartDoc rather than ModelDoc2. Think of IModelDoc2 as a container that can be used for all SOLIDWORKS files. It could be a part, an assembly or a drawing. There are many operations that are common across all file types in SOLIDWORKS such as adding a sketch, printing and saving. However, there are some operations that are specific to a file type. Material settings, for example, are only applied at the part level. Mates are only added at the assembly level. Views are only added to drawings. Since we are calling a function specific to a part, the IPartDoc interface is the appropriate reference. The challenge is that the ActiveDoc method returns an IModelDoc2 interface which could be an IPartDoc, an IAssemblyDoc or an IDrawingDoc. They are somewhat interchangeable. However, it is good practice to be explicit when you are trying to call a function that is unique to the file type. Explicit declaration also enables the correct IntelliSense information, making coding easier.

IPartDoc.SetMaterialPropertyName2 Method

The simplest way to set material properties is using the SOLIDWORKS materials library. SetMaterialPropertyName2 is a method of the IPartDoc interface and sets the material based on its configuration and a specific database.

IPartDoc.SetMaterialPropertyName2 (ConfigName, Database, Name)

- **ConfigName** is the name of the configuration to be used. Pass the name of a specific configuration as a string or use "" (an empty string) to set the material for the active configuration.
- **Database** is the path to the material database to use, such as *SOLIDWORKS Materials.sldmat*. If you enter "" (an empty string), it uses the default SOLIDWORKS material library. Use a fully qualified path if the library isn't working as expected.
- **Name** is the name of the material as it displays in the material library. If you misspell the material, nothing will be applied.

The macro is now fully operational. Try it out on any part.

Part 2: Working with Assemblies

You can now extend the functionality of this macro to assemblies.

Is the Active Document an Assembly?

To make this code universal for parts and assemblies, we need to know what document type is active. If the active document is an assembly, we need to do something to the selected component. If it is a part, we run the code we already have.

21. Add an If statement to check the active document type. The previous code has been moved inside this If statement (not bold).

```
If Result = DialogResult.OK Then
    Dim Model As ModelDoc2 = swApp.ActiveDoc
    If Model.GetType = swDocumentTypes_e.swDocPART Then
        'Assign the material to the part
```

```
    Dim Part As PartDoc = Model
    'Part = swApp.ActiveDoc
    Part.SetMaterialPropertyName2 ("Default", _
        "SOLIDWORKS Materials.sldmat", MyCombo.Text)
ElseIf Model.GetType = swDocumentTypes_e.swDocASSEMBLY Then
    Dim Assy As AssemblyDoc = Model
    'set materials on selected components

  End If
End If
```

Notice the interchange between ModelDoc2, IPartDoc and IAssemblyDoc. The declaration of Model has also been simplified. Rather than initializing the variable to Nothing like the previous examples, it is initialized directly to swApp.ActiveDoc. This is a shorthand way to declare the variable and set its value. When Part and Assy are declared, they are initialized to Model which is still a reference to the active document. However, since they are declared explicitly as IPartDoc and IAssemblyDoc, they inherit the document type specific capabilities of parts and assemblies.

Also, notice the use of the IModelDoc2.GetType method. GetType is used to return the type of ModelDoc that is currently active. This test is crucial before attempting to deal with specific IPartDoc and IAssemblyDoc methods. For example, if you use the general ModelDoc2 declaration and attach to the active document, and it is a part, any attempt to call an assembly API like AddMate will cause an exception or crash. The enumeration swDocumentTypes_e lists the possible types. When you typed in the code, you should have noticed the different document types show up in the IntelliSense pop-up.

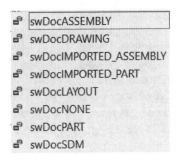

Selection Manager

In the Model Dimensions exercise we discussed basic selection methods. However, the code had to be specific. We had to pass the name of the component or a selection location, but that gets restrictive if you expect a user to interact with your macro. To get around those limitations you can employ a pre-selection method that is similar to most SOLIDWORKS features. You can require the user to pre-select the components he wishes to change prior to running the macro. Then write your macro to operate on each item the user selects. The Selection Manager interface makes this easy.

Connecting to the ISelectionMgr interface is similar to getting the IPartDoc (called Part). The Selection Manager is a child of IModelDoc2.

22. Add the following code inside the assembly section of the If statement to declare the Selection Manager and to attach to it.

```
ElseIf Model.GetType = swDocumentTypes_e.swDocASSEMBLY Then
    Dim Assy As AssemblyDoc = Model
    'set materials on selected components
    Dim SelMgr As SelectionMgr
    SelMgr = Model.SelectionManager

End If
...
```

From the Selection Manager, you can get to the selected object count, type, or even the xyz point in space where the object was selected. In this macro you will need to access the selected object count (number of items selected), and get to the components that were selected. Remember that components in SOLIDWORKS can be either parts or assemblies. Since we can only set density for parts, we will need to make sure the item selected is a part. For each item in the Selection Manager, get to IModelDoc2 and then set its material if it is a part.

23. Add the following code to set the material to all selected components.

```
ElseIf Model.GetType = swDocumentTypes_e.swDocASSEMBLY Then
    Dim Assy As AssemblyDoc = Model
    'set materials on selected components
    Dim SelMgr As SelectionMgr
    SelMgr = Model.SelectionManager

    Dim Comp As Component2
    Dim compModel As ModelDoc2
    For i As Integer = 1 To _
    SelMgr.GetSelectedObjectCount2 (-1)
        Comp = SelMgr.GetSelectedObjectsComponent4 (i, -1)
        compModel = Comp.GetModelDoc2
        If compModel.GetType = swDocumentTypes_e.swDocPART Then
            compModel.SetMaterialPropertyName2 ("Default", _
                "SOLIDWORKS Materials.sldmat", MyCombo.Text)
        End If
    Next
End If
...
```

For ... Next Statements and Loops

As was mentioned earlier, you want to set the material properties for each part that was selected by the user. What if the user has selected 500 parts? You certainly do not want to write 500 lines of code for each item selected. In many cases you will want to apply the same action to a variable number of items.

For ... Next statements allow you to repeat a section of code over as many iterations as you want. You just have to know how many times to loop through the code if you use a For ... Next statement.

```
For I As Integer = 0 To 10
    MsgBox ("You have clicked OK " & I & " times!")
Next I
```

Add this sample code to a procedure and then run. You get a message box stating how many times you have clicked OK. That is great if you know how many times the loop needs to process. In the macro, you do not know how many times to repeat the loop because you do not know how many parts the user might select. You can use ISelectionManager to help.

ISelectionMgr.GetSelectedObjectCount2

The number of selected items is retrieved by using GetSelectedObjectCount2. The argument passed is related to a selection Mark. A value of -1 indicates all selections will be counted, regardless of Mark. See the API Help for more information on marks. They're critical for features that require several distinct selection sets.

```
For i = 1 To SelMgr.GetSelectedObjectCount2(-1)
    '(loop code here)
Next i
```

The For loop starts with an initial i value of 1 so that it will only loop if the number of selected items is greater than zero. If nothing is selected, the method returns 0.

GetSelectedObjectsComponent4

The next step is to get the ModelDoc for each of the selected items. It requires a two-step process. The first gets the IComponent2 interface through the selection manager's GetSelectedObjectsComponent4 (*item number, Mark*) method. The Mark argument is again -1 to get the component regardless of selection Mark. The underlying ModelDoc is retrieved from IComponent2. Notice the declarations for compModel and Comp. They are specific to the type of SOLIDWORKS object we are accessing.

GetModelDoc2

IComponent2.GetModelDoc2 gives access to the underlying IModelDoc2. No arguments are required.

Now that you have the ModelDoc, you can use the same code from the part section to set the material properties after checking its type for parts.

Component vs. ModelDoc

If you have been wondering why we have to take the time to dig down to the IModelDoc2 interface of the Component, this discussion is for you. If not, and it all makes perfect sense, move on to the next subject.

Think of it this way – an IComponent2 interface understands information about the IModelDoc2 it references. It knows which configuration is showing, which instance it is, if it is hidden or suppressed, and even the component's rotational and translational position in the assembly. All of these can be accessed and changed using the IComponent2 interface. However, if you want to change something specific to the underlying part such as its material density, or to the underlying assembly such as its custom properties, then you must take the extra step of getting to the IModelDoc2 interface.

Verification and Error Handling

It's always a good practice to check the user's interactions to make sure they have done what you expected. After all, your macro may not have a user's guide. And even if it does, how many people really read that stuff? If you're reading this, you probably would. What about the other 95% of the population?

You should make sure the user is doing what you expect. First, define some criteria.

- Is the user in an assembly? The user must be in an assembly in our example to use the GetSelectedObjectsComponent4 method.

- If the active document is an assembly, has the user pre-selected at least one part? If not, they may assume they are applying material properties while nothing happens.

- Has the user selected items other than parts? If they select a plane or sketch, the macro may generate an exception or crash because there is no IModelDoc2 interface.

- Does the user even have a file open?

The only conditions left untested are the number of selections and if there is an active document.

24. Add the following immediately following the declaration of Model to check for an active document.

```
...
Dim Model As ModelDoc2 = swApp.ActiveDoc
If Model Is Nothing Then
  MsgBox ("You must first open a file.", MsgBoxStyle.Exclamation)
  Exit Sub
End If
...
```

25. Add the following to verify that the user has selected something in an assembly.

```
...
Dim Comp As Component2
Dim compModel As ModelDoc2
If SelMgr.GetSelectedObjectCount2 (-1) < 1 Then
  MsgBox ("You must select at least one component.", MsgBoxStyle.Exclamation)
  Exit Sub
End If
For i As Integer = 1 To SelMgr.GetSelectedObjectCount2 (-1)
  Comp = SelMgr.GetSelectedObjectsComponent3(i, -1)
...
```

If ... Then...Else Statements

If the active document is an assembly, you should check if the user has selected at least one component before continuing. Check GetSelectedObjectCount2 for a value less than one. If it is less than one the user has failed to select anything.

MsgBox

Make use of the Visual Basic MsgBox to give the user feedback. If you didn't already go through the Fundamentals chapter, this pre-defined dialog has an OK button by default, but it can have Yes and No buttons, OK and Cancel, or other combinations. If you use anything besides the default you can use the return value to determine which button the user selected, similar to using ShowDialog on the form.

The macro now gives the user better feedback. It makes good programming sense and is worth the extra effort to build good error handling into your macros. Users tend to quickly get frustrated when a tool crashes or generates undesired results.

26. Start the macro and test with both parts and selected components in assemblies. Verify the results by viewing the part and component assemblies.
27. Close Visual Studio.

Conclusion

Windows Forms are easy to design and are highly customizable. Get creative and ask for user feedback as you develop your own tools. The Selection Manager will also help you process most user selections. Finally, explore and experiment with IComponent2, IModelDoc2, IPartDoc, IAssemblyDoc and IDrawingDoc and their unique methods and properties, as well as their relationship to each other.

C# Example

```csharp
public void main()
{
    // Initialize the dialog
    var MyMaterials = new Dialog1();
    ComboBox MyCombo;
    MyCombo = MyMaterials.MaterialsCombo;

    // Set up materials list
    var MyProps = new string[] {"Alloy Steel", "6061 Alloy", "ABS PC"};
    MyCombo.Items.AddRange(MyProps);
    DialogResult Result;
    Result = MyMaterials.ShowDialog();
    if (Result == DialogResult.OK)
    {
        ModelDoc2 Model = (ModelDoc2)swApp.ActiveDoc;
        if (Model == null)
        {
            MessageBox.Show("You must first open a file.",
                "Materials", MessageBoxButtons.OK, MessageBoxIcon.Warning);
            return;
        }
```

```
if ((int)Model.GetType() == (int)swDocumentTypes_e.swDocPART)
{
    // Assign the material to the part
    PartDoc Part = (PartDoc)Model;
    // Part = swApp.ActiveDoc
    Part.SetMaterialPropertyName2("Default",
      "SOLIDWORKS Materials.sldmat", MyCombo.Text);
}
else if ((int)Model.GetType() ==
  (int)swDocumentTypes_e.swDocASSEMBLY)
{
    AssemblyDoc Assy = (AssemblyDoc)Model;
    // set materials on selected components
    SelectionMgr SelMgr;
    SelMgr = (SelectionMgr)Model.SelectionManager;
    Component2 Comp;
    ModelDoc2 compModel;
    if (SelMgr.GetSelectedObjectCount2(-1) < 1)
    {
        MessageBox.Show("You must select at least one component.",
          "Materials", MessageBoxButtons.OK, MessageBoxIcon.Warning);
        return;
    }

    for (int i = 1, loopTo = SelMgr.GetSelectedObjectCount2(-1);
      i <= loopTo; i++)
    {
        Comp = (Component2)SelMgr
          .GetSelectedObjectsComponent4(i, -1);
        compModel = (ModelDoc2)Comp.GetModelDoc2();
        if (compModel.GetType() == (int)swDocumentTypes_e.swDocPART)
        {
            PartDoc swPart = (PartDoc)compModel;
            swPart.SetMaterialPropertyName2("Default",
              "SOLIDWORKS Materials.sldmat", MyCombo.Text);
        }
    }
}
}
}
```

Custom Properties

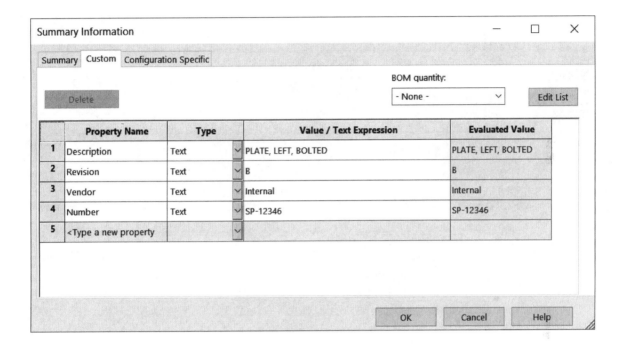

- **Setting Properties**

- **Modifying Properties**

- **Delete Properties**

- **Form Interactions**

- **Save and Copy Between Files**

Introduction

One of the most frequently automated areas of SOLIDWORKS is the custom property. They provide the basic storage for metadata or text in a SOLIDWORKS file. Until SOLIDWORKS developed the Property Tab Builder, users had to type in most custom properties unless they were linked to dimensions or equation results. If you wanted to provide a drop-down list for a property like Vendor, there was no pre-defined method other than SOLIDWORKS PDM datacards.

This example will enter custom properties from a macro as well as give you the tools to guide user input. The chapter will also introduce code organization through Modules and Classes as well as the programming design principle, Separation of Concerns (SoC).

The first step will create custom properties and add them to any SOLIDWORKS file. The methods used are valid for parts, assemblies or drawings.

Part 1: Adding Properties

Initial Code

The SOLIDWORKS macro recorder does not record entry or editing of File Properties, so the initial code will be written from scratch. The procedures have been broken out individually for easier use later in the chapter.

1. Start a new macro and save it as *properties.vbproj*.

2. Add the following code to the new macro inside the Partial Class SolidWorksMacro block.

```
Dim PropName() As String
Dim PropVal() As String
Dim PropMgr As CustomPropertyManager

Public Sub main()
  CreateTestProps()
  SetAllProps()
End Sub

Sub CreateTestProps()
  'Property names
  PropName = {"LastSavedBy", "CreatedOn", "Revision", "Material"}

  'Property values
  ReDim PropVal(3)
  PropVal(0) = "$PRP:" & Chr(34) & "SW-Last Saved By" & Chr(34)
  PropVal(1) = Date.Today
  PropVal(2) = "A"
  PropVal(3) = Chr(34) & "SW-Material" & Chr(34)
End Sub

Sub SetAllProps()
  'adds or sets all properties from PropName and PropVal arrays
  Dim Part As ModelDoc2 = swApp.ActiveDoc
```

```
    Dim PropMgr As CustomPropertyManager
    Dim value As Integer
    PropMgr = Part.Extension.CustomPropertyManager("")
    For m As Integer = 0 To PropName.Length - 1
      value = PropMgr.Add3(PropName(m), swCustomInfoType_e.swCustomInfoText, _
        PropVal(m), swCustomPropertyAddOption_e.swCustomPropertyReplaceValue)
    Next m

End Sub

''' <summary>
''' The SldWorks swApp variable is pre-assigned for you.
''' </summary>
Public swApp As SldWorks
```

Code Description

The macro will make use of arrays for property names and values similarly to the Material Properties macro. CreateTestProps is a procedure that populates two arrays with four values each. The first array, PropName, contains property names. The second, PropVal, contains property values. The arrays are populated with two different code techniques.

The SetAllProps procedure does the work of adding the properties to the active SOLIDWORKS document. The CustomPropertyManager interface will be introduced and is the best way to work with custom properties.

Class Variables and Scope

The two property arrays and a variable for the custom property interface have been declared outside of the main procedure block, but inside the class. The location of the variable declaration defines its scope. A variable declared within an If block or For loop is only available within that block. Variables declared within a procedure are only available to code in the procedure and following the declaration. But variables declared within a class are available to all procedures in the class. This concept will be discussed further when variables need to be available to the entire macro.

Dynamic Arrays

Notice another detail in the declaration of the arrays PropName() and PropVal(). The size argument of the array is left empty. By leaving out the size of the array, you have made it dynamic, enabling the array size to be modified as the macro runs. Dynamic arrays are a good practice since they are flexible.

Remember from a previous chapter, arrays can be populated with data by listing values within curly brackets { }. PropName is populated this way.

Reallocating Array Size

The size of a Dynamic Array can also change at runtime with the ReDim statement. This can be repeated anytime the array size needs to change. ReDim cannot change the data type of the array. Since arrays are zero-based, a size of 3 generates 4 array elements. PropVal is populated using individual values after a ReDim.

Either technique can be used in your code. Choose the one that best suits each situation.

Special SOLIDWORKS Property Values

SOLIDWORKS has several pre-defined properties you can reference, one of which is $PRP:"SW-Last Saved By." The property will always link to the Windows user name of who saved the SOLIDWORKS file last. The following list may be helpful in creating links to other predefined SOLIDWORKS properties.

All Documents	Parts Only
$PRP:"SW-Author"	"SW-Material"
$PRP:"SW-Comments"	"SW-Mass"
$PRP:"SW-Created Date"	"SW-Density"
$PRP:"SW-File Name"	"SW-Volume"
$PRP:"SW-Folder Name"	"SW-SurfaceArea"
$PRP:"SW-Keywords"	"SW-CalculatedCost"
$PRP:"SW-Last Saved By"	"SW-CenterofMassX"
$PRP:"SW-Last Saved Date"	"SW-CenterofMassY"
$PRP:"SW-Short Date"	"SW-CenterofMassZ"
$PRP:"SW-Subject"	"SW-Ix"

Drawings Only

$PRP:"SW-Current Sheet"	"SW-Iy"
	"SW-Iz"
$PRP:"SW-Sheet Format Size"	
$PRP:"SW-Sheet Name"	"SW-Px"
$PRP:"SW-Sheet Scale"	"SW-Py"
$PRP:"SW-Template Size"	"SW-Pz"
$PRP:"SW-Total Sheets"	"SW-Lxx"
	"SW-Lxy"
	"SW-Lxz"
	"SW-Lyx"
	"SW-Lyy"
	"SW-Lyz"
	"SW-Lzx"
	"SW-Lzy"
	"SW-Lzz"
	"SW-LinBlockTol1"
	"SW-LinBlockTol2"
	"SW-LinBlockTol3"
	"SW-LinBlockTol1Decimal"
	"SW-LinBlockTol2Decimal"
	"SW-LinBlockTol3Decimal"

Chr Function

In creating the link to $PRP:"SW-Last Saved By" and "SW-Material", quotation marks are needed within the property value. Use the Visual Basic character function with the code Chr(34) to add quotation marks within a string. Quotation marks in Visual Basic code have a specific meaning and cannot be used as a string character. They enclose and define a string. You cannot simply type "$PRP:"SW-Last Saved By"". That would be compiled as the string "$PRP:" followed by a function named SW-Last Saved By followed by an empty string. You must have quotes inside your string so that SOLIDWORKS knows how to handle the special property code. Chr(34) passes a quote character to the string without terminating the string definition. Other helpful ASCII characters are listed below.

Character Code	Return Value
Chr(8)	backspace
Chr(9)	tab
Chr(13)	carriage return
Chr(34)	"
Chr(38)	&
Chr(176)	°
Chr(177)	±
Chr(216)	Ø

As an alternative, Visual Basic.NET also allows you to enter a double-double quotation mark in a string. Not to be confused with a fast food cheese burger, the following two examples are identical. Even with the double-double quotation mark, don't forget to start and end any string with quotes.

```
PropVal(3) = Chr(34) & "SW-Material" & Chr(34)
PropVal(3) = """SW-Material"""
```

String Concatenation (&)

You have also used the character & to concatenate or combine the strings and Visual Basic characters. As shown earlier in the book, this is an effective way to build complex strings for a variety of functions.

Date.Today Function

To apply the current date in PropVal(1), the Visual Basic Date.Today function was used. This function returns the current date in mm/dd/yyyy format such as 12/31/2023.

Date.ToString()

If you prefer a different date format, you can make use of the ToString method. Most objects have a ToString method. The Date.ToString method has several possible arguments. One is simply the format of the date as a string. A Date can be transformed into your preferred string format using this technique.

```
PropVal(1) = Date.Today.ToString("MMMM dd, yyyy") 'December 31, 2023
```

To display the short date format in day/month/year, use the following.

```
PropVal(1) = Date.Today.ToString("dd/MM/yyyy") '31/12/2023
```

ICustomPropertyManager Interface

The SetAllProps procedure does the work of adding custom properties to the active SOLIDWORKS document by looping through the arrays. Back at the class level, a variable named PropMgr was declared as type CustomPropertyManager. The custom property manager interface has everything you need to work with SOLIDWORKS custom properties. Custom property manager interfaces are available for documents, configurations, and even for features. Since we are interested in setting properties to the general file properties for a part, assembly or drawing, the custom property manager interface was retrieved from the IModelDoc2.Extension (Part.Extension in our sample). An empty string was passed in the argument for configuration name to reference the general file properties. If you wanted to set configuration specific custom properties, you pass the name of the desired configuration as a string.

ICustomPropertyManager.Add3 Method

A For...Next loop fills in each property in the arrays using the CustomPropertyManager.Add3 method.

value = ICustomPropertyManager.Add3(FieldName, FieldType, FieldValue, OverwriteExisting)

- **FieldName** is a string for the name of the property. The first one in the example is "LastSavedBy."

- **FieldType** tells SOLIDWORKS if our value is text, number, date or yes/no. The macro uses the text type for all of our values (including the date). The only time you might want to use the other types is to limit the user input to certain types of values. The following is a list of field types from the swCustomInfoType_e enumeration.

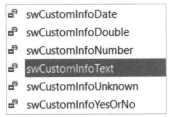

- **FieldValue** is the last argument required. Since FieldType was set to text, the input is the text from the array PropVal.

- **OverwriteExisting** enables the Add3 method to also overwrite existing property values. The enumeration options are as follows from swCustomPropertyAddOption_e.

The Add3 method will return an Integer value. If successful, it returns 0. Otherwise, it returns a value from swCustomInfoAddResult_e. The macro captures this returned value to the variable `value`. You could use the result to see if the method was successful or not.

Rather than hard-coding in a maximum value in the For loop, the variable m increases from 0 to `PropName.Length` - 1. An array's Length property returns its element count. Since the array is zero-based, the For loop has to stop at one less than its length.

Debug

3. Run the macro on a new part. It adds out the properties from the arrays along with their corresponding values. View them by selecting File, Properties from the SOLIDWORKS menu. Correct any errors prior to moving to the next section.

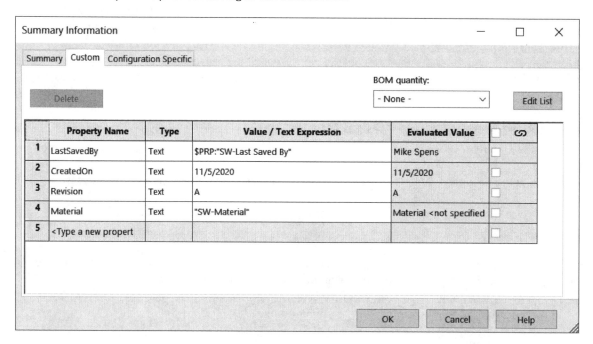

Part 2: Modifying Properties

Adding the Forms

Now that you understand how to set custom properties, it's time to give the user a simple way to control the value of properties entered through a user form.

4. Add a new dialog form by selecting Project, Add Windows Form…. Select the Dialog template and click Add.

5. Turn on the Toolbox from View, Toolbox. From the Common Controls group, add a ListBox and TextBox to the form and size them roughly as shown.

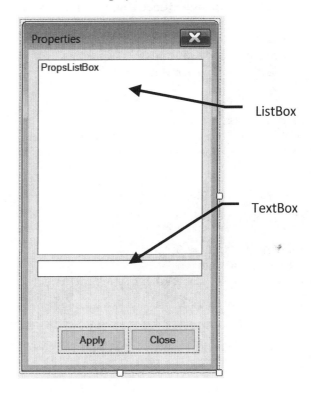

6. Modify the properties of the form and the newly created controls as follows.

Form Name = "PropsDialog"
Form Text = "Properties"
Form AcceptButton = (none)
ListBox Name = "PropsListBox"
ListBox Anchor = "Top, Bottom, Left, Right"
TextBox Name = "ValueTextBox"
TextBox Anchor = "Bottom, Left, Right"
OK Button Name = "ApplyButton"
OK Button Text = "Apply"
Cancel Button Text = "Close"

This macro will not be used to enter new properties yet. It will only change existing properties. The next section will add functionality to add and delete custom properties.

Classes and Modules

This is a good place for an explanation of classes and modules. Our current macro is built of two classes: SolidWorksMacro, the class created by default, and PropsDialog, the class created by adding a new form and modifying its name. There is an additional code container we have not used called a Module. The primary difference between classes and modules is how the code within them is referenced. To use code within a class, you must create a new instance of that class. Code and variables in a module are available without requiring an instance, depending on their declared scope. Generally, Modules are best used for the common variables, methods and procedures used by the entire application, or macro in our case. Classes are best for variables and code unique to an instance like a form or dialog.

As an example, if you had a function named MultiplyByTwo in a class named MathClass, you would need the following code to create and use MultiplyByTwo within a program (class) named MyApplication. Notice the use of the Public scope declaration of the function, making it visible to calling code outside the class.

```
Public Class MathClass
  Public Function MultiplyByTwo(ByVal num As Double) As Double
      MultiplyByTwo = num * 2
  End Function
End Class

Public Class MyApplication
Sub main()
  Dim MyNumber As Double = 5
  Dim mc As New MathClass
  Dim result As Double = mc.MultiplyByTwo(MyNumber)
End Sub
End Class
```

If MultiplyByTwo was in a Module named MathModule, you could call the method without declaring a named instance as follows.

```
Public Module MathModule
  Public Function MultiplyByTwo(ByVal num As Double) As Double
      MultiplyByTwo = num * 2
  End Function
End Module

Public Class MyApplication
Sub main()
  Dim MyNumber As Double = 5
  Dim result As Double = MathModule.MultiplyByTwo(MyNumber)
End Sub
End Class
```

Modules can greatly simplify the use of shared procedures and functions and provide a good location for application-level variables and procedures. If you need to access the same process from several classes, put that code into a module rather than in a class.

Separation of Concerns (SoC)

It is good design practice to keep your processing code separate from your user interface whenever possible. That means putting our property setting code outside of the button click event handler. It takes a little more effort, but it frees you to modify user interfaces without disturbing the underlying processing. It also keeps the processing code portable and reusable. If you need to use it in another application, it is not tied to specific controls or forms. There is much more to this design principle, but this is a good start. At this point, our macro has variables and processing code that will not be accessible to the form. A module will be used to keep our processing separate from our form as well as outside our base class SolidWorksMacro.

7. Add a new module by selecting Project, Add Module... Click Add on the Add New Item form to add Module1.vb and open it.
8. Switch back to the SolidWorksMacro.vb tab. Cut and paste the CreateTestProps and SetAllProps procedures into the Module1.vb tab within the Module Module1 section.
9. Cut and paste the declarations of PropName, PropVal and PropMgr from SolidWorksMacro.vb into Module1.vb as shown.

```
Module Module1
  Dim PropName() As String
  Dim PropVal() As String
  Dim PropMgr As CustomPropertyManager

  Sub CreateTestProps()
    'Property names
    PropName = {"LastSavedBy", "CreatedOn", "Revision", "Material"}

    'Property values
    ReDim PropVal(3)
    PropVal(0) = "$PRP:" & Chr(34) & "SW-Last Saved By" & Chr(34)
    PropVal(1) = Date.Today
    PropVal(2) = "A"
    PropVal(3) = Chr(34) & "SW-Material" & Chr(34)
  End Sub

  Sub SetAllProps()
    'adds or sets all properties from PropName and PropVal arrays
    Dim Part As ModelDoc2 = swApp.ActiveDoc
    Dim PropMgr As CustomPropertyManager
    Dim value As Integer
    PropMgr = Part.Extension.CustomPropertyManager("")
    For m As Integer = 0 To PropName.Length - 1
      value = PropMgr.Add3(PropName(m), swCustomInfoType_e.swCustomInfoText,
        PropVal(m), swCustomPropertyAddOption_e.swCustomPropertyReplaceValue)
    Next m
```

```
    End Sub

End Module
```

10. Add the Imports statements for the SOLIDWORKS libraries into Module1 as shown.

```
Imports SolidWorks.Interop.sldworks
Imports SolidWorks.Interop.swconst

Module Module1
  Dim PropName() As String
  Dim PropVal() As String
...
```

11. Change the scope of the two arrays to `Public`. This will make them accessible not only to procedures in this module, but also to any class in the macro.

```
Module Module1
  Public PropName() As String
  Public PropVal() As String
  Dim PropMgr As CustomPropertyManager
```

Sharing Data between Classes and Modules

There is one variable that still needs to be addressed. Notice the syntax error marking on the `swApp` variable in the `SetAllProps` procedure.

```
  Sub SetAllProps()
    'adds or sets all properties from PropName and PropVal arrays
    Dim Part As ModelDoc2 = swApp.ActiveDoc
```

Hovering your cursor over `swApp` displays a tooltip indicating that the name `swApp` is not declared. It is declared publicly in the SolidWorksMacro class. But recall that class variables are not accessible outside an instance of that class. The SolidWorksMacro class calls the module code, so we cannot create yet another instance of the same class in Module1 and get to its `swApp` interface. Each instance of a class is unique and so are its variables and interfaces.

When you need to get data from a class to another class or module, it is best to share a copy of that variable or interface. We need a new variable in Module1 to hold the copied reference.

12. Add the new public variable m_swApp to the Module1 variable declarations.

```
Module Module1
  Public PropName() As String
  Public PropVal() As String
  Dim PropMgr As CustomPropertyManager
  Public m_swApp As SldWorks
```

13. Change the variable used in SetAllProps to the newly declared module-level variable m_swApp.

```
Sub SetAllProps()
    'adds or sets all properties from PropName and PropVal arrays
    Dim Part As ModelDoc2 = m_swApp.ActiveDoc
    Dim PropMgr As CustomPropertyManager
    ...
```

Read Custom Properties

To make the form work for any file, custom properties need to be read out of the active document. The CreateTestProps procedure was only academic for our initial test. The macro needs a new procedure to read all properties from a part, assembly or drawing and fill out the PropName and PropVal arrays.

14. Add the following procedure to Module1 after SetAllProps to read all file properties from the active document.

```
Sub ReadFileProps()
  Dim Part As ModelDoc2 = m_swApp.ActiveDoc
  'make sure that Part is not nothing
  If Part Is Nothing Then
    MsgBox("Please open a file first.", MsgBoxStyle.Exclamation)
    Exit Sub
  End If

  'get the custom property manager
  PropMgr = Part.Extension.CustomPropertyManager("")

    'resize the PropVal array
    Dim propCt As Integer = PropMgr.Count
    ReDim PropVal(propCt - 1)
    ReDim PropName(propCt - 1)
    'fill in the array of properties
    For k As Integer = 0 To propCt - 1
    PropName(k) = PropMgr.GetNames(k)
    PropMgr.Get6(PropName(k), False, PropVal(k), "", Nothing, False)
  Next

End Sub
```

Notice the similarity between SetAllProps and ReadFileProps. The same process is used to get the active document and its custom property manager. A quick quality check was added to ensure that the Part variable is connected to a document. If the user tries to run the macro before opening a file, they'll get a message to open a file first and the procedure will exit.

The arrays are resized with the Redim statement, using the variable propCt set to the CustomPropertyManager.Count property. It returns the number of properties in that custom property instance, in this case, the number of file properties. The arrays are zero-based, so the size is reduced by 1. The same method is used in the For loop to count array elements.

The CustomPropertyManager.GetNames method gets the names of all properties for that Custom Property Manager instance. The integer argument defines which custom property you want. Without this integer (k), the method would return an array of property names rather than individual property names. As an alternative, you could move the population of the PropName array to just before the For loop using the following syntax.

```
PropName = PropMgr.GetNames
For k As Integer = 0 To propCt - 1
  PropMgr.Get6(PropName(k), False, PropVal(k), "", Nothing, False)
Next
```

As long as the arrays have the same size, you can simply set one to the other.

Within the loop, the next line of code populates each element of the PropVal array. The method used to get custom property values is CustomPropertyManager.Get6. This method expects several arguments, some are inputs and some are outputs.

value = ICustomPropertyManager.Get6(FieldName, UseCached, ValOut, ResolvedValOut, WasResolved, LinkToProperty)

- **FieldName** is the name of the custom property to get.

- **UseCached** is a Boolean. If set to True, a previously stored value is returned. If False, the value is re-calculated which can take extra processing time, but will be more consistently accurate. For speed, custom properties are cached in memory for configurations that are not active. This can return unexpected results if the property value is linked to dimensions, materials, mass or equations that haven't been rebuilt.

- **ValOut** returns the value of the custom property, but not its evaluated or resolved value. For example, if the property were linked to a dimension, this value would return the name of the dimension rather than its value.

- **ResolvedValOut** returns the evaluated or resolved value. For example, it would return the dimension value rather than the dimension name if it is linked.

- **WasResolved** will return a True or False to confirm whether or not the custom property value was not current based on unactivated configurations.

- **LinkToProperty** is used to link properties between derived parts. Set the value to True to preserve a link to the parent part and False to disable the link. Property linking is uncommon and is False in this example.

- **Value** is an integer value from the enumeration swCustomInfoGetResult_e. It allows you to determine if the returned value was fully resolved or simply a cached value from an unactivated configuration.

Since this method returns four different values, you must create variables prior to the call as containers for the results. This is true for ValOut and ResolvedValOut. If you are not interested in the return of one of these, pass an empty string "". To ignore the WasResolved return value, pass the Nothing keyword. We have also ignored the integer return value by not setting the result of the method to a variable.

Rather than entering a specific property name in the Get5 call, notice how you can directly use an array element of PropName to retrieve the corresponding property value.

```
PropMgr.Get6(PropName(k), False, PropVal(k), "", Nothing, False)
```

The arrays are now populated with custom property names and values. The next step will be to load the form with values and display it. In keeping with our SoC design technique, another procedure will be added to Module1 to do the work. This will be the procedure we call from the SolidWorksMacro class main procedure, so this will also be the mechanism to copy the SOLIDWORKS application interface into the m_swApp module variable.

15. Add a new procedure named LoadForm to Module1 as shown.

```
Sub LoadForm(ByVal swapp As SldWorks)
  'set the local SolidWorks variable to the running SolidWorks instance
  m_swApp = swapp
  'read the properties from the file
  ReadFileProps()

End Sub
```

By adding an argument variable requiring an instance of the SOLIDWORKS application, we can pass the SolidWorksMacro class SOLIDWORKS instance to the module. It is copied into the module-level variable m_swApp by simply setting them equal. The procedure then calls the ReadFileProps to load the arrays. Any call to ReadFileProps before the SOLIDWORKS application instance is passed into the module would throw the following error.

16. Switch to the *SolidWorksMacro.vb* tab and edit the main procedure as follows to call LoadForm, passing the SOLIDWORKS application. Delete all other procedure calls in Sub main.

```
Partial Class SolidWorksMacro

  Public Sub main()
    LoadForm(swApp)
  End Sub
```

The code still needs to show the form. First, an additional Imports statement is needed to reference the Windows Forms namespace.

17. Switch to the *Module1.vb* tab and add the following Imports statement.

```
Imports SolidWorks.Interop.sldworks
Imports SolidWorks.Interop.swconst
Imports System.Windows.Forms
```

18. Add the following code to the LoadForm procedure that creates an instance of the PropsDialog form, loads the data from the arrays, and displays the form.

```
Sub LoadForm(ByVal swapp As SldWorks)
  'set the local SolidWorks variable to the running SolidWorks instance
  m_swApp = swapp
  'read the properties from the file
  ReadFileProps()

  'initialize the new form
  Dim PropDia As New PropsDialog
  'if there are properties, fill in the controls
  If PropName.Length > 0 Then
    'load the listbox with the property names
    PropDia.PropsListBox.Items.AddRange(PropName)
    'set the first item in the list to be active
    PropDia.PropsListBox.SelectedItem = 0
    'show the first item's value in the text box
    PropDia.ValueTextBox.Text = PropVal(0)
  End If
  'show the form to the user
  Dim DiaRes As DialogResult
  DiaRes = PropDia.ShowDialog
End Sub
```

New Form Instance

Before filling the ListBox control, the form PropsDialog must be initialized to a variable using the New keyword.

Filling the ListBox Control

An If statement checks for an empty properties array before filling the ListBox and TextBox. This check eliminates errors caused by populating the ListBox with nothing.

The ListBox control has many of its own methods and properties. Filling the ListBox is done in the example using ListBox.Items.AddRange. When using AddRange you must pass an array of values, typically strings. AddRange uses the same element order as the incoming array. This will help us match property names to property values later.

SelectedItem Property

The SelectedItem property of the ListBox can be used to determine what is selected by getting the value, or to force a selection by setting it. In this case, it is used to select the first property name in the list by setting it to 0.

```
PropDia.PropsListBox.SelectedItem = 0
```

TextBox.Text

To show the value of the selected property, you have set the TextBox control's Text property to the value of the corresponding array value. Since you have selected the first item at location zero, set the TextBox.Text property to the same index in the PropVal array.

UserForm.ShowDialog

The ShowDialog method of the form completes the interaction by making it visible to the user and waits for the user's response. Since we used the Dialog template for the form, the Close button is already set to close the dialog.

If the user selects Apply, the arrays of property names and values will eventually be written back to the file.

User Interactions

You now have the framework for the tool, but nothing happens if the user selects a different property in the ListBox. The user must also be able to make changes to the property values that will then be stored back to the SOLIDWORKS file. Additional form control interactions will be needed to make this work.

The first step is to associate the user interactions in the ListBox with the value displayed in the TextBox. When the user selects a property from the ListBox, the TextBox should display that property's value. We have the names and values in arrays. Now we need the right event handler.

19. Switch to the view of *Dialog1.vb* form by double-clicking on it in the Solution Explorer or selecting the Dialog1.vb [Design] tab.

Double-click on the ListBox control to view the code for its default event, a procedure named PropsListBox_SelectedIndexChanged.

SelectedIndexChanged Procedure

The SelectedIndexChanged procedure for the ListBox control was automatically created when you double-clicked on the ListBox control. The two pull-down lists at the top of the code window can be used to generate procedures for additional user interactions for each control on the form (event handlers).

20. Add the following code to the PropsListBox_SelectedIndexChanged procedure.

```
Private Sub PropsListBox_SelectedIndexChanged (
    ByVal sender As System.Object, ByVal e As System.EventArgs) _
    Handles PropsListBox.SelectedIndexChanged

    ValueTextBox.Text = PropVal(PropsListBox.SelectedIndex)

End Sub
```

When the user clicks in the ListBox control it changes its SelectedIndex property. This line of code ensures that the TextBox control will always display the value from PropVal that corresponds to our ListBox when the user makes a selection.

21. Test your macros ability to display property values. Run the macro and click through some of the properties in the ListBox control. Make sure that the correct property value is displayed in the TextBox when an item is selected from the list. Make sure there are no errors before moving on to the next section.

Making Things Stick

Editing capability still needs to be added to make the macro useful. If the user types a new value into the TextBox, the array value will need to change. Code behind the Apply button will be used to change the array as well as write the property back to the active part, assembly or drawing.

Find the ApplyButton click event handler procedure named OK_Button_Click. The default procedure for a Button control is its Click event. *(Note: even though the procedure is named OK_Button_Click, it handles the ApplyButton.Click event. The Handles statement to the right of the procedure definition indicates what the procedure actually handles, not its name.)*

22. Change the code as follows to modify the array and save all properties to the file. Don't forget to remove the existing code that came from the Dialog template. It is not needed for this macro and would cause the form to close if the user clicked Apply.

```
Private Sub OK_Button_Click(ByVal sender As System.Object, _
ByVal e As System.EventArgs) Handles ApplyButton.Click
    'set the property value
    PropVal(PropsListBox.SelectedIndex) = ValueTextBox.Text
    SetAllProps()
End Sub
```

When the Apply button is clicked, one of the PropVal elements is set to the Text property of the TextBox control named ValueTextBox. The SelectedIndex property of the ListBox control is again used to make sure the correct element of the PropVal array is changed.

A call to SetAllProps updates all file properties. Only one property may have changed, but rather than writing something unique, this simply ensures all array values are updated in the file.

Debugging Tips

Your macro is now ready to be used to modify custom properties. Test your code before you move on to the next section and correct any errors.

Writing code requires attention to detail. One extra or missing character can give you serious headaches. VSTA is pretty good at alerting you to problems by underlining errors and giving tooltip feedback for what the problems might be. It will do some auto-correcting like adding parentheses if you forget them. Fix errors when they appear. It will save time in the long run.

Debugging skills will make or break you as a programmer. Hopefully some of my favorite tips and tools will help you find the problems before things get unwieldy.

Breakpoints

A breakpoint will force code execution to pause on the selected line. This is extremely helpful when you want to check the contents of a variable during execution. To add a breakpoint you can either use the F9 key or click in the narrow, grey column on the left side of the code window. To turn off a breakpoint, repeat the same process.

23. Test the For loop criteria in SetAllProps by adding the breakpoint shown in the image below.

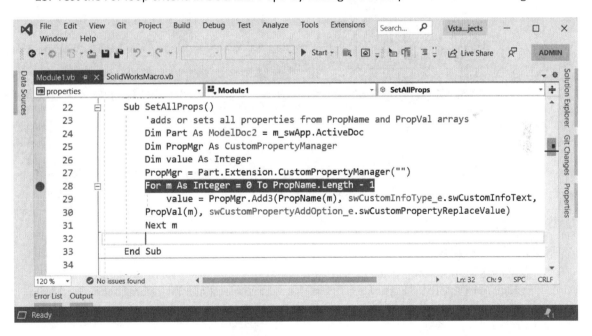

24. After setting the breakpoint, run the macro. Edit a property value and click Apply. The code will pause on the breakpoint as shown with the yellow highlighting and arrow.

```
1 reference
Sub SetAllProps()
    'adds or sets all properties from PropName and PropVal arrays
    Dim Part As ModelDoc2 = m_swApp.ActiveDoc
    Dim PropMgr As CustomPropertyManager
    Dim value As Integer
    PropMgr = Part.Extension.CustomPropertyManager("")
    For m As Integer = 0 To PropName.Length - 1
        value = PropMgr.Add3(PropName(m), swCustomInfoType_e.swCustomInfoText,
        PropVal(m), swCustomPropertyAddOption_e.swCustomPropertyReplaceValue)
    Next m

End Sub
```

25. Hover your cursor over the text PropName.Length to display a tooltip showing the current Length value.

26. Highlight only PropName and then hover your mouse over the highlight to see properties of the array. If you hover over the ▷ symbol, you will see the expanded array elements. Notice that the maximum array index is always one less than the Length.

```
Dim value As Integer
PropMgr = Part.Extension.CustomPropertyManager("")
For m As Integer = 0 To PropName.Length - 1
    value = PropMgr.Add3(Pro ◄ ⚙ PropName {Length=4} ⊟ ype_e.swCustomInfoText,
    PropVal(m), swCustomProper ● (0) ⚲ ▾ "LastSavedBy" stomPropertyReplaceValue)
Next m                        ● (1) ⚲ ▾ "CreatedOn"
                             ● (2) ⚲ ▾ "Revision"
End Sub                       ● (3) ⚲ ▾ "Material"
```

While paused at a breakpoint, you can review any variable that is already assigned a value. To continue running the rest of the code, click the Continue ▶ button or F5 on the keyboard.

Debug Stepping

While paused, you can step through your code one line at a time by clicking Step Into ⬇, typing F11, or selecting Debug, Step Into. Step Over ↻ (F10) and Step Out ↑ (Shift+F11) are similarly useful debugging functions. Try them while your code is paused to understand the different behavior. Step Over is helpful when you would like your code to run through a sub routine without going in one line at a time.

Run To Cursor

Another quick way to make your code pause at a specific location is to right-click on a line and select Run To Cursor (Ctrl+F10). You can use this method as an alternative to a breakpoint. Once the code is paused, you can use this method repeatedly to run through a block of code and pause again without having to set or remove breakpoints.

Watch

Add Watch and QuickWatch are also great ways to explore variables during code execution. If you add a watch, the Watch window appears. An added watch will remain in the project until you remove it – even if you close and re-open the project. A Quick Watch is only available when code is paused and will show the structure of the variable in a separate window.

27. Stop code execution by clicking Stop Debugging 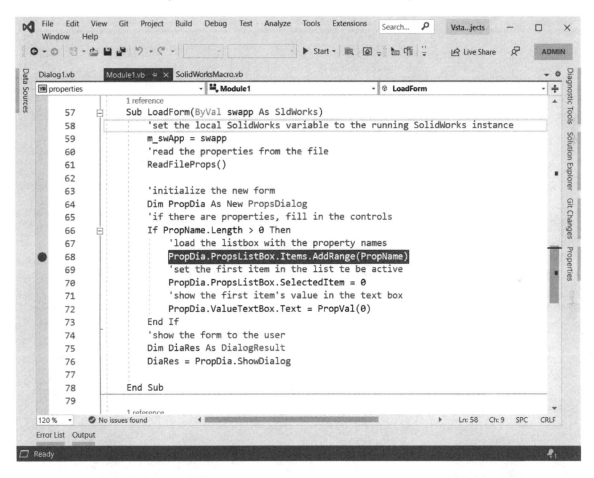.

28. Find the LoadForm procedure in Module1.vb and create a breakpoint on the line shown.

29. Test QuickWatch by running your code to the breakpoint. Once the code has paused, right-click on PropDia and select QuickWatch. Every property of the form is visible in a separate window.

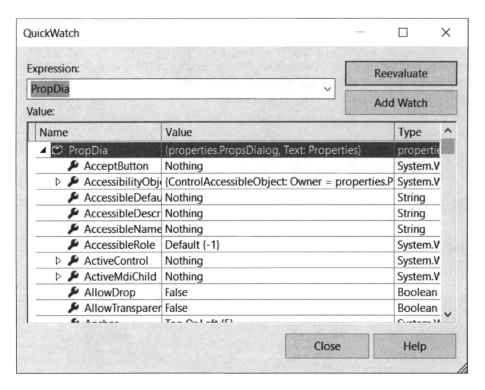

Some property and variable values can even be changed on the fly if you need to test different behavior while debugging.

30. Close the QuickWatch dialog and stop code execution by clicking .

31. Clear all breakpoints by selecting Debug, Delete All Breakpoints or typing Ctrl+Shift+F9.

Part 3: Add and Delete Properties

Let's continue to expand the properties tool and explore additional related topics.

Modifying the UI

Additional controls are needed on the form to add and delete properties.

32. Open Dialog1.vb [Design] and add two more Buttons to the form.

33. Change the properties as shown.

    Button1 (Name) = "AddButton"
    Button1 Text = "Add"
    Button2 (Name) = "DeleteButton"
    Button2 Text = "Delete"

Your form could look something like the image shown after making the changes.

Additional Procedures

The two arrays, PropName and PropVal, are the macro's core data. To enable adding properties, both arrays need to have elements added. To enable deletion of properties, the arrays need to have elements removed and associated properties need to be removed. To keep things organized, these operations will again be built in new procedures.

34. Switch back to Module1.vb and add the following procedure to add new properties.

```
Sub AddProperty(ByVal Name As String, ByVal Value As String)
    PropMgr.Add3(Name, swCustomInfoType_e.swCustomInfoText, Value, _
    swCustomPropertyAddOption_e.swCustomPropertyReplaceValue)

    're-read the properties arrays
    ReadFileProps()
End Sub
```

The first line of code should look similar to what was used in SetAllProps. The new property is added using the custom property manager's Add3 method. Instead of adding the individual elements to the property arrays, a call to ReadFileProps will reload them and will include the newly added property.

35. Add the following procedure below the End Sub statement of the AddProperty method to be used to delete properties.

```
Sub DeleteProperty(ByVal Name As String)
    PropMgr.Delete2(Name)

    're-read the properties arrays
    ReadFileProps()
End Sub
```

Deleting a property is simple when using the custom property manager. Simply pass the name of the property to delete to the Delete method.

Now that we have the two procedures, they need to be called from the corresponding form buttons where we'll determine the name and value to add or the name to delete.

AddButton Code

36. Create the click event handler for the Add button by switching back to the Dialog1.vb [Design] tab and double-clicking on the Add button. As an alternative, from the Dialog1.vb code window, select AddButton from the Class Name drop-down at the top-middle of the code window, and then select the Click event from the Method Name drop-down on the top-right.

37. Add the following code to the AddButton button's click event handler.

```
Private Sub AddButton_Click(ByVal sender As System.Object, _
ByVal e As System.EventArgs) Handles AddButton.Click
    'get the property to add
    Dim NewName As String
    Dim NewVal As String
    NewName = InputBox ("Enter new property name:")
    NewVal = InputBox ("Enter property value for " & NewName & ":")

    'add it to the file and update the array
    AddProperty(NewName, NewVal)
End Sub
```

InputBox

The use of the InputBox method is a quick and dirty way to receive information from the user without the need to create forms. This Visual Basic function (not SOLIDWORKS) displays a dialog box on the screen allowing the user to type in any value and then returns that value to the variables NewName and NewVal. The InputBox argument is the prompt that displays on the InputBox dialog when it is shown. Concatenating the NewName variable in the second InputBox reminds the user of the new property name.

Finally, the property is added to the file by calling the newly created AddProperty method.

DeleteButton Code

38. Create the click event handler for the Delete button by double-clicking on the button on the form.

39. Add the following code to the Delete button's click event handler.

```
Private Sub DeleteButton_Click(ByVal sender As System.Object, _
ByVal e As System.EventArgs) Handles DeleteButton.Click
    'get the list's selected item to know what to delete
    Dim PropToDelete As String
    PropToDelete = PropsListBox.SelectedItem

    'verify deletion from the user
    Dim answer As MsgBoxResult = MsgBox("Delete " & PropToDelete & "?", _
    MsgBoxStyle.YesNo + MsgBoxStyle.Question)
    If answer = MsgBoxResult.Yes Then
        'Delete the property from the file
        DeleteProperty(PropToDelete)
    End If
End Sub
```

Before you delete a selected item out of the ListBox control and the SOLIDWORKS file, it is customary to ask the user if they are absolutely sure. I have seen some creative comments in these subtle, yet often ignored reminders.

Additional MsgBox Settings

When you run, you will notice that the Message Box has a question mark icon and a yes and no button. The change is made by passing the added Visual Basic constants MsgBoxStyle.YesNo and MsgBoxStyle.Question just after the message text. As you typed the constants you likely saw several other styles available for use with a Message Box. These style options can be added together to enable special buttons along with typical icons. The two constants are added together to form the complete Message Box style argument. (*Hint: for quick access to help on Visual Basic functions to learn more about them, simply select the function name in your code and type F1 on the keyboard.*)

The MsgBox function also returns values as described in the first chapter. This is important when the user has the option to select from more than one button. The macro needs to know which one was selected. From a yes/no Message Box, clicking No returns the Visual Basic constant MsgBoxResult.No while clicking yes returns MsgBoxResult.Yes.

Refreshing the Form

After deleting or adding a property to the file, the ListBox needs to be refreshed with the updated file properties.

40. Add a new procedure above or below the DeleteButton_Click procedure within the PropsDialog class to refresh the form with the updated array values.

```
Sub RefreshForm()
    'reset list of properties displayed to the user
    PropsListBox.Items.Clear()
    PropsListBox.Items.AddRange(PropName)
    PropsListBox.SelectedIndex = PropName.Length - 1
    Me.Refresh()
End Sub
```

The ListBox is reset by clearing and then re-adding items from the PropName array and selecting the last property by its index. You could replace this with a zero if you prefer to return to the first item in the list.

In Visual Basic, Me will always refer to the class. In this case, Me is the form itself. A form's Refresh method will simply refresh the graphical display of the form.

41. Call RefreshForm at the end of both the AddButton_Click and DeleteButton_Click procedures as shown.

```
Private Sub AddButton_Click(ByVal sender As System.Object, _
    ByVal e As System.EventArgs) Handles AddButton.Click
    'get the property to add
    Dim NewName As String
    Dim NewVal As String
    NewName = InputBox("Enter new property name:")
    NewVal = InputBox("Enter property value for " & NewName & ":")

    'add it to the file and update the array
    AddProperty(NewName, NewVal)
    RefreshForm()
End Sub
Private Sub DeleteButton_Click(ByVal sender As System.Object, _
    ByVal e As System.EventArgs) Handles DeleteButton.Click
    'get the list's selected item to know what to delete
    Dim PropToDelete As String
    PropToDelete = PropsListBox.SelectedItem

    'verify deletion from the user
    Dim answer As MsgBoxResult = MsgBox("Delete " & PropToDelete & "?", _
    MsgBoxStyle.YesNo + MsgBoxStyle.Question)
    If answer = MsgBoxResult.Yes Then
        'Delete the property from the file
        DeleteProperty(PropToDelete)
```

```
  End If

  RefreshForm()
End Sub
```

Debug

You now have a fully functional custom property tool. Take the time to test your macro and to debug any errors before you move onto the next section.

Part 4: Save and Copy Between Files

So far, the macro does not do anything that SOLIDWORKS cannot already do. It was an academic exercise. The next step is to add functionality that does not exist in SOLIDWORKS. You will also learn about file input and output (or IO) with basic text file reading and writing.

While you are running your macro, all of the arrays and variables are stored in RAM. When the macro is closed, all of that valuable information is tossed out the window. In many situations it is helpful to store data permanently to a file or database so it can be accessed the next time you run your macro. Visual Basic can create, read from, write to and destroy files almost as easily as it saves variables to RAM.

Saving Text to a File Using System.IO

42. Add another Button control to your form.

43. Change its Text property to Save and its Name property to SaveButton.

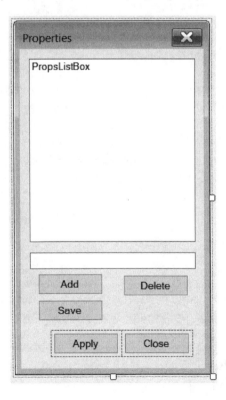

44. Double-click the new Save button to access its click event handler `SaveButton_Click`. Add the following code to this procedure.

```
Private Sub SaveButton_Click(ByVal sender As System.Object, _
ByVal e As System.EventArgs) Handles SaveButton.Click
    Dim FileName As String = "savedprops.txt"
    Try
        Dim sr As New StreamWriter(FileName)
        For i As Integer = 0 To PropName.Length - 1
            sr.WriteLine(PropName(i))
            sr.WriteLine(PropVal(i))
        Next
        sr.Close()
    Catch ex As Exception
        MsgBox(ex.Message, MsgBoxStyle.Exclamation)
        Exit Sub
    End Try
End Sub
```

45. Add the following Imports statement for System.IO to the top of the *Dialog1.vb* code window.

```
Imports System.Windows.Forms
Imports System.IO
```

```
        Dim sr As New StreamWriter(FileName)
        For i As Integer  ⑨ ▼  1      Type 'StreamWriter' is not defined.
            sr.WriteLine(PropNam
            sr.WriteLine(PropVal  Show potential fixes (Ctrl+.)
        Next
```

Hint: hover your cursor over undefined object types, StreamWriter in this case. Click the lightbulb and select Imports.System.IO from the list of potential fixes to quickly add the required Imports statement.

System.IO

The System.IO namespace contains many tools for working with files including the Path class previously discussed. The StreamWriter class provides an easy way to create and write to text files. For example, StreamWriter.WriteLine is used to write an entire string as a line in the text file. A return character is automatically appended to the end of the line. StreamWriter.Close simply closes the file so that other applications can access it.

StreamWriter

When the variable sr is declared as a New instance of the StreamWriter class, it can simply be passed the name of the file to write to as an argument. The file is created if it does not exist already. (*Hint: The full path to the file can be substituted for the name if the file is in a location other than the macro location.*)

Try ... Catch Statements

A Try ... Catch statement was used in this example to handle errors or exceptions. Any problems need to be caught while writing to the file to avoid a macro exit or crash. For example, an exception might occur if the file path is read only or if the file is already open by another application. The structure is pretty simple. If anything in the Try block causes an exception or code failure, processing jumps straight into the Catch block. In this example, the message from the exception is passed to the user in a simple message box with the exclamation mark icon. If the Try block is successful, the Catch block is skipped. I use Try ... Catch blocks regularly to trap errors and handle the result cleanly. Use them anywhere there may be failure points in your code projects.

Reading Text from a File Using System.IO

The object of reading data from a file will be to populate another file with a predefined set of custom properties. That means you will need to first read the information from the file into the PropName and PropVal arrays. Then add the custom properties to the active SOLIDWORKS file and refresh the ListBox control.

46. Add another Button control to your form. You can copy and paste an existing button.

47. Change its Text to "Load" and its Name to "LoadButton".

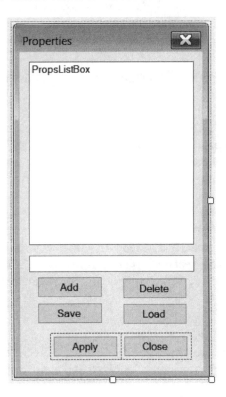

48. Double-click on the Load button and add the following code to read the array values from the file, then write them to the SOLIDWORKS properties and refresh the form.

```
Private Sub LoadButton_Click(ByVal sender As System.Object, _
ByVal e As System.EventArgs) Handles LoadButton.Click
    Dim FileName As String = "savedprops.txt"
    Dim sw As New StreamReader(FileName)
    Dim i As Integer = 0
    Do While Not sw.EndOfStream
```

```
        ReDim Preserve PropName(i)
        ReDim Preserve PropVal(i)
        PropName(i) = sw.ReadLine
        PropVal(i) = sw.ReadLine
        i = i + 1
    Loop

    'write the properties to the part
    SetAllProps()
    RefreshForm()
End Sub
```

StreamReader

The StreamReader is the complement to StreamWriter. It provides easy methods to read from a text file. You create a new instance of StreamReader in the same way you did with StreamWriter. You pass an argument for the file path to read. Again, if your file is not in the macro's directory, pass the full file path, including its folder, rather than only the file name.

Do Loops

Do loops are a perfect way to loop until a condition is met. Or, in this example, no longer met. They are similar to For loops. But instead of counting until a number is reached, the Do loop keeps processing until the While condition is False. Since it does not increment a counter, you must create the code to increment any counters needed. Be careful with Do loops. It is very easy to make a Do loop repeat infinitely. It is not a bad idea to build in a safety net. You could use the following If statement to exit the loop if more than 1000 lines were read in the file.

```
If i > 1000 Then Exit Do
```

StreamReader.EndOfStream Property

The StreamReader.EndOfStream property is used in the macro as an exit point for the loop. It simply returns False until the end of a file is reached. Then it returns True. Since the Do loop's While statement exits when it is False, the Not statement before StreamReader.EndOfStream reverses the Boolean value to its opposite.

StreamReader.ReadLine

The ReadLine function reads each successive line in the text file. A line must be terminated with a return character. It returns the line as a string. In this example, the result is passed to the corresponding array.

49. Test the macro on some existing files. Use the Save button to save properties from one file and the Load button to load them into another. Verify the results by reviewing the File Properties after running the macro.

50. Open the file *savedprops.txt*, saved in the macro *bin* folder. We have used a simple text format as follows.

```
Property Name 1
```

```
Property Value 1
Property Name 2
Property Value 2
...
```

51. Close Visual Studio.

Using the new macro, you could fill the text file with property names and values from one part, drawing or assembly, then load them into another part, drawing or assembly.

Conclusion

This exercise was intended to help you understand how to read and write to SOLIDWORKS custom properties. Hopefully some of the additions of using forms, modules and classes have also helped better understand the .NET framework. And now you have a macro that does something that SOLIDWORKS does not.

Utilize the Separation of Concerns (SoP) design principle regularly in your macros to make them easier to develop, debug and re-use.

An effective user interface can often make or break a macro. If you spend the time to think it out carefully and most of all, follow standards, your macros can be easy to use and will not frustrate the user or get in his way.

There are many other controls available to build tables, common dialog controls, timers and almost anything else you can think of. Browse through the Visual Studio Toolbox to see them. Test them by dragging them onto a form and reviewing their properties. Each control is well documented in the Visual Studio help. The internet is a great source of examples of how to use each one.

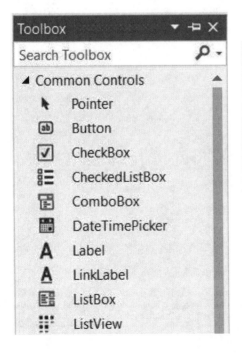

C# Example

The entire C# properties source is available in the downloadable example files referenced in the introduction.

```csharp
public static string[] PropName;
public static string[] PropVal;
private static CustomPropertyManager PropMgr;
public static SldWorks m_swApp;

public static void ReadFileProps()
{
    ModelDoc2 Part = (ModelDoc2) m_swApp.ActiveDoc;
    // make sure that Part is not nothing
    if (Part == null)
    {
        MessageBox.Show("Please open a file first.",  "Properties",
            MessageBoxButtons.OK,
            System.Windows.Forms.MessageBoxIcon.Exclamation);
        return;
    }

    // get the custom property manager
    PropMgr = Part.Extension.get_CustomPropertyManager("");

    // resize the PropVal array
    int propCt = PropMgr.Count;
    PropVal = new string[propCt];
    PropName = new string[propCt];
    string[] propNames = (string[])PropMgr.GetNames();
    // fill in the array of properties
    for (int k = 0; k < propCt; k++)
    {
        PropName[k] = propNames[k];
        bool resolved;
        bool linked;
        string resolvedVal;
        PropMgr.Get6(PropName[k], false,out PropVal[k],out resolvedVal,
            out resolved,out linked);
    }

}
```

Model Creation

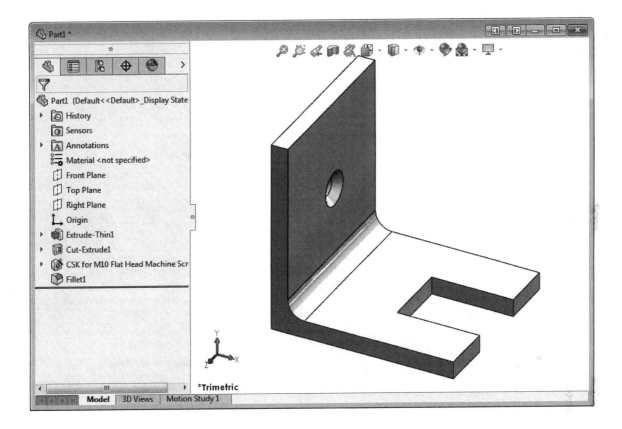

- **Geometry Creation**

- **Thin Feature Extrusions**

- **Extruded Cuts**

- **Hole Wizard**

- **Adding Fillets**

- **Feature Traversal and Editing**

Introduction

Whenever you need to automate any type of geometry creation, the best place to start is with a good recording of the process. With careful planning you can create these types of macros with very little manual programming. And if you ask me, the less you have to type, the better.

The first half of this exercise will examine several API elements used in the creation of the simple part shown here. Sketching, sketch dimensions, extrusions, thin features, hole wizard features and fillets are all pieces that make up the model. Follow these instructions exactly to learn the process. Once you are comfortable with the techniques, try it on a part you would like to automate.

The second half will teach you how to edit existing geometry beyond simply changing dimensions. You will learn how to traverse the FeatureManager and collect information about each feature in the tree. You will also learn the API calls required to modify those features. You can quickly build power editing utilities by employing these methods.

Part 1: Geometry Creation

Macro Recording Process

As you may have already learned, SOLIDWORKS records nearly every action you make in a macro. As you record the following steps, be overly cautious about avoiding extra clicks, selections or even panning, zooming and rotating. This will keep the recording clean and simple. There will be dozens of lines of code as a result of the steps below. It will be much more difficult to edit if there are hundreds. I will also try to point out the important code so you will be able to remove unnecessary lines if they are recorded.

1. Start the process by making sure you have nothing open in SOLIDWORKS. This will eliminate extraneous code that you would have to delete anyway. The simpler the better.

2. From the Tools menu, select Macros, Record or click ▮▮● .

3. Click the New button ▯ or File, New and select the generic Part template from the Templates tab. You can use a different template if you wish, but the resulting code may vary.

4. Make sure the unit system of this part is MMGS by going to Tools, Options, selecting the Document Properties tab and selecting Units from the list. Or choose MMGS from the quick access menu in the bottom right corner of the status bar.

5. Select the Front plane from the Feature Manager and start a sketch by clicking ▦.

6. Sketch a vertical line from the origin with a length of roughly 100mm. Draw in the positive Y direction.

7. Add a dimension to the sketch line to a length of 100mm using the Smart Dimension tool. Make sure to select the line for the dimension and not its endpoints to avoid extra lines of code.

8. Sketch a horizontal line from the origin with a length of roughly 100mm in the positive X direction.

9. Add a dimension to the sketch line to a length of 100mm using Smart Dimension. Your resulting sketch should look like the following.

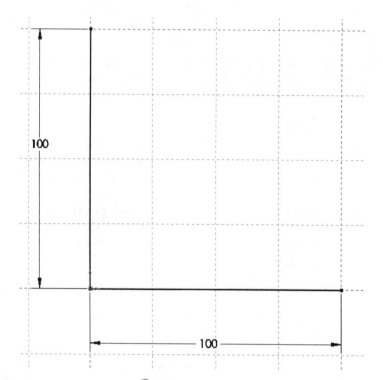

10. Click the Extruded Boss\Base tool . Set the depth to 100mm in Direction 1. Set the Thin Feature thickness to 10mm. Click ✔.

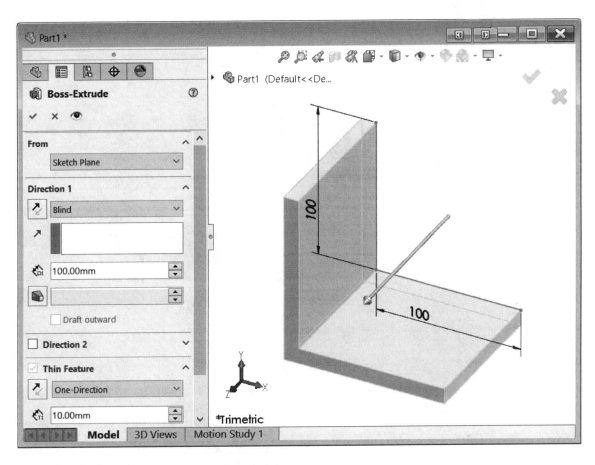

11. Select the face shown and start a new sketch.

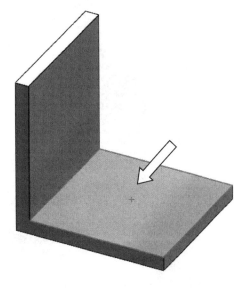

12. Sketch a rectangle roughly as shown.

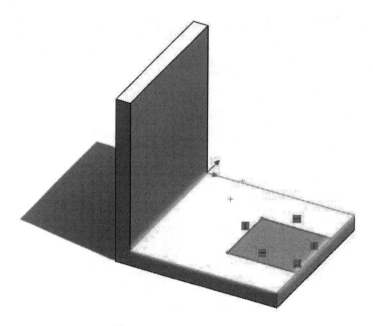

13. Click on the Extruded Cut tool 📦 and set the end condition to "Through All" and click ✔.

14. Select the upper face near the middle as shown and select the Hole Wizard 📦. Use the same settings as shown in the image (Countersink, ANSI Metric, Flat Head Screw, M10, Through All). Click ✔ to complete the hole in its selected location.

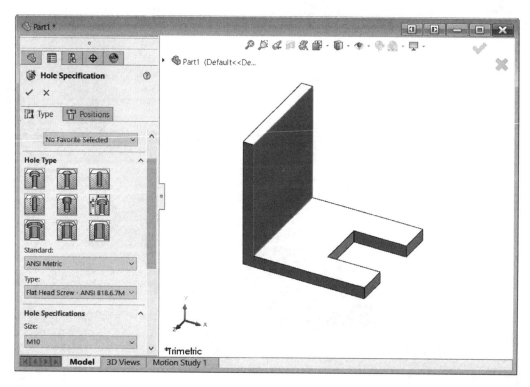

15. Select the edge between the two previously selected faces and create a 10mm radius Fillet as shown. Click OK ✔ to complete the fillet.

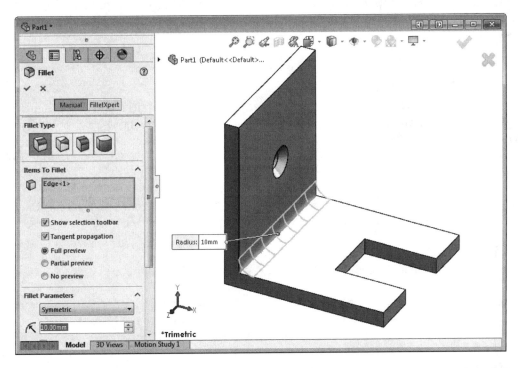

16. Stop the macro by selecting Tools, Macro, Stop and save it with the name *PartModel.vbproj*.

You have just recorded the basics for several feature types. You have also recorded two different sketching methods: using lines and rectangles. You have recorded selection of faces and edges. We will examine each of these steps and the related methods.

Creating a New Part

Edit the macro *PartModel.vbproj* and you will see the following code. Your macro will be much longer. But to make it easier to describe, it will be broken into small pieces here. All references to ModelView have been removed along with some additional, unnecessary code.

```
Public Sub main()

    Dim swDoc As ModelDoc2 = Nothing
    Dim swPart As PartDoc = Nothing
    Dim swDrawing As DrawingDoc = Nothing
    Dim swAssembly As AssemblyDoc = Nothing
    Dim boolstatus As Boolean = False
    Dim longstatus As Integer = 0
    Dim longwarnings As Integer = 0
    '
    'New Document
    Dim swSheetWidth As Double
    swSheetWidth = 0
```

```
Dim swSheetHeight As Double
swSheetHeight = 0
swDoc = CType(swApp.NewDocument("C:\...\templates\Part.prtdot", _
    0, swSheetWidth, swSheetHeight), ModelDoc2)
swPart = swDoc
```
...

ISldWorks.NewDocument

The first code section beyond the variable declarations uses swApp.NewDocument. This is a method of the SOLIDWORKS application and is a simple way to make a new part, drawing or assembly. All you need to pass is a string representing the path and name of a template to use. The three arguments after the template path are only meaningful if creating a new drawing from a template that doesn't have a defined sheet format size.

If you recall from earlier in the book, CType is a Visual Basic function that converts or verifies the return type. You could simplify the code line to create a new document by removing the CType function from the call as follows.

```
swDoc = swApp.NewDocument ("C:\...\Templates\Part.prtdot", 0, 0, 0)
```

If you replace the last arguments with zeros, you can also remove the variable declarations for swSheetWidth and swSheetHeight since they don't apply to parts. Do not forget to enter a real path to a template if you are copying the code here in the book rather than a recording.

Creating a Sketch

The next section of the macro illustrates how to select a plane, insert a sketch, add sketch lines and dimensions. Line wrapping is shown here to fit the page and numeric value have been simplified.

```
boolstatus = swDoc.Extension.SelectByID2("Front Plane", "PLANE", 0, 0, 0, _
    False, 0, Nothing, 0)
swDoc.SketchManager.InsertSketch(True)
swDoc.ClearSelection2(True)
Dim skSegment As SketchSegment = Nothing
skSegment = CType(swDoc.SketchManager.CreateLine(0R, 0R, 0R, 0R, _
    0.087032R, 0R), SketchSegment)
swDoc.ClearSelection2(True)
boolstatus = swDoc.Extension.SelectByID2("Line1", "SKETCHSEGMENT", _
    -0.000699R, 0.04445R, 0, False, 0, Nothing, 0)
Dim myDisplayDim As DisplayDimension = Nothing
myDisplayDim = CType(swDoc.AddDimension2(-0.0335R, _
    0.0439R, 0), DisplayDimension)
swDoc.ClearSelection2(True)
Dim myDimension As Dimension = Nothing
myDimension = CType(swDoc.Parameter("D1@Sketch1"), Dimension)
myDimension.SystemValue = 0.1R
swDoc.ClearSelection2(True)
skSegment = CType(swDoc.SketchManager.CreateLine(0R, 0R, 0R, 0.105348R, _
    0R, 0R), SketchSegment)
swDoc.ClearSelection2(True)
```

```
boolstatus = swDoc.Extension.SelectByID2("Line2", "SKETCHSEGMENT", 0.0252R, _
    0.0007156R, 0, False, 0, Nothing, 0)
myDisplayDim = CType(swDoc.AddDimension2(0.0524R, _
    -0.02474R, 0), DisplayDimension)
swDoc.ClearSelection2(True)
myDimension = CType(swDoc.Parameter("D2@Sketch1"), Dimension)
myDimension.SystemValue = 0.1R
...
```

SelectByID2

After creating a new part, the selection of the Front Plane was recorded using the IModelDocExtension.SelectByID2 method as discussed in the Model Dimensions exercise.

ISketchManager

You might notice that the call to insert a sketch, described below, is in the form IModelDoc2.SketchManager.InsertSketch recalling that the variable swDoc is declared as ModelDoc2. Based on the structure of the call, you can tell that the SketchManager interface is accessed from IModelDoc2. The SketchManager gives access to all of the sketching methods and properties for any type of model, whether it be a part, assembly or drawing. There are some obsolete sketching methods of the IModelDoc2 interface itself. The SketchManager interface should be used for the best performance and the most capability. All of the recorded sketch calls are made by this two-layer reference to the SketchManager rather than setting the model's SketchManager to an additional variable. If you want to simplify your code, you could add something like the following and replacing all references to swDoc.SketchManager with the variable SketchMan.

```
Dim SketchMan As SketchManager = swDoc.SketchManager
SketchMan.InsertSketch(True)
...
```

InsertSketch

The SketchManager.InsertSketch method simply starts a sketch on the currently selected plane (or face). A Boolean value of True was passed to this method to tell it to rebuild the model with any changes made by this sketch. In rare circumstances you might want to pass False, but most users want their models to actually change after a sketch is changed. If you don't pre-select a sketch plane or flat face, a new sketch will be added on the first plane in the model, typically the Front Plane. InsertSketch will also be used to exit the sketch.

CreateLine

The sketching of the vertical and horizontal lines is done by the call SketchManager.CreateLine. This method requires six input values representing x, y and z for the line's start point (all zeros in this example for both lines) and then the x, y and z values representing the line's end point in meters. The x, y and z coordinates are in the local sketch coordinates. Until 3D sketching is introduced, the z values are meaningless.

Visual Basic Literals

You might be wondering why the numeric values in the recorded code are followed by R. Similar to the concept of CType, R forces the numeric value to be a Double data type. It is not a necessary programming element, but like CType, it helps validate your input.

ClearSelection2

IModelDoc2.ClearSelection2 is recorded repeatedly since SOLIDWORKS clears selections prior to selecting another tool. The macro can be made more efficient by deleting or commenting out these extra actions. The code will be modified later to remove redundant ClearSelection2 and SelectByID2 methods.

AddDimension2

The next lines of code add the first dimension to the vertical sketch line. The call to the IModelDoc2 method AddDimension2 requires something to be pre-selected. That is why the previous SelectByID2 call was made. AddDimension2 requires three arguments, but none of them are the actual value of the dimension. They represent the x, y and z position of the dimension text. Since you are in a 2D sketch, the z value will always be zero as you can see in the recorded code. AddDimension2 returns the dimension (DisplayDimension interface) that was just created.

Again, to simplify the code, the CType statement and literal R could be removed as follows. Do not change the recorded numeric argument values.

```
Dim myDisplayDim As DisplayDimension = Nothing
myDisplayDim = swDoc.AddDimension2(-0.03748, 0.0410061, 0)
```

SystemValue

The next lines should look familiar from the Model Dimensions chapter. swDoc.Parameter("D1@Sketch1") and myDimension.SystemValue were used to control dimensions from Excel. The same method has been recorded to set the value of the dimension to 100mm (or 0.100 meters). The same process is then repeated for the second line.

Code Simplification

At this point, your code can be simplified to remove the redundant ClearSelection2 followed by SelectByID2 methods. This is possible because of the recorded order of operations. This would not be as simple had you recorded sketching the two lines before adding dimensions.

17. Change the code around the CreateLine and AddDimension2 methods as follows by removing the redundant selections. The CType statements and literals have also been removed from this code for simplicity. Line lengths have been set to exactly 0.100 meters in the CreateLine call.

```
Dim skSegment As SketchSegment = Nothing
skSegment = swDoc.SketchManager.CreateLine(0, 0, 0, 0, 0.1, 0)
Dim myDisplayDim As DisplayDimension = Nothing
myDisplayDim = swDoc.AddDimension2(-0.03748, 0.04100, 0)
Dim myDimension As Dimension = Nothing
myDimension = swDoc.Parameter("D1@Sketch1")
myDimension.SystemValue = 0.1
```

```
skSegment = swDoc.SketchManager.CreateLine(0, 0, 0, 0.1, 0, 0)
myDisplayDim = swDoc.AddDimension2(0.033165, -0.02890, 0)
myDimension = swDoc.Parameter("D2@Sketch1")
myDimension.SystemValue = 0.1
```

Creating Features

After creating the sketch, the first feature is created using the following code. Unnecessary code has been removed and line continuations are shown.

```
'
'Named View
swDoc.ShowNamedView2("*Trimetric", 8)
swDoc.ViewZoomtofit2()
Dim myFeature As Feature = Nothing
myFeature = swDoc.FeatureManager.FeatureExtrusionThin2(True, _
    False, False, 0, 0, 0.1R, 0.01R, False, False, False, False, _
    0.017453292R, 0.0174532R, False, False, False, False, True, _
    0.01R, 0.01R, 0.01R, 0, 0, False, 0.005R, True, True, 0, 0, _
    False)
swDoc.ISelectionManager.EnableContourSelection = False
```

ShowNamedView2

When you need to show a specific view in your model, use the ShowNamedView2 method of IModelDoc2 as recorded here. The first argument is the name of the view as a string. The second argument is the view ID from swStandardViews_e as shown below.

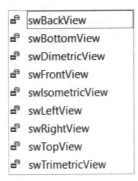

If you chose to use the view ID, you can pass an empty string to the view name. If the two conflict, the view ID will take precedence. This is recorded when SOLIDWORKS rotates the model to the Trimetric view when creating the first feature.

ViewZoomtofit2

As you might guess from the name, ViewZoomtofit2 is the Zoom to Fit action. It belongs to the IModelDoc2 interface so it works for parts, assemblies and drawings. Use this API call at the end of macros to give your users a good resulting view.

FeatureExtrusionThin2

The next lines of code represent the actual extrusion. A variable name myFeature is declared as a Feature. The IFeature interface is used to reference any item that displays in the FeatureManager tree in SOLIDWORKS. This can include sketches, folders as well as features like extrudes, cuts and fillets.

The IFeatureManager interface is used in a similar way to the SketchManager. Notice that the call to FeatureExtusionThin2 is a member of FeatureManager which is a member of swDoc which is a reference to IModelDoc2. This may seem like a complicated path to get to the point of adding a feature to the model. But like the SketchManager, the FeatureManager interface provides a comprehensive set of methods and properties for creating SOLIDWORKS features.

Take a quick look at all the arguments passed to the call swDoc.FeatureManager.FeatureExtrusionThin2, ignoring the surrounding CType function. There are thirty of them! Only three of these values are important to the thin feature you recorded. The remaining twenty-seven are still required. They simply become placeholders for this feature. I will discuss the three that make a difference to the macro and you can review the rest through the API Help by going to FeatureExtrusionThin2 in the index.

value = IFeatureManager.FeatureExtrusionThin2 (Sd, Flip, Dir, **T1**, T2, **D1**, D2, Dchk1, Dchk2, Ddir1, Ddir2, Dang1, Dang2, OffsetReverse1, OffsetReverse2, TranslateSurface1, TranslateSurface2, Merge, **Thk1**, Thk2, EndThk, RevThinDir, CapEnds, AddBends, BendRad, UseFeatScope, UseAutoSelect, T0, StartOffset, FlipStartOffset)

- **T1** is a long integer that indicates the end condition. You have a blind extrusion represented by 0 in the recorded code. The different end conditions are enumerated in swEndConditions_e and are as follows.

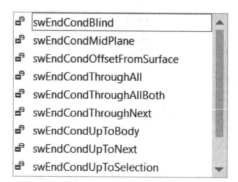

- **D1** sets the distance in meters for the first direction. Since you extruded the feature 100mm the recorded value is 0.100 (meters).

- **Thk1** sets the thickness of the feature in meters. Your macro should show a value of 0.010 (meters) since the thickness was set at 10mm.

Simplify Debugging

There is still a lot of code to review in this macro. It is easy to lose motivation to continue clean-up. Rather than taking on an entire recording at once, simplify your macro by commenting out large code blocks.

18. Select all code starting after the `swDoc.ISelectionManager.EnableContourSelection` line to the end of the procedure, excluding the End Sub line. Comment out the entire selection by clicking 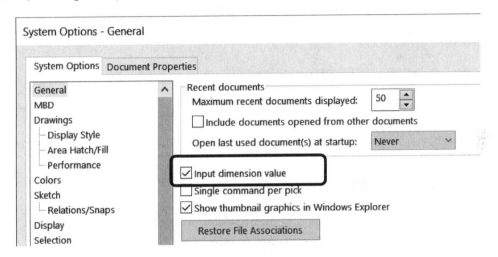 or Edit, Advanced, Comment Selection.

19. Close all parts currently open in SOLIDWORKS and run the macro by clicking ▶ or Debug, Start.

Rather than running to completion, every time a dimension is created, the Dimension Modify box displays and the macro pauses. If you click ✔ each time, the macro will complete. For true automation, how might the problem be solved?

In the General category of the SOLIDWORKS System Options you will find the "Input dimension value" setting. If unchecked, the Dimension Modify box will not display when the macro is run. If selected, it will and will require user input every time. Take the extra step of finding out if the user has this setting on or off, turn it off, add the dimensions and then turn the setting back to where the user had it. This will be similar to the code used in the Controlling Options chapter.

20. Add code to store and change the Input dimension value setting. Place the code immediately before the call to select the Front Plane.

```
'Store the user's setting
Dim usersSetting As Boolean
usersSetting = swApp.GetUserPreferenceToggle _
  (swUserPreferenceToggle_e.swInputDimValOnCreate)
swApp.SetUserPreferenceToggle _
  (swUserPreferenceToggle_e.swInputDimValOnCreate, False)

'create sketch
boolstatus = swDoc.Extension.SelectByID2("Front Plane", _
  "PLANE", 0, 0, 0, False, 0, Nothing, 0)
swDoc.SketchManager.InsertSketch(True)
...
```

21. Add code to restore the user's previous settings after completing the sketch, just before the swDoc.ShowNamedView2 line.

```
...
myDisplayDim = swDoc.AddDimension2(0.03316549917093, -0.02890194508009, 0)
myDimension = swDoc.Parameter("D2@Sketch1")
myDimension.SystemValue = 0.1

'Restore the user's setting
swApp.SetUserPreferenceToggle _
  (swUserPreferenceToggle_e.swInputDimValOnCreate, usersSetting)

'

'Named View
swDoc.ShowNamedView2("*Trimetric", 8)
```

22. Test the new macro after making the changes. It should create the part without showing the Dimension Modify dialog boxes.

23. Close all parts that were created by the macro.

Intelligent Dimension Placement

At this point the sketch is functional. But it is not as flexible as you might need. The lengths and positions of the dimensions are hard-coded. Make the following changes to prompt the user for size and automatically locate the dimensions relative to the sketch lines.

24. Modify the code section for the sketch, lines and dimensions as shown in bold.

```
'create sketch
Dim length As Double
Dim xLoc As Double
Dim yLoc As Double
length = CType(InputBox("Enter the line length in meters:"), Double)

boolstatus = swDoc.Extension.SelectByID2("Front", _
  "PLANE", 0, 0, 0, False, 0, Nothing, 0)
swDoc.SketchManager.InsertSketch(True)
swDoc.ClearSelection2(True)
Dim skSegment As SketchSegment = Nothing
skSegment = swDoc.SketchManager.CreateLine(0, 0, 0, 0, length, 0)
yLoc = length / 2
xLoc = -(length / 2)
Dim myDisplayDim As DisplayDimension = Nothing
myDisplayDim = swDoc.AddDimension2(xLoc, yLoc, 0)
Dim myDimension As Dimension = Nothing
myDimension = swDoc.Parameter("D1@Sketch1")
myDimension.SystemValue = length
skSegment = swDoc.SketchManager.CreateLine(0, 0, 0, length, 0, 0)
xLoc = length / 2
```

```
yLoc = -(length / 2)
myDisplayDim = swDoc.AddDimension2(xLoc, yLoc, 0)
myDimension = swDoc.Parameter("D2@Sketch1")
myDimension.SystemValue = length
...
```

First, variables are declared for the length of the sketch lines as well as the x and y placement location for the dimension text. Notice the replacement of sketch points in each CreateLine method. The first uses the `length` variable to set the second endpoint y value. In the second, `length` is used for the second endpoint x value. Before adding the dimensions, the text x and y locations are derived from half of the `length` value. The first dimension is placed to the left of the vertical line by making the x value negative. The second is placed below the horizontal by making the y value negative.

An InputBox is used again for simple user input. Use caution with the InputBox. It returns the user-entered value as a string. That means your code needs to handle all user entry validation.

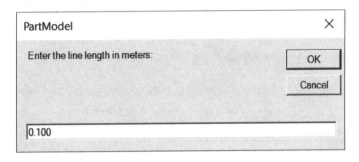

In this example, we make use of the CType function seen throughout the recorded macros. CType converts the first argument into the data type of the second. In this example, it converts the returned String from the InputBox into a Double data type. However, if any string values are typed into the InputBox, CType will fail. For example, if the user entered "100mm," you would get the following error.

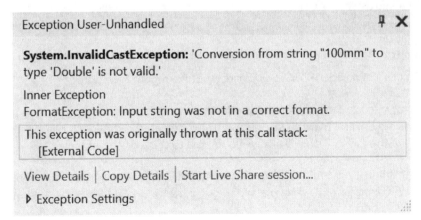

Additional Error Handling

The Visual Basic IsNumeric function comes to the rescue to trap the possible user error.

25. Modify the code to store the user input in a String variable and then make sure the value is a number.

```
'create sketch
Dim length As Double
Dim xLoc As Double
Dim yLoc As Double
Dim userLength As String
userLength = InputBox("Enter the line length in meters:")
If Not IsNumeric(userLength) Then
  MsgBox("Please enter only values in meters.", MsgBoxStyle.Exclamation)
  Exit Sub
End If
length = CType(userLength, Double)
```

If the user enters non-numeric characters, the code will stop and let them know. Since IsNumeric returns False if there are string characters, the Not keyword is used to invert the logic. If a problem is caught with the logic, a message box is displayed to the user with a warning exclamation icon and then the main procedure is ended early with the Exit Sub statement. Otherwise, it continues and converts the user input to a double value.

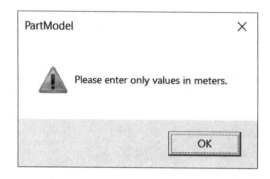

Extruded cuts

Now that the first extrusion is complete, you can uncomment the next block of code for cleanup and testing.

26. Uncomment the code block for the extruded cut. Select all commented code up to the Hole Wizard code, then click ⚏ or Edit, Advanced, Uncomment Selection.

This is the code for the cut. First, the face was selected, a sketch was added again using InsertSketch, the rectangle was drawn, and the cut-extrude feature was created. The sketch rectangle should always start at the edge of the face, which will always be related to the length variable.

27. Make the following modifications to this section of code shown in bold, commenting out unnecessary lines and replacing the first point of the CreateCornerRectangle method with the length variable.

```
'sketch for cut extrude
boolstatus = swDoc.Extension.SelectByRay(0.071, 0, 0.036, _
  -0.4, -0.515, -0.758, 0.0005, 2, False, 0, 0)
swDoc.SketchManager.InsertSketch(True)
'swDoc.ClearSelection2(True)
```

```
'boolstatus = swDoc.Extension.SetUserPreferenceToggle(...
'boolstatus = swDoc.Extension.SetUserPreferenceToggle(...
Dim vSkLines As Array = Nothing
vSkLines = swDoc.SketchManager.CreateCornerRectangle _
  (length, -0.0748, 0, 0.0549, -0.0244, 0)
myFeature = swDoc.FeatureManager.FeatureCut4(True, False, _
  False, 1, 0, 0.01, 0.01, False, False, False, False, _
  0.017, 0.017, False, False, False, False, False, True, _
  True, True, True, False, 0, 0, False, False)
'swDoc.ISelectionManager.EnableContourSelection = False
```

SetUserPreferenceToggle

Two lines of SetUserPreferenceToggle code were commented out and simplified in the example above. They record setting the behavior of the sketch rectangle tool for centerline display. They are not necessary since centerline use is not important to this macro. Refer back to the section on controlling options along with the API help to review the options for using rectangle centerlines if needed.

CreateCornerRectangle

After selecting the face, a new sketch was inserted. The rectangle was created by the command SketchManager.CreateCornerRectangle. The CreateCornerRectangle method of the SketchManager requires a sketch to be active and takes six arguments. It returns an array of sketch segments, specifically lines.

value = ISketchManager.CreateCornerRectangle(X1, Y1, Z1, X2, Y2, Z2)

The first three values are for the first corner point of the rectangle. The next three are for the second corner point. All arguments require double values in meters. Only the X and Y values are meaningful since the sketch is 2D. The X and Y locations are also in the local sketch coordinate system, not the model coordinate system. The returned value is an array of SketchSegment objects representing the individual sketch lines of the rectangle. The array will contain four elements unless you include centerlines in the rectangle.

Hint: If you intend to add dimensions to the resulting rectangle, your code may be more understandable if you use individual lines, immediately followed by dimensions in the same order as the first sketch. Only the first horizontal and vertical sketch line would need dimensions.

FeatureCut

The second to last call is the creation of the cut feature using FeatureManager.FeatureCut4. Just like the FeatureExtrusionThin method, FeatureCut4 requires dozens of arguments. Take a moment to reference the API Help to learn more about the 27 settings available to the command. I will discuss those that are most important to the feature you created here.

value = IFeatureManager.FeatureCut4(Sd, Flip, Dir, **T1**, T2, D1, D2, Dchk1, Dchk2, Ddir1, Ddir2, Dang1, Dang2, OffsetReverse1, OffsetReverse2, TranslateSurface1, TranslateSurface2, NormalCut, UseFeatScope, UseAutoSelect, AssemblyFeatureScope, AutoSelectComponents, PropogateFeatureToParts, T0, StartOffset, FlipStartOffset, OptimizeGeometry)

Only the fourth argument **T1** is relevant to this cut. It represents the end condition as a long integer from swEndConditions_e. You made the cut "Through All" represented by the integer 1 in the code. The same list of end conditions discussed for a thin feature extrusion would also apply to an extruded cut.

28. Optionally, delete the commented code lines and run a debug test.

HoleWizard5

Immediately following the FeatureCut operation you will find the code that handles creation of the counterbore hole using the Hole Wizard. The recording provides over 26 lines of code! Luckily, most of that code is not needed for a single hole.

29. Select and uncomment the code from the SelectByRay line through the line calling HoleWizard5.

```
boolstatus = swDoc.Extension.SelectByRay(0, 0.0514, 0.0485, _
  -0.4, -0.515, -0.758, 0.0, 2, False, 0, 0)
'
'Hole Wizard
Dim swHoleFeature As Object = Nothing
swHoleFeature = swDoc.FeatureManager.HoleWizard5(1, 1, 36, _
  "M10", 1, 0.011, 0.01, -1, 0.02, 1.571, 0, 1, 0, 0, 0, _
  0, 0, -1, -1, -1, "", False, True, True, True, True, False)
```

The simplest way to automate a single Hole Wizard hole is to pre-select a face prior to calling the HoleWizard5 method. Even though this is not a requirement in the SOLIDWORKS interface, many API features require a selection prior to the call being made. The hole will be created at the selection point by default. If you want to create multiple holes, you must pre-select several sketch points prior to calling HoleWizard5. This is another method you might want to review through the API Help for additional detail.

value = IFeatureManager.HoleWizard5(GenericHoleType, StandardIndex, FastenerTypeIndex, SSize, EndType, Diameter, Depth, Length, Value1, Value2, Value3, Value4, Value5, Value6, Value7, Value8, Value9, Value10, Value11, Value12, ThreadClass, RevDir, FeatureScope, AutoSelect, AssemblyFeatureScope, AutoSelectComponents, PropagateFeatureToParts**)**

Out of the first five arguments, four are long integer enumerations from SolidWorks.Interop.swconst. The next two are the diameter and depth of the primary hole as double values. Length is only used for slots. Value1 through Value12 are unique to the type of hole and are double values. Not all twelve are used for all holes. ThreadClass is used only for ANSI standard and is either "1B", "2B" or "3B". The rest of the arguments are Boolean value options and should typically be left as recorded.

The following table lists the SOLIDWORKS enumerations for GenericHoleType, StandardIndex and some for the FastenerTypeIndex. The possible values for EndType are the same as those for extrusions discussed earlier.

GenericHoleType	**swWzdGeneralHoleTypes_e** swWzdCounterBore swWzdCounterBoreSlot swWzdCounterSink swWzdCounterSinkSlot swWzdHole swWzdHoleSlot swWzdLegacy swWzdPipeTap swWzdTap
StandardIndex	**swWzdHoleStandards_e** swStandardAnsiInch swStandardAnsiMetric swStandardAS swStandardBSI swStandardDIN swStandardDME swStandardGB swStandardHascoMetric swStandardHelicoilInch swStandardHelicoilMetric swStandardIS swStandardISO swStandardJIS swStandardKS swStandardPCS swStandardPEMInch swStandardPEMMetric swStandardProgressive swStandardSuperior
FastenerTypeIndex	**swWzdHoleStandardFastenerTypes_e** **(ANSI Inch countersink holes)** swStandardAnsiInchFlatSocket82 swStandardAnsiInchFlatHead100 swStandardAnsiInchFlatHead82 swStandardAnsiInchOval

Error Corrections

The macro recorder included a lengthy section of code that is intended to give control over the hole location after it is created. It comprises the next 22 lines of code and is not needed if you provide a precise location for the hole by the selection location. In addition, as of publication, there are errors recorded in the same code due to recording bugs (SPRs).

30. Delete the commented code from the line below the HoleWizard5 call to the line before the next call to SelectByRay.

Adding Fillets

The remaining code is used to create a fillet along the selected edge.

31. Uncomment the remaining commented lines of code for the fillet feature beginning at the SelectByRay line.

There are some recorded problems with the code that need to be fixed. The recorded code is also much more complicated than it needs to be for a simple edge fillet.

32. Correct the recorded code errors and simplify the code shown in bold. Comment out or delete the unnecessary lines noted below. Numeric values have been simplified and line continuations are shown for clarity.

```
'create fillet
swDoc.ClearSelection2(True)
boolstatus = swDoc.Extension.SelectByRay(0.000462, -0.0003591, 0.06352, _
  -0.4000, -0.51503, -0.7580, 0.001405, 1, False, 1, 0)
Dim swFeatData As SimpleFilletFeatureData2 = Nothing
swFeatData = CType(swDoc.FeatureManager.CreateDefinition _
  (swFeatureNameID_e.swFmFillet), SimpleFilletFeatureData2)
'
swFeatData.Initialize(swSimpleFilletType_e.swConstRadiusFillet)
'
'Dim swEdge As Object = Nothing
'Dim edgesArray() As Object = Nothing
'swEdge = CType(swDoc.ISelectionManager.GetSelectedObject6(1, 1), Object)
'Dim edgesVar As Object
'swFeatData.Edges = edgesVar
'
'swFeatData.AsymmetricFillet = False
swFeatData.DefaultRadius = 0.01R
'swFeatData.ConicTypeForCrossSectionProfile = swFeatureFilletCircular
'swFeatData.CurvatureContinuous = False
'swFeatData.ConstantWidth = 0.01R
'swFeatData.IsMultipleRadius = False
'swFeatData.OverflowType = swFilletOverFlowType_Default
''
Dim swFeature As Feature = Nothing
swFeature = CType(swDoc.FeatureManager.CreateFeature(swFeatData), Feature)
```

Selection Marks

After clearing any existing selections, the edge is selected using SelectByRay. To make sure the edge is selected, the 8th argument is 1. It is important to note that a Mark of 1 (second from the last argument) was used with this selection. Some features allow multiple selection types. SOLIDWORKS uses the Mark value to indicate which selection is used for each part of the feature. For example, if you were to create a sweep, you would need to pre-select the profile and path using the API. The sweep feature knows how to use each selection by its Mark. A fillet requires a Mark value of 1 for the selected edge.

CreateFeature

The SOLIDWORKS API has two different methods for creating fillets. The FeatureManager.CreateFeature method is recorded and shown here. CreateFeature can create several different feature types and has a different structure than FeatureFillet3. FeatureFillet3 is more like FeatureExtrusionThin2 and FeatureCut4 and has 16 arguments. Explore IFeatureManager.FeatureFillet3 in the API help for more detail.

CreateFeature only requires one argument of a FeatureData type. It also assumes a pre-selected edge, but is not limited to pre-selection. It returns a Feature interface as well. The call to create the fillet is the last line of code in the procedure.

```
Dim swFeature As Feature = Nothing
swFeature = CType(swDoc.FeatureManager.CreateFeature(swFeatData), Feature)
```

All code above that call creates a FeatureData interface that carries the definition of the fillet, beginning with the call to swDoc.FeatureManager.CreateDefinition.

CreateDefinition

The CreateDefinition method of IFeatureManager can be used to create many feature types, but primarily patterns, sweeps and fillets. Take a look at the API help for more detail and a full list of feature types and requirements. CreateDefinition only requires passing the type of feature based on the swFeatureNameID_e enumeration. swFmFillet is the type recorded in the code. The error in the recorded code was that it was missing the enumeration the type came from.

CreateDefinition returns a specific type of FeatureData based on the type. The swFmFillet type returns a SimpleFilletFeatureData2 interface. The CType detail is removed below to clarify the code.

```
Dim swFeatData As SimpleFilletFeatureData2 = Nothing
swFeatData = swDoc.FeatureManager.CreateDefinition _
  (swFeatureNameID_e.swFmFillet)
```

Once you have the new SimpleFilletFeatureData2 interface, you must first initialize it with the fillet type to be created. That could be a face fillet, full round fillet, or constant radius fillet. The Initialize method expects the fillet type as an argument based on the swSimpleFilletType_e enumeration. The recorded code uses the swConstRadiusFillet option.

```
swFeatData.Initialize(swSimpleFilletType_e.swConstRadiusFillet)
```

Finally, properties like DefaultRadius can be set for the SimpleFilletFeatureData2 interface.

```
swFeatData.DefaultRadius = 0.01R
```

Debug

After modifying the macro, test it to make sure everything works correctly. The code will create a new part each time you run. Try adjusting some of the input values to get different size models.

Improving Performance

As you run the macro, each action takes time as if the user were performing the steps. View orientations and previews each add graphics processing overhead to display information on the screen. If you are truly trying to automate the procedure, speed is the driving factor rather than seeing the build operation. There is a simple way to disable the graphical updating of the screen while your code runs.

33. Modify your code as shown immediately following the NewDocument method to disable graphics updating while the part is being built.

```
...
swDoc = CType(swApp.NewDocument("C:\...\Templates\Part.prtdot", _
  0, 0, 0), ModelDoc2)

'disable graphics updating for this part
swDoc.ActiveView.EnableGraphicsUpdate = False

'Store the user's setting
Dim usersSetting As Boolean
usersSetting = swApp.GetUserPreferenceToggle _
  (swUserPreferenceToggle_e.swInputDimValOnCreate)
swApp.SetUserPreferenceToggle _
  (swUserPreferenceToggle_e.swInputDimValOnCreate, False)
...
```

EnableGraphicsUpdate

EnableGraphicsUpdate is a property of the IModelView interface. Setting EnableGraphicsUpdate to False turns off all graphical refreshing of the model. Be aware that if your macro crashes or is stopped after making this call, graphics refreshing will still be disabled for the model. Save, close and re-open the file to re-enable graphics updating manually.

Calls to ModelView are regularly recorded in your macros but we have been removing them. ModelView can be accessed from the IModelDoc2 interface using the ActiveView method. The code is kept simple in this example by using the combined code. If you prefer something more explicit, you could replace the one line with the following.

```
'disable graphics updating for this part
Dim MyView As ModelView = Nothing
MyView = swDoc.ActiveView
MyView.EnableGraphicsUpdate = False
```

34. Finish your macro with the following code to enable graphics updating after the model is completed. Place the code immediately above End Sub.

```
...
    swDoc.ActiveView.EnableGraphicsUpdate = True
End Sub
```

Run another test of your modified code. You should notice a significant difference in the performance as well as the graphical behavior.

35. Close Visual Studio.

Part 2: Feature Editing and Traversal

Now that you have learned to create basic features in parts, the next logical step is to explore how to navigate and edit existing features. Once a feature has been created, all of its feature information can be accessed and modified as if you were editing the feature through the user interface. As an example, you may need to enable draft on an extruded boss or cut. You may wish to find all features of a given type in the FeatureManager. Understanding how to traverse the FeatureManager will help greatly. A more detailed feature traversal approach is explained in the Favorite Code Examples chapter.

Feature Editing

Unfortunately, feature editing operations are not recorded in macros. So rather than recording a macro to begin this example, a new macro will be created.

36. Open the part named *Block.sldprt*. *(Hint: you can open any simple part with a cut named "Cut-Extrude1" and several fillet features.)*

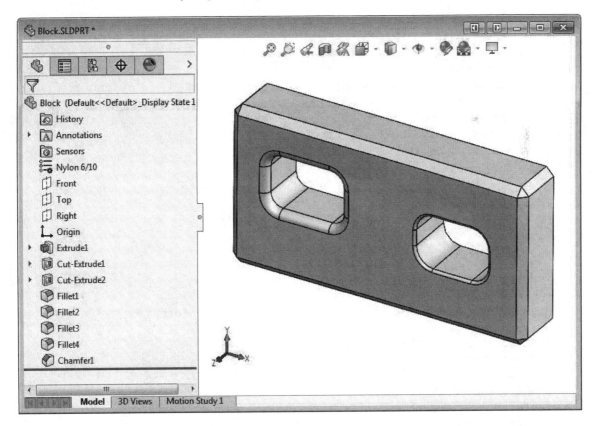

37. Start a new macro named *FeatureEdit.vbproj*.

38. Add the following code to turn on 2° draft for *Cut-Extrude1* and to change its end condition to Through All.

```
Sub main()
    Dim Part As PartDoc
    Dim MyFeature As Feature
    Dim featureDef As ExtrudeFeatureData2
    Dim retval As Boolean
    Dim message As String

    Part = swApp.ActiveDoc

    MyFeature = Part.FeatureByName("Cut-Extrude1")
    featureDef = MyFeature.GetDefinition

    'get some settings from the feature
    If featureDef.GetDraftWhileExtruding(True) = False Then
        message = "The selected feature has no draft." & vbCr
    End If

    Select Case featureDef.GetEndCondition(True)
        Case swEndConditions_e.swEndCondBlind
            message += "Blind"
        Case swEndConditions_e.swEndCondThroughAll
            message += "Through All"
        Case swEndConditions_e.swEndCondThroughAllBoth
            message += "Through All Both"
        Case swEndConditions_e.swEndCondUpToSurface
            message += "Up To Surface"
        Case swEndConditions_e.swEndCondMidPlane
            message += "Mid Plane"
        Case swEndConditions_e.swEndCondOffsetFromSurface
            message += "Offset From Surface"
        Case swEndConditions_e.swEndCondThroughNext
            message += "Up To Next"
        Case swEndConditions_e.swEndCondUpToBody
            message += "Up To Body"
    End Select

    MsgBox(message & " end condition.", MsgBoxStyle.Information)

    'rollback to edit the feature
    retval = featureDef.AccessSelections(Part, Nothing)

    'modify some feature values
    featureDef.SetEndCondition(True, _
      swEndConditions_e.swEndCondThroughAll)
    featureDef.SetDraftWhileExtruding(True, True)
    featureDef.SetDraftAngle(True, 2 * Math.PI / 180)
```

```
    'complete the edit operation
    retval = MyFeature.ModifyDefinition(featureDef, Part, Nothing)

    'in case the modification failed
    If retval = False Then
        featureDef.ReleaseSelectionAccess()
    End If
End Sub
```

FeatureByName

First, notice the use of the PartDoc data type. Since all actions will focus on editing part features, explicitly using the PartDoc type provides direct access to part methods and properties. After the declarations and connecting to the active document, the FeatureByName method of IPartDoc is used to get the Cut-Extrude1 feature in the part. This code makes the assumption that you know the name of the feature you want to work with. If you do, the process of getting to the feature is easy.

value = IPartDoc.FeatureByName(Name)

39. **Name** is the only argument required for this. It is simply the name of the feature as a string.

40. **Value** is the return value and is an IFeature interface.

The Feature interface can be compared loosely to ModelDoc2. Feature contains the general methods and properties for all features in the FeatureManager like name, suppression state, as well as type. Each specific feature type has its own unique properties and methods and can be derived from the Feature interface like PartDoc or AssemblyDoc can be derived from ModelDoc2.

Other methods are available to get a feature rather than having to know its name. The SelectionManager interface could be used if you expect the user to pre-select. This method was shown in the Material Properties exercise to get to selected parts in an assembly.

Also consider using FeatureByName to help select a feature as an alternative to SelectByID2. Once you have it, use the feature interface's Select2 method. For example, use the following code to get and select the Front plane, assuming the variable Part is a PartDoc data type.

```
Dim MyFeature As Feature
MyFeature = Part.FeatureByName("Front")
MyFeature.Select2(False, -1)
```

More about the Select2 method later in this chapter.

GetDefinition

After getting to a specific feature you must get its feature definition interface using the method IFeature.GetDefinition. The feature's settings are then available to be evaluated or changed.

GetDraftWhileExtruding and GetEndCondition

Each feature has many methods available through the feature definition interface. Since you have used GetDefinition on a cut-extrude feature, the return interface is IExtrudeFeatureData2. Two of the available methods are shown in the code to get the draft setting and the end condition. You can find a list of all the possible members of this interface by looking up IExtrudeFeatureData2 in the index of the API Help.

value = IExtrudeFeatureData2.GetDraftWhileExtruding(Forward)

- **Forward** is a Boolean value that sets whether to get the draft setting for direction 1 or direction 2. True represents direction 1. Use False to get direction 2.

- **value** is returned as another Boolean value. If True is returned, draft is turned on in the requested direction.

value = IExtrudeFeatureData2.GetEndCondition(Forward)

- **Forward** is again a Boolean value that sets whether to get direction 1 (True) or direction 2 (False).

- **value** is returned as a constant from swEndConditions_e. Any of the following values can be returned depending on the setting of the IExtrudeFeatureData2 interface.

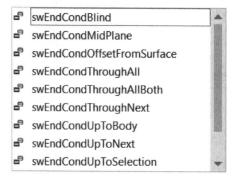

Select Case

The Visual Basic Select Case block is an effective way to check a list of conditions as long as you are checking only one data type. Imagine writing the same check with an If statement. It certainly could be done, but the Select Case technique makes it easier since you do not need to build each comparison expression. A single variable or expression is listed on the first line. Then on each Case line there is a variable or constant to compare. Case Else statements can be used if no other condition is met. If a Case value matches the Select Case comparison, the block of code just under the Case line is executed and the code continues after End Select. Select Case blocks are compared in order of the Case statements.

In the following example, any code under the Case "A" statement would run. String comparisons are case sensitive, so the line Case "a" would be evaluated to False.

```
Dim val As String = "A"
Select Case val
  Case "B"
    'it is B
```

```
  Case "a"
    'it is a
  Case "A"
    'it is A
  Case Else
    'it is none of the other cases
End Select
```

AccessSelections

Before a feature can be modified you must rollback the feature history to the point where the feature was built. SOLIDWORKS does this automatically when you select Edit Feature. However, if you are editing programmatically you must make the call to the AccessSelections method of the specific feature data interface to get the same results.

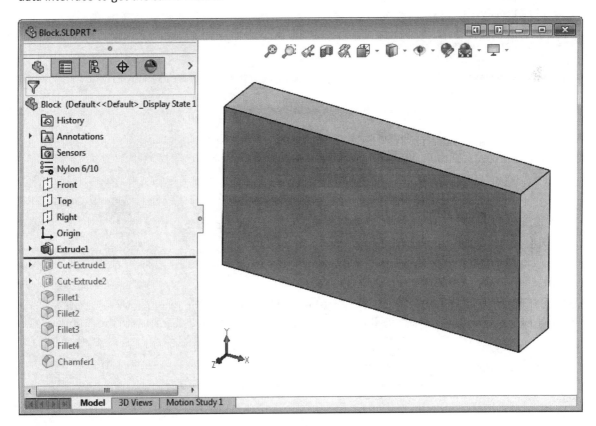

value = IFeatureData.AccessSelections(TopDoc, Component)

- **TopDoc** must be the top level document interface. If you are editing a feature from a part in an assembly, the IAssemblyDoc interface must be passed to TopDoc.

- **Component** is the IComponent interface of the part to be edited if an assembly is the TopDoc. If TopDoc is a part you can pass Nothing (an empty object).

- **value** will be True if successful. Check this value before you try to change the feature data. You will get errors if you attempt to modify an invalid feature data definition.

- **FeatureData** must be a valid feature data interface. For a list of those available, type AccessSelections into the index of the SOLIDWORKS API Help.

Modifying the Feature Data

Just as there are many methods used to get feature settings from the feature data interface, there are also many methods used to modify them. Three are used in this example. Review the API Help for a full list of methods available to each feature type.

IExtrudeFeatureData2.SetEndCondition(Forward, EndCondition)

- **Forward** should be set to True if you wish to set the end condition for direction 1. Set it to False for direction 2.

- **EndCondition** must be passed a constant from swEndConditions_e.

IExtrudeFeatureData2.SetDraftWhileExtruding (Forward, DraftWhileExtrude)

- **Forward** is again True if you wish to turn draft on for direction 1.

- **DraftWhileExtrude** must be True if you want to turn on draft in the selected direction. Pass False if you wish to turn draft off.

IExtrudeFeatureData2.SetDraftAngle (Forward, DraftAngle)

- **Forward** is set to True if you wish to set the draft angle for direction 1.

- **DraftAngle** requires a double value in radians for the draft angle. Do not forget to do any conversion necessary if you have a value desired in degrees.

You can think of this operation as if changing settings in the Property Manager dialogs. It is as if you first selected Through All, then clicked the Draft On/Off button and typed in 2 degrees into the draft angle box.

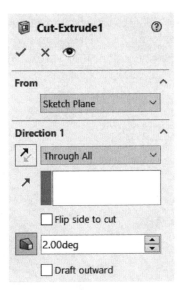

ModifyDefinition

Now that the new settings have been entered into the FeatureData interface, you must actually apply them and rebuild the model. Compare this to clicking ✔ on the Cut-Extrude1 Property Manager.

value = IFeature.ModifyDefinition (Definition, TopDoc, Component)

- **Definition** must be passed the same feature data definition that has just been modified.

- **TopDoc** is again the IAssemblyDoc interface if you are editing a part in an assembly or the IPartDoc interface if you are simply editing a part.

- **Component** is again the IComponent interface if editing a part in an assembly and Nothing if you are working on a part only.

- **value** returns True if the operation was successful. Check this value to determine whether you need to use ReleaseSelectionAccess as described below.

ReleaseSelectionAccess

If for some reason the edit operation failed, you should return the model back to its previous state. This is only important if you have made a call to AccessSelections. The call required is simply IFeatureData.ReleaseSelectionAccess.

41. Save and test the macro on the currently open part and verify that the *Cut-Extrude1* feature has been modified.

42. Close Visual Studio.

Feature Traversal

The macro you have built is quite limited. Everything is hard coded. There is no user interface or user interaction of any type. What if you want to modify every cut-extrude feature in a part and add two degrees of draft? You would need the ability to traverse the FeatureManager to find all cut-extrude features. If you wanted to build a tool to automatically suppress or edit all fillets in a model, you would need something similar. The following example illustrates how to find fillet features by their feature type.

Initial Code

43. Start a new macro named *FilletEdit.vbproj*.

44. Add the following code to traverse all fillets and display their names to the user.

```
Public Sub main()
    Dim swDoc As ModelDoc2
    Dim MyFeature As Feature
    Dim retval As Boolean

    swDoc = swApp.ActiveDoc

    'get the first feature
    MyFeature = swDoc.FirstFeature

    'loop through remaining features
    Do While MyFeature IsNot Nothing
        If MyFeature.GetTypeName2 = "Fillet" Then
            MsgBox("Found: " & MyFeature.Name)
        End If
        MyFeature = MyFeature.GetNextFeature
    Loop
End Sub
```

Running this macro will display a message with the name of each fillet feature in a part. It is not exactly the enhancement to SOLIDWORKS every user has been screaming for, but it illustrates some important methods.

FirstFeature and GetNextFeature

The first, next method of iterating through elements in the FeatureManager is referred to as traversal. The first Feature is retrieved using the call IModelDoc2.FirstFeature. A Visual Basic Do Loop is used to traverse each subsequent feature using the IFeature.GetNextFeature method. At some point, GetNextFeature will hit the end of the FeatureManager and will return Nothing (an empty object pointer).

Do While Loop

From the Material Properties chapter you learned about For loops that iterate over a known count. Traversal requires a different kind of loop – one that combines looping with a logical statement. The loop should exit when the MyFeature variable is set to Nothing at the end of the FeatureManager. A Do While loop does exactly that. It begins with the logical test at Do While. If the statement is True, the block code within the loop is run. This loop checks whether MyFeature is pointing to an object interface, written in code as `MyFeature IsNot Nothing`. End a Do While loop with the Loop statement. If you need an additional logical test within a Do While loop that will exit the loop, add an Exit Do statement anywhere within the Do While block.

GetTypeName2

The loop filters features using IFeature.GetTypeName2. This method returns a string representing the type of feature by name. The following list shows many of the common feature type names.

"Chamfer"	"Fillet"
"Cavity"	"Draft"
"MirrorSolid"	"CirPattern"
"LPattern"	"MirrorPattern"
"Shell"	"Extrusion"
"Cut"	"RefCurve"
"Revolution"	"RevCut"
"Sweep"	"SweepCut"
"SurfCut"	"Thicken"
"ThickenCut"	"VarFillet"
"HoleWzd"	"Imported"
"DerivedLPattern"	"CosmeticThread"

Name

The user is presented with the name of each of the fillets in the model using the IFeature.Name method. It simply returns the name of the feature seen in the FeatureManager.

To better understand feature names and feature type names, try the following.

45. Modify the loop as shown below, commenting out the original If statement and adding a new message.

```
'loop through remaining features
Do While MyFeature IsNot Nothing
    'If MyFeature.GetTypeName2 = "Fillet" Then
    '    MsgBox("Found: " & MyFeature.Name)
    'End If
    MsgBox("Feature: " & MyFeature.Name & vbCrLf & _
      "Feature Type: " & MyFeature.GetTypeName2)
    MyFeature = MyFeature.GetNextFeature
Loop
```

46. Run the macro to display each feature name and type name.

47. Stop the macro, remove the comments around the If statement and comment out the new message as shown.

```
'loop through remaining features
Do While MyFeature IsNot Nothing
    If MyFeature.GetTypeName2 = "Fillet" Then
        MsgBox("Found: " & MyFeature.Name)
    End If
    'MsgBox("Feature: " & MyFeature.Name & vbCrLf & _
    '  "Feature Type: " & MyFeature.GetTypeName2)
    MyFeature = MyFeature.GetNextFeature
Loop
```

Code Changes

Now make the macro do something useful. There are only a few lines of code required to find all fillets that match a specification. You can then perform an appropriate operation on them. In this example, fillets under a specific size will be suppressed.

48. Modify the existing loop in the macro to find and suppress all fillets having a default radius less than or equal to a value input by the user.

```
'loop through remaining features
Dim radius As String
Dim featureDef As Object
radius = InputBox("Suppress all fillets < or = (in meters)")
If Not IsNumeric(radius) Then
  MsgBox("Please enter only numeric values.", MsgBoxStyle.Exclamation)
  Exit Sub
End If
Do While MyFeature IsNot Nothing
    If MyFeature.GetTypeName = "Fillet" Then
    '    MsgBox("Found" & MyFeature.Name)
        featureDef = MyFeature.GetDefinition
        If featureDef.DefaultRadius <= CDbl(radius) Then
            MyFeature.Select2(False, 0)
            swDoc.EditSuppress2()
        End If
    End If
    MyFeature = MyFeature.GetNextFeature
Loop
```

SimpleFilletFeatureData2 and DefaultRadius

When GetDefinition is used on a fillet, the returned feature data definition is the ISimpleFilletFeatureData2 interface with several unique methods and properties. DefaultRadius is the method used in this example to get the radius of the fillet. For a complete list of methods and properties available to ISimpleFilletFeatureData2, explore the API Help.

Error Checking

As with the previous section, when using the InputBox function to get a numeric value, the IsNumeric function will help validate the user input. Converting radius to a double gives you an "apples to apples" comparison to the DefaultRadius value which is also a double data type.

Select2

To suppress the feature, it has to be selected, just like a user would work in SOLIDWORKS. As described earlier, the Select2 method of the IFeature interface is a quick way to select the feature when you already have its interface. Compare this to SelectByID2. For SelectByID2 you need the name of the feature and its type. You could use IFeature.Name to first get the name and then use SelectByID2 to select the feature. Feature.Select2 is simpler.

value = IFeature.Select2 (Append, Mark)

- **Append** requires a Boolean value. If it is True then the item is added to the current selection. If it is False, the current selection is cleared and this new item is selected.
- **Mark** marks the selection with an integer required for some methods. Feature suppression does not require a mark so it does not matter what value you use in this example.
- **value** will be either True or False to indicate whether the selection was successful or not.

EditSuppress2

Once a feature is selected, a quick call to EditSuppress2 from ModelDoc2 will suppress the feature.

49. Test your macro. All fillets with radii less than the value entered into the InputBox should be suppressed. The following value would suppress Fillet1, Fillet3 and Fillet4.

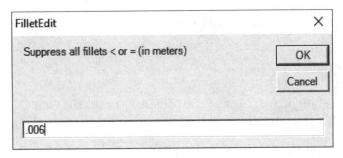

50. Close Visual Studio.

Conclusion

At this point, the possibilities for part automation are nearly endless. You could add code to the first macro using the InputBox method for additional dimension values. You could query an Excel spreadsheet or a database for values. A user form could be created that could check or restrict user input if you are trying to build a knowledge-based design tool. This same record-and-edit technique can be used to simplify the automation of most geometry while the traversal technique is effective for any post-creation editing.

If you intend to build a design-to-order tool, you may get better performance and simpler code if you start from an existing model and modify it rather than building each new model from an empty template. Use IModelDocExtension.SaveAs to generate new files after modifying a master model.

C# Examples

Feature Edit

```csharp
public void Main()
{
    PartDoc Part;
    Feature MyFeature;
    ExtrudeFeatureData2 featureDef;
    bool retval;
    string message = "";

    Part = (PartDoc)swApp.ActiveDoc;

    MyFeature = (Feature)Part.FeatureByName("Cut-Extrude1");
    featureDef = (ExtrudeFeatureData2)MyFeature.GetDefinition();

    //get some settings from the feature
    if (!featureDef.GetDraftWhileExtruding(true))
    {
        message = "The selected feature has no draft.\n";
    }

    switch (featureDef.GetEndCondition(true))
    {
        case (int)swEndConditions_e.swEndCondBlind:
            message += "Blind";
            break;
        case (int)swEndConditions_e.swEndCondThroughAll:
            message += "Through All";
            break;
        case (int)swEndConditions_e.swEndCondThroughAllBoth:
            message += "Through All Both";
            break;
        case (int)swEndConditions_e.swEndCondUpToSurface:
            message += "Up To Surface";
            break;
        case (int)swEndConditions_e.swEndCondMidPlane:
```

```
            message += "Mid Plane";
            break;
        case (int)swEndConditions_e.swEndCondOffsetFromSurface:
            message += "Offset From Surface";
            break;
        case (int)swEndConditions_e.swEndCondThroughNext:
            message += "Through Next";
            break;
        case (int)swEndConditions_e.swEndCondUpToBody:
            message += "Up To Body";
            break;
        default:
            break;
    }

    MessageBox.Show(message + " end condition.");

    //rollback to edit the feature
    retval = featureDef.AccessSelections(Part, null);
    //modify some feature values
    featureDef.SetEndCondition(true,
        (int)swEndConditions_e.swEndCondThroughAll);
    featureDef.SetDraftWhileExtruding(true, true);
    featureDef.SetDraftAngle(true, 2 * Math.PI / 180);

    //complete the edit operation
    retval = MyFeature.ModifyDefinition(featureDef, Part, null);

    //in case the modification failed
    if(!retval){
        featureDef.ReleaseSelectionAccess();
    }

    return;
}
```

Feature Traversal

```
public void Main()
{
    ModelDoc2 swDoc;
    Feature MyFeature;
    bool retval;

    swDoc = (ModelDoc2)swApp.ActiveDoc;

    //get the first feature
    MyFeature = (Feature)swDoc.FirstFeature();
```

```
        //loop through remaining features
        string radius;
        object featureDef;
        radius = Interaction.InputBox("Suppress all fillets < or = (in meters)");
        if(!IsNumeric(radius)){
            MessageBox.Show("Please enter only numeric values.", "Feature Edit",
                MessageBoxButtons.OK, MessageBoxIcon.Exclamation);
            return;
        }

        while (MyFeature != null) {
            if (MyFeature.GetTypeName2() == "Fillet")
            {
                //MessageBox.Show("Found: " + MyFeature.Name);
                featureDef = (SimpleFilletFeatureData2)MyFeature.GetDefinition();
                SimpleFilletFeatureData2 filletfeat =
                    (SimpleFilletFeatureData2)featureDef;
                if (filletfeat.DefaultRadius <= Convert.ToDouble(radius))
                {
                    MyFeature.Select2(false, 0);
                    swDoc.EditSuppress2();
                }
            }
            //MessageBox.Show("Feature: " + MyFeature.Name
            // + System.Environment.NewLine
            //+ "Feature Type: " + MyFeature.GetTypeName2());
            MyFeature = (Feature)MyFeature.GetNextFeature();
        }

        return;
}

//helper function to check string for numeric (double) value
public bool IsNumeric(string value)
{
    Double num = 0;
    bool isDouble = false;

    // Check for empty string.
    if (string.IsNullOrEmpty(value))
    {
        return false;
    }
    isDouble = Double.TryParse(value, out num);
    return isDouble;
}
```

Data Import and Export

	A	B	C
1	1.0000	0.0000	0.0000
2	0.9659	-0.2588	0.0417
3	0.8660	-0.5000	0.0833
4	0.7071	-0.7071	0.1250
5	0.5000	-0.8660	0.1667
6	0.2588	-0.9659	0.2083
7	0.0000	-1.0000	0.2500
8	-0.2588	-0.9659	0.2917
9	-0.5000	-0.8660	0.3333
10	-0.7071	-0.7071	0.3750
11	-0.8660	-0.5000	0.4167
12	-0.9659	-0.2588	0.4583
13	-1.0000	0.0000	
14	-0.9659	0.2588	
15	-0.8660	0.5000	0.5833
16	-0.7071	0.7071	0.6250
17	-0.5000	0.8660	0.6667
18	-0.2588	0.9659	0.7083
19	0.0000	1.0000	0.7500
20	0.2588	0.9659	0.7917
21	0.5000	0.8660	0.8333
22	0.7071	0.7071	0.8750
23	0.8660	0.5000	0.9167
24	0.9659	0.2588	0.9583
25	1.0000	0.0000	1.0000

- 3D Points

- Code Modules, Classes and Portability

- 3D Curves

- Export Points

- Export Sheet Metal Flat Patterns

Introduction

In this chapter, you will create a macro that imports 3D point data and 3D curve data from Microsoft Excel. This comes in handy if you have design data generated by other programs that you must use during the automation process. We will also explore exporting points as well as a common request, sheet metal flat patterns.

3D Points

The 3D point macro is quite simple. It makes use of the 3D sketch capabilities of SOLIDWORKS and automates creation of sketch points. Follow these instructions to record the initial code.

1. Create a new part in SOLIDWORKS.

2. Start recording a new macro by selecting Tools, Macro, Record or clicking **II●** on the macro toolbar.

3. Start a 3D sketch by selecting Insert, 3D Sketch or by clicking [3D].

4. Select the Point sketch tool ▪ and create a few points on the screen. It does not matter exactly where you place them.

5. Exit the sketch.

6. Stop the macro recorder and save the new macro as *3DPoints.vbproj*.

7. Edit the new macro and review the code. It should look something like the following. All CType statements have been removed from this sample for clarity.

```
Public Sub main()

    Dim swDoc As ModelDoc2 = Nothing
    Dim swPart As PartDoc = Nothing
    Dim swDrawing As DrawingDoc = Nothing
    Dim swAssembly As AssemblyDoc = Nothing
    Dim boolstatus As Boolean = False
    Dim longstatus As Integer = 0
    Dim longwarnings As Integer = 0
    swDoc = swApp.ActiveDoc
    swDoc.SketchManager.Insert3DSketch(True)
    Dim skPoint As SketchPoint = Nothing
    skPoint = swDoc.SketchManager.CreatePoint(-0.025697, 0.038464, 0)
    skPoint = swDoc.SketchManager.CreatePoint(-0.048349, -0.028166, 0)
    skPoint = swDoc.SketchManager.CreatePoint(0.033022, -0.019803, 0)
    swDoc.ClearSelection2(True)
    swDoc.SketchManager.InsertSketch(True)
End Sub
```

Insert3DSketch

The only two new API calls used in this code are ISketchManager.Insert3DSketch and ISketchManager.CreatePoint. The Insert3DSketch method is straightforward – it starts a 3D sketch. Unlike InsertSketch which starts a 2D sketch, Insert3DSketch does not expect or require a plane to be selected first. The Boolean argument passed to the call causes the model to rebuild when you exit the sketch if set to True. To close or exit a 3D Sketch, call ISketchManager.InsertSketch as was done with 2D sketches.

CreatePoint

ISketchManager.CreatePoint requires three arguments: the x, y and z location of the newly created point in meters. It returns a reference to a ISketchPoint interface. Just like all other SOLIDWORKS API interfaces, sketch points also have their own properties and methods.

Code Changes

So how do you make this a useful tool? It would be tedious to code every point into a macro if you are trying to import a 3D point cloud. Instead, make the utility read the point values from an Excel spreadsheet.

The code should also work for any number of 3D points. We will assume that there is an open Excel spreadsheet with point locations in three columns. The first row should contain the first point's x, y and z location, the second row contains the next point and so on. The following image is an example of the expected format.

	A	B	C
1	1.0000	0.0000	0.0000
2	0.9659	-0.2588	0.0417
3	0.8660	-0.5000	0.0833
4	0.7071	-0.7071	0.1250
5	0.5000	-0.8660	0.1667
6	0.2588	-0.9659	0.2083
7	0.0000	-1.0000	0.2500
8	-0.2588	-0.9659	0.2917
9	-0.5000	-0.8660	0.3333
10	-0.7071	-0.7071	0.3750
11	-0.8660	-0.5000	0.4167
12	-0.9659	-0.2588	0.4583
13	-1.0000	0.0000	0.5000
14	-0.9659	0.2588	0.5417
15	-0.8660	0.5000	0.5833
16	-0.7071	0.7071	0.6250
17	-0.5000	0.8660	0.6667
18	-0.2588	0.9659	0.7083
19	0.0000	1.0000	0.7500
20	0.2588	0.9659	0.7917
21	0.5000	0.8660	0.8333
22	0.7071	0.7071	0.8750
23	0.8660	0.5000	0.9167
24	0.9659	0.2588	0.9583
25	1.0000	0.0000	1.0000

Single Turn Helix Points

When the macro reaches an empty cell, the end of the data has been reached and code should stop. That sounds like a perfect place for a loop as well, but should it be a Do While or For loop?

Also, this is code that would be helpful in multiple applications. This is the perfect time to use a separate code file.

Code Files and Portability

8. Create a new code module by selecting Project, Add Module. Name the module ExcelDataReader.vb.

9. Create a new function in the new module as shown below.

```
Function getExcelPoints() As List(Of Double())
    Dim Excel As Object = GetObject(, "Excel.Application")

End Function
```

Functions or Procedures

At this point it is worth taking a minute to remind you of a distinction. Functions and procedures are different in one aspect. A procedure operates on data and can accept as many input arguments as needed. However, it does not return anything to the code that calls it. Functions can still have arguments, but they also return data back to the calling code.

As an example, the CreatePoint method discussed above returns a SketchPoint interface. It is a function. Insert3DSketch starts a sketch, but does not return a value. It could be considered a procedure or method. Whenever you create code outside of the main procedure, the first thing to decide is if it will be a procedure or a function. If you need something returned by the code block, declare it as a function.

List(Of Type)

This new function will return a List of Double values. Lists are defined in Visual Basic as an expandable collection of items. They have more flexibility than arrays. You can add to, remove from, find in and clear Lists with built-in commands. You can also use For Each loops rather than looping by index. Lists give enough flexibility that they are often preferred to arrays whenever you need flexibility.

In this example, the Function returns a List of Arrays of Double values. It has to return the x, y and z value for each point. So each array of Doubles will have those three values.

Working with Excel

In an earlier chapter you learned how to use the Excel VBA interface to control SOLIDWORKS. The new code shows how to use the SOLIDWORKS VSTA interface to attach to the currently running Excel application and gather information from cells in the active spreadsheet.

New variables were added – an Object type variable used to capture the Excel application, an integer variable to be used as a counter for the loop and three double values to store the x, y and z location of each point. Late binding is being used in this example to avoid adding a reference to the Microsoft Excel Object Library.

GetObject vs. CreateObject

To capture the currently running Excel interface, the GetObject method is used. This is a Visual Basic method that captures a currently running application. CreateObject is another Visual Basic method that connects to an application, but CreateObject launches another instance of the application rather than capturing the instance currently running.

GetObject was used in the Model Dimensions chapter to connect to SOLIDWORKS from Excel. In this example, we're connecting to Excel from a SOLIDWORKS macro. The structure of the call is the same. Late binding is used to simplify code and leave out the reference to the Microsoft Office type library.

10. Add the following code to read cells with values from the first three columns of an open Excel file.

```
Function getExcelPoints() As List(Of Double())
    Dim Excel As Object = GetObject(, "Excel.Application")
    Dim i As Integer = 1
    Dim xPt As Double = 0
```

```
    Dim yPt As Double = 0
    Dim zPt As Double = 0
    Dim points As New List(Of Double())
    Do While Excel.Cells(i, 1).Value IsNot Nothing
        xPt = Excel.Cells(i, 1).value
        yPt = Excel.Cells(i, 2).value
        zPt = Excel.Cells(i, 3).value
        Dim p As Double() = {xPt, yPt, zPt}
        points.Add(p)
        i += 1
    Loop
    Return points
End Function
```

Excel.Cells

In a previous exercise, you used the Excel.Range method and passed the cell name in the "A1" format. The Range method is great if you know exactly which cell you want by its name. An alternative technique is used here to read Excel cell values. The cells need to be read incrementally, one at a time, three across and then down to the next row until you find an empty cell. The Excel.Cells method works best here and retrieves a single Cell interface. It expects two integers as arguments. The first is for the row index and the second is for the column index (the opposite of the "A1" convention where the column is listed before the row). The Value property of the Cell interface returns cell's underlying value, not just the displayed text. For example, a cell may have a date or currency formatting or even truncated decimal display. If you need to retrieve the displayed text, use the Cell's Text property instead.

Loop Logic

A Do While loop does the job of testing whether the upcoming row's column 1 cell has a value or is Nothing (an empty cell). Notice the variable i is declared and set to 1 initially to make the loop begin with the first row. If the Cell.Value property IsNot Nothing (it has a value), the three variables, xpt, ypt and zpt are populated with the Excel cell values from the row. Those three values are put into an array of Doubles p.

The points variable will be the value returned by the function. That means it must have the same type declaration. It has been declared as a List(Of Double()), or a List of arrays of Doubles.

Lists have an Add method that adds an element to the list. For each row in Excel, the array of three Doubles is added to the points List.

Our counter, i, is increased by one using the special += expression. This expression will add the right side of the expression to the variable if it is numeric, or concatenate if it is a string.

Another pass is made to the loop to test whether cell "A2" (Excel.Cells(2,1)) IsNot Nothing. If it is not empty, the code block continues. When it finds an empty cell, the loop exits and the points are returned to the method that called the Function.

With the new module and function, the main procedure can be simplified significantly.

11. Modify the recorded code as shown below in bold.

```
Public Sub main()

    Dim swDoc As ModelDoc2 = Nothing
    Dim swPart As PartDoc = Nothing
    Dim swDrawing As DrawingDoc = Nothing
    Dim swAssembly As AssemblyDoc = Nothing
    Dim boolstatus As Boolean = False
    Dim longstatus As Integer = 0
    Dim longwarnings As Integer = 0
    swDoc = CType(swApp.ActiveDoc, ModelDoc2)
    swDoc.SketchManager.Insert3DSketch(True)

    swDoc.SketchManager.AddToDB() = True

    For Each pt As Double() In getExcelPoints()
        swDoc.SketchManager.CreatePoint(pt(0), pt(1), pt(2))
    Next

    swDoc.SketchManager.AddToDB() = False
    swDoc.ClearSelection2(True)
    swDoc.SketchManager.InsertSketch(True)
End Sub
```

For Each ... Next Loop

The For Each, Next block in the middle of the macro does the work of looping through the List returned from getExcelPoints. But this example is different from previous For loops. When working with Lists and Arrays, the For Each method simplifies the loop. The structure is as follows. Replace the code in [ ] with variables, arrays and processing code.

```
For Each [looping variable declaration] In [Array, List or Collection]
    [processing code here]
Next
```

The new For Each loop will iterate through each item in the List, an array of three Doubles in our example, and create the sketch point for each using their respective array elements 0, 1 and 2. The variable pt must be the same datatype as the List. No indexing is needed and no separate variable declarations are needed.

SketchManager.AddToDB

The AddToDB property of the ISketchManager interface can be invaluable to accurately create sketch geometry. This was not necessary when sketching was introduced in the model creation chapter. However, if you try creating sketch geometry that is very small compared to your zoom scale, you will run into unexpected behavior. Just like the user interface, there is a snapping zone even for the API. When two points are drawn close enough to one another, they merge automatically. AddToDB essentially creates sketch geometry without solving. The API Help explains it as adding sketch geometry directly to the database. No sketch relations are created either. Setting this property to True enables sketch geometry to be created consistently and quickly. Just be aware, you must set the property back to False before exiting your code or SOLIDWORKS sketches will not behave as expected. When sketching highly detailed or very small sketches through the API, setting AddToDB to true will give your macros accuracy and a significant performance boost.

Classes versus Modules

You have explored the basic differences of classes and modules. But when should you use one instead of the other? Macros always start from a class named SolidWorksMacro. Every form and dialog is also a class. You just used a module for a common Excel operation. This slightly more advanced topic will show you how to use a class in place of the ExcelDataReader module. This is not a necessary step for the macro but it is a good place to introduce the topic.

Point Class

Start by adding a custom Point class to your macro.

12. From the menu, select Project, Add Class. Name the class Point.vb.

13. Add the following public variables to the new class.

```
Public Class Point
    Public X As Double
    Public Y As Double
    Public Z As Double
End Class
```

ExcelDataReader Class

Next, create an ExcelDataReader2 class. The class will be nearly identical to the module with a minor difference.

14. Select Project, Add Class. Name the new class ExcelDataReader2.vb.

15. Add the following code to the class.

```
Public Class ExcelDataReader2
    Dim Excel As Object
    Public Sub New()
        Try
            Excel = GetObject(, "Excel.Application")
        Catch ex As Exception
            MsgBox("Please open an Excel Workbook first.",
                    MsgBoxStyle.Critical)
        End Try
    End Sub
End Class
```

The New() method is run each time a new instance of this class is created. If an Excel Workbook is open, the variable Excel is connected to the running session using GetObject. Otherwise, a message is sent to the user to open a workbook.

16. Copy the getExcelPoints function from the module to the class as shown below.

```
Public Class ExcelDataReader2
    Dim Excel As Object
    Public Sub New()
        Try
            Excel = GetObject(, "Excel.Application")
        Catch ex As Exception
            MsgBox("Please open an Excel Workbook first.",
                    MsgBoxStyle.Critical)
        End Try
    End Sub

    Function getExcelPoints() As List(Of Double())
        Dim Excel As Object = GetObject(, "Excel.Application")
        Dim i As Integer = 1
        Dim xPt As Double = 0
        Dim yPt As Double = 0
        Dim zPt As Double = 0
        Dim points As New List(Of Double())
        Do While Excel.Cells(i, 1).Value IsNot Nothing
            xPt = Excel.Cells(i, 1).value
            yPt = Excel.Cells(i, 2).value
            zPt = Excel.Cells(i, 3).value
            Dim p As Double() = {xPt, yPt, zPt}
            points.Add(p)
            i += 1
        Loop
        Return points
    End Function
End Class
```

17. Make the following modifications to the getExcelPoints function. Be sure to delete the line that uses GetObject to connect to the Excel Application. This has already been done when the class was instantiated (when a new instance is created).

```
Public Function getExcelPoints() As List(Of Point)
    Dim i As Integer = 1
    Dim points As New List(Of Point)
    Do While Excel.Cells(i, 1).Value IsNot Nothing
        Dim p As New Point
        p.X = Excel.Cells(i, 1).value
        p.Y = Excel.Cells(i, 2).value
        p.Z = Excel.Cells(i, 3).value
        points.Add(p)
```

```
        i += 1
    Loop
    Return points
End Function
```

The first change is the scope of the function. Public makes it available from outside of the class. The function's return type has been changed to a List of the new Point class. The two variables p and points are also using the new Point class. Notice that the variable p has also been moved into the While loop.

Instead of creating an array of Doubles, the While loop now creates a new instance of the Point class. The X, Y and Z variables of that class are set to the Excel cell values.

See how the Point class makes the code more readable.

18. Switch back to SolidWorksMacro.vb and change the code in main() as shown, using the new classes.

```
swDoc = CType(swApp.ActiveDoc, ModelDoc2)
swDoc.SketchManager.Insert3DSketch(True)

swDoc.SketchManager.AddToDB() = True

Dim edr As New ExcelDataReader2
Dim points As List(Of Point) = edr.getExcelPoints()
For Each pt As Point In points
    swDoc.SketchManager.CreatePoint(pt.X, pt.Y, pt.Z)
Next

swDoc.SketchManager.AddToDB() = False
swDoc.ClearSelection2(True)
swDoc.SketchManager.InsertSketch(True)
```

Notice the difference using the class versus the module. The variable edr is used to store a new instance of the class, like you have seen with forms. The Excel application is connected when the class is created, so there is no need to set it or see it from the main code. And finally, the X, Y and Z values from each Point in the List are used in the CreatePoint call rather than array elements.

Using classes can take a little more planning and effort, but they make your code more portable and readable in the long run.

Unit Conversion Functions

An assumption has been made to this point. We assume that the values passed to SOLIDWORKS are all in meters. What if the spreadsheet data is in millimeters, inches, or some other unit? The following function can make unit conversion a snap.

```
Function inchToMeters(InchVal As Double) As Double
    Return InchVal * 0.0254
```

```
End Function

Function mmToMeters(mmVal As Double) As Double
    Return mmVal / 1000
End Function
```

These quick functions return the input argument, multiplied or divided by a factor to get the expected result. Functions can be as simple or as complex as needed, but when a Return statement is hit, the value of the Return statement line is passed back to the code that called the function and no additional function code is evaluated.

If needed, a function can have multiple Return statements. Consider the following example.

```
Function IsGreaterThanOne(value as Double) as Boolean
    If value > 1 Then
        Return True
    End If
    Return False
End Function
```

If the argument passed to the function is greater than one, the If statement evaluates to True and the Return True statement sends a True Boolean value back to the calling code. Functions always exit at a Return statement so the Return False statement never runs. If the argument is equal to or less than one, the code in the If statement is skipped and the Return False line is run, passing a False Boolean value back to the calling code.

19. Add another code module to your macro by selecting Project, Add Module. Or as a shortcut, right-click the project name in the Solution Explorer and select Add, Module. Name the new module *Utilities.vb* and click Add.

20. Add the code for the inchToMeters and mmToMeters functions as shown below. They are declared Public to make them available anywhere in the macro code.

```
Module Utilities
    Public Function inchToMeters(ByVal InchVal As Double) As Double
        Return InchVal * 0.0254
    End Function

    Public Function mmToMeters(mmVal As Double) As Double
        Return mmVal / 1000
    End Function
End Module
```

Where is the right place to use these conversion functions? You could convert the point values when each sketch point is created, or you could convert the values before they are stored in the Point class. The choice is up to you as you build your projects. In this example, we will convert the values to meters when they are read from the Excel workbook into the Point class.

21. Go to the ExcelDataReader2.vb tab. Add the following Imports statement.

```
Imports SolidWorks.Interop.swconst
Public Class ExcelDataReader2
    Dim Excel As Object
```

22. Modify the getExcelPoints function as shown below, including the new units parameter.

```
Public Function getExcelPoints(units As swLengthUnit_e) As List(Of Point)
    Dim i As Integer = 1
    Dim points As New List(Of Point)
    Dim multiplier As Double = 1
    If units = swLengthUnit_e.swINCHES Then multiplier = inchToMeters(1)
    If units = swLengthUnit_e.swMM Then multiplier = mmToMeters(1)
    Do While Excel.Cells(i, 1).Value IsNot Nothing
        Dim p As New Point
        p.X = Excel.Cells(i, 1).value * multiplier
        p.Y = Excel.Cells(i, 2).value * multiplier
        p.Z = Excel.Cells(i, 3).value * multiplier
        points.Add(p)
        i += 1
    Loop
    Return points
End Function
```

SOLIDWORKS already has a length unit enumeration named swLengthUnit_e in swConst type library. The units parameter in the function allows your calling code to specify the incoming units in inches or millimeters. You could continue this logic with any units you might need. Don't forget to add conversion functions for additional units in the Utilities.vb module.

Notice the shorthand format for the two If statements. You can write the logical check and the Then result in one line, ignoring the End If statement for simplicity.

It might seem strange that we have not used the conversion functions directly on the Excel cell values. Instead, a multiplier variable has been created that multiplies each value.

Consider the logic and code required to handle the conversion of each cell value. You would need something like the following. Imagine how this code would look if you wanted to handle more units options!

```
Do While Excel.Cells(i, 1).Value IsNot Nothing
    Dim p As New Point
    If units = swLengthUnit_e.swINCHES Then
        p.X = Utilities.inchToMeters(Excel.Cells(i, 1).Value)
        p.Y = Utilities.inchToMeters(Excel.Cells(i, 2).Value)
        p.Z = Utilities.inchToMeters(Excel.Cells(i, 3).Value)
    End If
    If units = swLengthUnit_e.swMM Then
        p.X = Utilities.mmToMeters(Excel.Cells(i, 1).Value)
        p.Y = Utilities.mmToMeters(Excel.Cells(i, 2).Value)
        p.Z = Utilities.mmToMeters(Excel.Cells(i, 3).Value)
    End If
    points.Add(p)
    i += 1
Loop
```

23. Switch back to the SolidWorksMacro.vb tab and add the units parameter for swINCHES to getExcelPoints.

```
Dim edr As New ExcelDataReader2
Dim points As List(Of Point) = edr.getExcelPoints(swLengthUnit_e.swINCHES)
For Each pt As Point In points
    swDoc.SketchManager.CreatePoint(pt.X, pt.Y, pt.Z)
Next
```

24. Open the sample Excel spreadsheet named *Helix Table.xls* from the downloaded examples. This spreadsheet has been set up to build points to create a single-turn helix. Alternatively, create an Excel spreadsheet with several rows of x, y and z values beginning at cell A1.

25. With the spreadsheet open, start a new part in SOLIDWORKS and then run your macro to create the results shown here.

26. Close Visual Studio.

3DCurves

Points are great, but it is typically more useful to build 3D curves in SOLIDWORKS rather than just points. Recording a macro using the Curve Through XYZ Points tool generates code similar to the following.

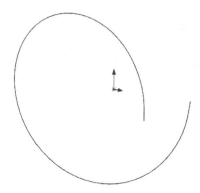

```
swDoc.InsertCurveFileBegin()
swDoc.InsertCurveFilePoint(-0.1265384925976, 0.01485370121131, 0)
swDoc.InsertCurveFilePoint(-0.1122488559892, 0.0554663526245, 0)
swDoc.InsertCurveFilePoint(-0.1231541049798, -0.01259744279946, 0)
swDoc.InsertCurveFileEnd()
```

InsertCurveFileBegin and InsertCurveFileEnd

There are three calls involved in building a curve from data points. The first and last simply tell SOLIDWORKS that you are starting and finishing the creation of a curve. The first of these is IModelDoc2.InsertCurveFileBegin. When you are done creating points for the curve you must call IModelDoc2.InsertCurveFileEnd. The variable named swDoc, in this example code, must be a pointer to an IModelDoc2 interface. However, this process is only valid for parts and assemblies, not drawings since they are 3D.

InsertCurveFilePoint

The real action happens through IModelDoc2.InsertCurveFilePoint. This call, like sketch points, requires the x, y and z location of the point to be created in, you guessed it, meters.

Sample Code

You can reuse the same structure used in the 3D points macro to gather the data from Excel. The following code shows the use of the InsertCurveFile methods. It is identical to the 3D points macro other than the changes in bold.

27. Copy the 3DPoints macro folder and its contents. Rename the folder 3DCurve. This will not rename the macro files, but will create a new macro with all of the same code.

28. Edit the new macro from the copied folder and change the code as shown in bold. The unused variable declarations have been removed for simplicity.

```
Public Sub main()
    Dim swDoc As ModelDoc2 = Nothing
    swDoc = CType(swApp.ActiveDoc, ModelDoc2)
    swDoc.InsertCurveFileBegin()

    Dim edr As New ExcelDataReader2
    Dim points As List(Of Point) = _
```

```
        edr.getExcelPoints(swLengthUnit_e.swINCHES)
    For Each pt As Point In points
        swDoc.InsertCurveFilePoint(pt.X, pt.Y, pt.Z)
    Next
    swDoc.InsertCurveFileEnd()
    swDoc.ViewZoomtofit2()
End Sub
```

29. Test and debug the new 3DCurves macro as necessary.

30. Close Visual Studio.

Exporting Point Data

You may also find a need to reverse the process and export point data. The ISketchManager interface can be used to access SketchPoints. Each SketchPoint interface will contain its coordinates. The following code illustrates a simple technique.

31. Open a new or existing SOLIDWORKS part. If it is a new part, save it with any name before continuing to avoid empty file path errors.

32. Create a new 3D Sketch and sketch a variety of points and lines. Keep the sketch active – don't exit.

33. Create a new macro by selecting Tools, Macro, New. Save the macro as *PointExport.vbproj*.

34. Add the following code to the main procedure to export the X, Y, Z coordinates of all points in the active sketch to a tab-delimited text file.

```
Public Sub main()
    'assume the desired sketch is currently in edit mode
    Dim model As ModelDoc2 = swApp.ActiveDoc
    Dim skMgr As SketchManager = model.SketchManager
    Dim ptSketch As Sketch = skMgr.ActiveSketch
    If ptSketch Is Nothing Then
        MsgBox("Please edit the desired sketch before running.", _
            MsgBoxStyle.Exclamation)
        Exit Sub
    Else
        'initialize a string for the output file
        Dim outputString As String
        'it will be tab-delimited
        Dim headerRow As String = "X" & vbTab & "Y" & vbTab & "Z"
        outputString = headerRow

        'gather all sketch points
        Dim points As Object
        points = ptSketch.GetSketchPoints2
        For Each skPoint As SketchPoint In points
            Dim x As Double = skPoint.X
```

```
        Dim y As Double = skPoint.Y
        Dim z As Double = skPoint.Z

        'add the text to the output string
        outputString += & vbCrLf & x.ToString & vbTab _
           & y.ToString & vbTab & z.ToString
    Next

        'save the resulting data to a file
        Dim filePath As String = _
          IO.Path.ChangeExtension(model.GetPathName, ".txt")
        My.Computer.FileSystem.WriteAllText (filePath, outputString, False)
        'open the text file
        Process.Start(filePath)
    End If

  End Sub
```

ActiveSketch

The code needs to be told which sketch to process. There are a few different ways to get to a specific sketch in SOLIDWORKS. You can get a feature by name directly through the IPartDoc or IAssemblyDoc interface. You can traverse the FeatureManager from top to bottom, finding a sketch by some criteria. Alternatively, you can simply get the active sketch.

This example assumes the sketch is in its edit state. The SketchManager's ActiveSketch method returns that sketch.

Some error checking verifies that there was an active sketch. If the returned sketch is nothing, an empty result, the user is presented with a message to edit a sketch before running.

Tab-delimited Output

The macro creates a tab-delimited text file that could be opened in Excel or used for input to other applications. It uses a strategy of populating a string variable named outputstring with the entire content that needs to be written to the file. After declaring the variable it is populated with header values of X, Y and Z separated by tab characters, represented by the Visual Basic constant vbTab.

The first time through the loop, outputString is "X[tab]Y[tab]Z" where [tab] is an actual tab character. There is no new line or carriage return character. To make a multi-lined output file, a carriage return is added using the Visual Basic constant vbCrLf. This represents both a carriage return and a line feed character as described in the first chapter. The first set of x, y and z values are added, separated by tab characters. The Visual Basic.NET ToString method is used to turn the Double values into strings so they can be appended to outputString. As in the case of converting strings to double values using CType, ToString ensures the proper data type before combining. The loop continues until all points have been processed.

WriteAllText

This is an additional technique for writing text to a file. From the Custom Properties chapter, StreamWriter is best suited for writing an indeterminate number of lines of text, written one line at a time. The WriteAllText method of My.Computer.FileSystem writes all text from a string into a file, with only one line of code. The first argument required is the full path to the file to be created or edited. Don't forget to include the extension. This macro makes use of IO.Path.ChangeExtension introduced in the PDF chapter. The second argument of WriteAllText is the string of text to write. The third is a Boolean value telling the method to either overwrite or append to existing files. If set to False, any existing content in the file will be overwritten. Pass True to append new content to an existing file. If no file exists, you can pass True or False with the same result.

Code Snippets

Writing text to a file is extremely common. So are many other typical file and math operations. Code snippets are a great way to generate commonly used code blocks without having to remember the method and syntax. Right-click in the code window and select Insert Snippet to access the library. Writing text to a file can be found under fundamentals, filesystem code snippet grouping.

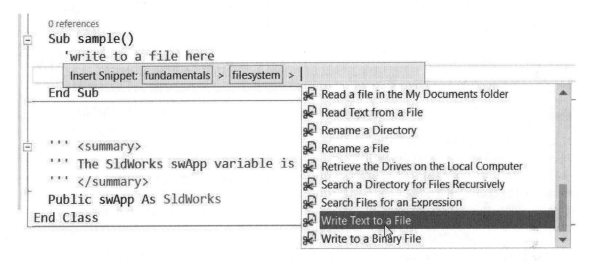

The snippet builds the following line of code. The full example macro replaces the hard-coded path and string input with variables. The snippet highlights the elements you should change in light yellow as a reminder. Use the Tab key to step between the highlighted elements.

```
My.Computer.FileSystem.WriteAllText("C:\Test.txt", "Text", True)
```

Another snippet was used to open the text file at the end of the macro. From os, Process, select Run a Program Associated with a File Type. This generates the code below.

```
Process.Start("C:\Test.txt")
```

The string "C:\Test.txt" was then replaced with filePath to open the newly created text file.

Sketch.GetSketchPoints2

The GetSketchPoints2 method of the sketch returns an array of SketchPoints to be evaluated through the For Each loop. This includes individual sketch points like you would see when using the Hole Wizard, along with the sketch points at each end of sketch segments. The array of points returned by GetSketchPoints2 is defined in the API help as the generic Object type. Remember that the Object type can contain any data type, including an array. You can still loop through the array stored in the Object using a For Each block.

SketchPoint.X, Y and Z

The X, Y and Z properties of a SketchPoint return double values representing the coordinate location of the point. The loop adds the coordinates, converted to string values, along with tab and return characters to the `outputString` variable for each sketch point. Appending to an existing string is done by setting `outputString` equal to its current value, then appending the additional values as discussed above.

If you needed to segregate individual sketch points from the end points of sketch segments, use SketchPoint.Type. It will return an integer based on the enumeration swSketchPointType_e. For example, `swSketchPointType_e.swSketchPointType_Internal` represents the endpoint of a sketch line or other sketch segment while `swSketchPointType_e.swSketchPointType_User` is an individual sketch point.

35. Test and debug the macro as needed. Then close Visual Studio.

Saving Sheet Metal Flat Patterns

No discussion of exporting from SOLIDWORKS would be complete without sheet metal flat patterns to DWG or DXF. The API makes this one easy with one call. The process is similar to saving a PDF from a drawing. You first need to decide on an output name for the DWG or DXF. In the example below, we will simply change the extension of the open file to DXF. The resulting flat pattern will be saved to the same folder with the same name as the originating part. You could take it a step further and use batch processing to generate flat patterns from all parts in a folder. For each part in a folder, you could check to see if a flat pattern feature exists by traversing the feature tree, and then export the flat pattern. That technique is explained in the Favorite Code Examples chapter.

Since the SOLIDWORKS macro recorder creates an unusually complicated process, we will start with a new macro.

36. Open or create a sheet metal part in SOLIDWORKS. If it is a new part, save it before continuing.

37. Create a new macro named *ExportFlatPattern.vbproj* and add the following code to the main procedure.

```
Public Sub main()
    'get the active document
    Dim swDoc As ModelDoc2 = swApp.ActiveDoc
    If swDoc.GetType <> swDocumentTypes_e.swDocPART Then
        MsgBox("This macro is for parts only.", MsgBoxStyle.Exclamation)
        Exit Sub
    End If
    'get the part interface
    Dim swPart As PartDoc = swDoc
    'set a new path for the flat pattern
    Dim flatPatternPath As String
    Dim ext As String = ".DXF"
```

```
'could also use .DWG if prefered
flatPatternPath = IO.Path.ChangeExtension(swDoc.GetPathName, ext)
'export the flat pattern with bend lines
Dim bendSetting As Long
bendSetting = 2 ^ 0 + 2 ^ 2
'bendSetting = 2 ^ 0   'use this setting for no bend lines
swPart.ExportToDWG2(flatPatternPath, swDoc.GetTitle,
                    swExportToDWG_e.swExportToDWG_ExportSheetMetal,
                    True, Nothing, False, False, bendSetting, Nothing)
End Sub
```

The general process (or algorithm) is to 1) connect to the active document, 2) connect the IModelDoc2 interface to the more specific IPartDoc interface, and 3) export the flat pattern view. The macro includes additional details to make it robust and easier to read. For example, IModelDoc2.GetType is used to make sure the active document is a part. IO.Path.ChangeExtension is used to modify the file path.

IPartDoc.ExportToDWG2

The IPartDoc interface's ExportToDWG2 method is the call that does all of the work. This is the same method you would use to export a selected face to DXF or DWG (not a common action). The method requires the path of the DXF or DWG to be created along with several options for the output alignment and style.

value = IPartDoc.ExportToDWG2(FilePath, ModelName, Action, ExportToSingleFile, Alignment, IsXDirFlipped, IsYDirFlipped, SheetMetalOptions, Views)

- **FilePath** is a string value with a complete file path to the output DXF or DWG file. It should include the new file name and extension.

- **ModelName** is the name of the active part to export as a string.

- **Action** is defined in swExportToDWG_e. Always use swExportToDWG_ExportSheetMetal for sheet metal flat patterns.

> swExportToDWG_ExportAnnotationViews
> swExportToDWG_ExportSelectedFacesOrLoops
> swExportToDWG_ExportSheetMetal

- **ExportToSingleFile** should always be True for sheet metal.

- **Alignment** is an array of 12 Double values related to output alignment. To maintain default alignment, pass Nothing.

- **IsXDirFlipped** is True to flip the output on the X axis, False to maintain the default.

- **IsYDirFlipped** is True to flip the output on the Y axis, False to maintain the default.

- **SheetMetalOptions** is a Bitmask of sheet metal output options. This value is only evaluated if the Action is swExportToDWG_ExportSheetMetal. More information below.

- **Views** is an array of annotation view names to export and is only used if the Action is swExportToDWG_ExportAnnotationViews. This does not apply to sheet metal flat patterns.

The SheetMetalOptions are called a Bitmask. A Bitmask is a way to pack several options into one value. There are 8 unique output options in SheetMetalOptions. Instead of creating ExportToDWG2 with 16 parameters, the 8 options are combined. The following table shows the options and their corresponding Bit and Long value.

Bit	Action if set to 1	Long Value
Bit #1	Export the flat pattern geometry	1
Bit #2	Show hidden edges	2
Bit #3	Show bend lines	4
Bit #4	Show sketches	8
Bit #5	Merge coplanar faces	16
Bit #6	Show Library Features	32
Bit #7	Show Forming Tools	64
Bit #12	Show the bounding box	2048

To set a specific bit, add its Long value to the Bitmask. For example, flat pattern geometry, bend lines and sketches, use the following.

```
bendsetting = 1 + 4 + 8
```

As another way to create the Bitmask, add 2 ^ n where n is the index of the Bit. They are zero-based, so Bit #1 is 0, Bit #5 is 4, etc.

```
bendsetting = 2 ^ 0 + 2 ^ 2 + 2 ^ 3
```

38. Test and debug the macro. Try different sheet metal parts to make sure the output file path is working as expected. Try different output options with DWG and DXF as well as flat patterns with and without bend lines, sketches, etc.

39. Close Visual Studio.

Note: ExportToDWG2 requires a body to be pre-selected if the sheet metal part has multiple sheet metal bodies.

Conclusion

Though these API methods are simple, they can add a tremendous amount of power to a customized application to generate and export geometry. Curves and points can be used to generate flow paths, input scanned data or create mathematical curves. Exported data can be used in other applications or for laser or plasma cutting of sheet metal parts.

Practice using separate code modules and classes to add flexibility and code portability to your macros and applications.

C# Examples

3D Curve

```csharp
public void Main()
{
    ModelDoc2 swDoc = null;
    swDoc = ((ModelDoc2)(swApp.ActiveDoc));
    swDoc.SketchManager.Insert3DSketch(true);
    swDoc.SketchManager.AddToDB = true;

    ExcelDataReader edr = new ExcelDataReader();
    foreach (Point p in
edr.getExcelPoints(SolidWorks.Interop.swconst.swLengthUnit_e.swINCHES))
    {
        swDoc.SketchManager.CreatePoint(p.X, p.Y, p.Z);
    }

    swDoc.SketchManager.InsertSketch(true);
    swDoc.SketchManager.AddToDB = false;
    swDoc.ClearSelection2(true);
    swDoc.ViewZoomtofit2();

}
```

Utilities Class

```csharp
static class Utilities
{
    public static double inchToMeters(double InchVal)
    {
        return InchVal * 0.0254;
    }
}
```

```
    public static double mmToMeters(double mmVal)
    {
        return mmVal / 1000;
    }
}
```

ExcelDataReader Class

```
public class ExcelDataReader
{
    dynamic excel;
    public ExcelDataReader()
    {
        excel = Marshal.GetActiveObject("Excel.Application");
    }

    public List<Point> getExcelPoints(swLengthUnit_e units)
    {
        int i = 1;
        double multiplier = 1;
        List<Point> points = new List<Point>();
        if (units == swLengthUnit_e.swINCHES) { multiplier =
Utilities.inchToMeters(1); }
        if (units == swLengthUnit_e.swMM) { multiplier =
Utilities.mmToMeters(1); }
        while (excel.Cells(i, 1).Value != null)
        {
            Point p = new Point();
            p.X = excel.Cells(i, 1).Value * multiplier;
            p.Y = excel.Cells(i, 2).Value * multiplier;
            p.Z = excel.Cells(i, 3).Value * multiplier;
            points.Add(p);
            i++;
        }
        return points;
    }

}
```

ExportFlatPattern

```
public void Main()
{
    ModelDoc2 swDoc = null;
    PartDoc swPart = null;

    //get the active document
    swDoc = (ModelDoc2)swApp.ActiveDoc;
    //exit if not a part
    if (swDoc.GetType() != (int)swDocumentTypes_e.swDocPART)
```

```
    {
        swApp.SendMsgToUser2("This macro is for parts only.",
            (int)swMessageBoxIcon_e.swMbWarning,
            (int)swMessageBoxBtn_e.swMbOk);
        return;
    }
    //get the part interface
    swPart = (PartDoc)swDoc;
    //set a new path for the flat pattern
    string FlatPatternPath;
    string ext = ".DXF";
    //could also use .DWG if prefered
    FlatPatternPath = System.IO.Path.ChangeExtension(swDoc.GetPathName(),
                                                     ext);

    //export the flat pattern with bend lines
    int bendSetting = 2 ^ 0 + 2 ^ 2;
    swPart.ExportToDWG2(FlatPatternPath, swDoc.GetTitle(),
        (int)swExportToDWG_e.swExportToDWG_ExportSheetMetal,
        true, null, false, false, bendSetting, null);
    return;
}
```

Drawing Automation

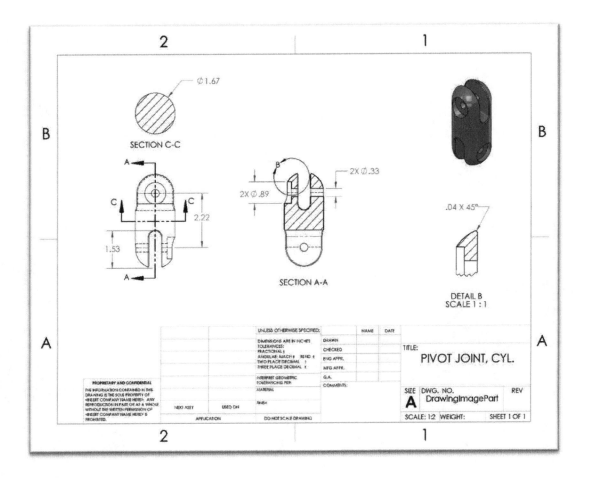

- **Creating Drawing Views**

- **Inserting Dimensions**

- **Batch Creation**

Introduction

Drawings lend themselves to automation in many ways. Just look at all the redundant work often done while creating drawings. This exercise will use drawing view creation and inserting of dimensions as a way to examine some of the drawing specific API calls in SOLIDWORKS. The goal of the chapter will be to build a macro that creates drawings for every part in a given directory. To get there, basic Windows directory and file traversal will be covered.

Creating Drawings

Typical drawing view operations record very well. The initial code will be built by recording the insertion of standard three views, an isometric or named view, dimensions in each view and setting the view's display style. Since the macro will be expected to build multiple drawings for the user, it will be useful to also record the action of creating a new drawing.

1. Open the part *PlanetGear.sldprt*. Another part will work as well, assuming it has features and dimensions. Just watch out for path mismatches.

2. Start recording a macro.

3. Create a new drawing by selecting File, New or clicking [icon]. Select any drawing template as a base for the drawing. The code below will show selection of the generic drawing template and selection of a sheet format.

4. Cancel any automatic Model View creation.

5. Insert three standard views by selecting Insert, Drawing View, Standard 3 View or by clicking [icon] , selecting PlanetGear and clicking OK [icon] .

6. Insert an isometric view by selecting Insert, Drawing View, Model... or by clicking [icon]. Select the open part *PlanetGear*, select Next [icon] and then select the "*Isometric" view by selecting [icon] from the Model View PropertyManager, placing the view in the top-right corner of the drawing.

7. Insert all model dimensions by selecting Insert, Model Items or by clicking [icon]. Select Entire model as the Source and check Import items into all views as shown below.

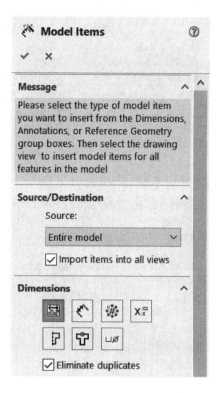

8. Save the drawing in the same location as *PlanetGear.sldprt* with the name *PlanetGear.slddrw*.

9. Stop recording the macro and save it with the name *AutoDrawing.vbproj*.

10. Edit the new macro.

Code Description

The recorded code will be similar to that below. Your path to the drawing template and part will be different. All CType statements have been removed for simplicity. Redundant calls along with IModelView calls have also been removed.

```
Public Sub main()

  Dim swDoc As ModelDoc2 = Nothing
  Dim swPart As PartDoc = Nothing
  Dim swDrawing As DrawingDoc = Nothing
  Dim swAssembly As AssemblyDoc = Nothing
  Dim boolstatus As Boolean = False
  Dim longstatus As Integer = 0
  Dim longwarnings As Integer = 0
  swDoc = swApp.ActiveDoc
  '
  'New Document
  Dim swSheetWidth As Double
  swSheetWidth = 0.2794
  Dim swSheetHeight As Double
```

```
swSheetHeight = 0.21589999999999998
swDoc = swApp.NewDocument("C:\...\Drawing.drwdot", _
   12, swSheetWidth, swSheetHeight)
Dim swDrawing As Object = Nothing
swDrawing = swDoc
Dim swSheet As Object = Nothing
swSheet = swDrawing.GetCurrentSheet()
swSheet.SetProperties2(12, 12, 1, 1, False, swSheetWidth, _
   swSheetHeight, True)
swSheet.SetTemplateName("C:\...\a - landscape.slddrt")
swSheet.ReloadTemplate(True)
boolstatus = swDrawing.Create3rdAngleViews("C:\...\PlanetGear.SLDPRT")
Dim myView As View = Nothing
myView = swDrawing.CreateDrawViewFromModelView3( _
   "C:\...\PlanetGear.SLDPRT", "*Isometric", 0.22256684234752588, _
   0.15014366743383195, 0)
Dim vAnnotations As Array = Nothing
vAnnotations = swDrawing.InsertModelAnnotations3(0, 32776, True, _
   True, False, True)
'
'Zoom To Fit
swDoc.ViewZoomtofit2()
'
'Save As
longstatus = swDoc.SaveAs3("C:\...\PlanetGear.SLDDRW", 0, 2)
End Sub
```

NewDocument

The NewDocument method of ISldWorks was introduced briefly in the Model Creation chapter. However, there are some additional arguments that are useful when creating a new drawing.

value = ISldWorks.NewDocument(TemplateName, PaperSize, Width, Height)

This method of the SOLIDWORKS interface creates a new part, assembly or drawing based on a named template. When used to create a drawing, you may need to provide valid arguments for PaperSize, Width and Height. If you use a template that has sheet size already defined (*.drwdot), the arguments PaperSize, Width and Height are ignored. The method requires you to pass values for these arguments, so you can leave them at whatever value was recorded or change them to zeros. If you always use the generic SOLIDWORKS drawing template and select a Sheet Format, you will need to pass the appropriate settings for the last three arguments.

- **PaperSize** must be passed one of the values from swDwgPaperSizes_e as shown below. The recorded code shows a value of 12, or swDwgPapersUserDefined, since the recorded code used a generic template.

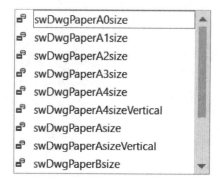

- **Height** and **Width** are simply the size of the sheet in meters. These arguments are only read if you chose swDwgPapersUserDefined.

Setting Sheet Properties

The macro recording includes code that sets up the new drawing sheet format. Since formats can be unique to each sheet, the first step is to get the ISheet interface by calling IDrawing.GetCurrentSheet. The swSheet variable has been declared generically as Object, but could be modified to use the specific Sheet data type.

The next call sets up the scale and other various settings of the current drawing sheet using ISheet.SetProperties2. The example code specifies a custom paper size, a user-defined sheet format. The drawing sheet is full scale (1:1) and uses third angle projections.

ISheet.SetProperties2(PaperSz, Templ, Scale1, Scale2, FirstAngle, Width, Height, SameCustomPropAsSheetInDocProp)

- **PaperSz** defines the sheet size from swDwgPaperSizes_e shown below.

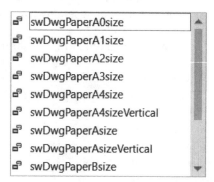

- **Templ** defines the sheet format used, but only if using one of the standard sheet formats. These are defined in swDwgTemplates_e shown below.

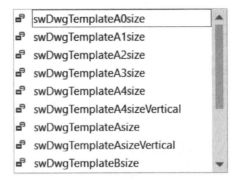

- **Scale1** is the sheet scale numerator as an integer.

- **Scale2** is the sheet scale denominator as an integer.

- **FirstAngle** is a Boolean value. If True, the drawing will be set up for first angle projections and third angle if False.

- **Width** is the width of the sheet in meters if a custom paper size is specified in the first argument.

- **Height** is the height of the sheet in meters if a custom paper size is specified.

- **SameCustomPropAsSheetInDocProp** is a Boolean and defines where to read the property links for that sheet in notes. See the API help for more detail.

Modifying the Sheet Format

The next two calls define the specific sheet format (*.slddrt) to be used. The first call to ISheet.SetTemplateName defines the sheet format to use. The only argument is the full path to a valid sheet format as a string. The second call refreshes the sheet display by using ISheet.ReloadTemplate. The Boolean argument is set to True if you want to keep any note changes that have been applied to the sheet format.

If you will be using a drawing template (*.drwdot) that includes a pre-defined sheet format, and I would certainly recommend it, the code could be simplified significantly as shown below.

```
'
'New Document
swDoc = swApp.NewDocument("C:\...\draw.drwdot", 0, 0, 0)
Dim swDrawing As DrawingDoc = Nothing
swDrawing = swDoc, DrawingDoc
boolstatus = swDrawing.Create3rdAngleViews("C:\...\PlanetGear.SLDPRT")
...
```

Create3rdAngleViews

The next line creates the standard three views in a third angle projection on the active drawing from the file path as an argument. IDrawingDoc.Create3rdAngleViews(ModelName). This method will create third angle views of either parts or assemblies. The current API call is Create3rdAngleViews2; however, there is no difference in the argument or return value. If you prefer the views to display in first angle projection, your recorded code would have then shown the call Create1stAngleViews2.

This method automatically positions the three views on the drawing sheet and automatically scales the sheet if you have the system option "Automatically scale new drawing views" turned on under the Drawings category. If you wish to locate the views by an x, y, z position, use the next method and create the views one at a time.

CreateDrawViewFromModelView3

The next lines declare a variable for the new drawing view and create a named view at a specific x, y, z sheet location.

value = IDrawingDoc.CreateDrawViewFromModelView3(ModelName, ViewName, LocX, LocY, LocZ)

- **ModelName**, as you can see in the recorded code, is the path to the part to insert.

- **ViewName** is a string and must be a named view in the model. This example uses the "*Isometric" view name. However, this could be replaced with any other view name such as "*Trimetric", "*Front" or any user specified name.

- **LocX**, **LocY** and **LocZ** variables are double values that represent the x, y and z location of the center of the view with respect to the sheet in meters. Do not ask me why they require you to pass a Z value, just leave it as 0.

InsertModelAnnotations3

After building the desired views, you have finally recorded the steps of inserting all dimensions into the drawing and getting them back in an array of annotations.

value = IDrawingDoc.InsertModelAnnotations3(Option, Types, AllViews, DuplicateDims, HiddenFeatureDims, UsePlacementInSketch)

- **Option** is a value from swImportModelItemsSource_e that allows you to specify whether dimensions are inserted for entire views, selected components, selected features or from only an assembly. The recorded example uses 0 representing swImportModelItemsFromEntireModel from the enumeration.

⊞ swImportModelItemsFromAssemblyOnly
⊞ swImportModelItemsFromEntireModel
⊞ swImportModelItemsFromSelectedComponent
⊞ swImportModelItemsFromSelectedFeature

- **Types** is a long value where you pass which types of annotations to show. The possible types from swInsertAnnotation_e are listed below. You can combine multiple types by simply adding them together. For example: swInsertCThreads + swInsertDatums + swInsertNotes.

        swInsertCThreads
        swInsertDatums

```
swInsertDatumTargets
swInsertDimensions
swInsertInstanceCounts
swInsertGTols
swInsertNotes
swInsertSFSymbols
swInsertWelds
swInsertWeldBeads
swInsertWeldBeads_ET
swInsertSketches
swInsertAxes
swInsertCurves
swInsertPlanes
swInsertSurfaces
swInsertPoints
swInsertOrigins
swInsertDimensionsMarkedForDrawing
swInsertDimensionsNotMarkedForDrawing
swInsertHoleWizardProfileDimensions
swInsertHoleWizardLocationDimensions
swInsertDimensionsNotMarkedForDrawing
swInsertholeCallout
```

- **AllViews** should be set to True to get all dimensions into all views on the drawing and False otherwise.

- **DuplicateDims** is a Boolean that when set to False, eliminates duplicate dimensions.

- **HiddenFeatureDims** is a Boolean that when set to True, adds dimensions for hidden features.

- **UsePlacementInSketch** is a Boolean that when set to True, places the dimensions as they were placed in the sketch.

SaveAs3

The macro is finished with a call to IModelDoc2.SaveAs3. As discussed in the PDF saving chapter, the recorded method is an obsolete method, but its simplicity is preferred for this example. See IModelDocExtension.SaveAs3 in the API Help if you would like to explore the latest save method.

Batch Processing Files in a Folder

The goal is to create a drawing for every part or assembly in a directory. But the methods used to insert views into drawing sheets require a full path string. So how do you get that for every file in a directory? We will again use the System.IO namespace. This namespace provides a variety of standard file and folder operations, including getting all files in a folder.

11. Add the following code to the top of the code window to import the System.IO and System.Diagnostics namespaces. System.Diagnostics will be explained later.

```
Imports SolidWorks.Interop.sldworks
Imports SolidWorks.Interop.swconst
```

```
Imports System
Imports System.IO
Imports System.Diagnostics
```

Divide Code into Procedures

This is another opportunity to separate the code into re-usable blocks. The operations that will be introduced in this section are helpful for a variety of other macros. Begin by making the existing code independent of the main procedure rather than simply adding to it.

12. Rename the main procedure to DrawingFromModel and add two arguments named drawingTemplate and modelPath, both as strings.

```
Public Sub DrawingFromModel(drawingTemplate As String, _
 modelPath As String)
  Dim boolstatus As Boolean
  Dim longstatus As Long
...
```

13. Replace all hard coded path strings in the newly named DrawingFromModel procedure with the two new variables for drawingTemplate and modelPath. The code below assumes the drawing template has a pre-defined sheet format so the extra sheet format code has been removed. You can continue with included sheet properties and formatting if you choose. The output drawing path is also automated using IO.Path.ChangeExtension as was shown in previous chapters.

```
Public Sub DrawingFromModel(ByVal drawingTemplate As String, _
 ByVal modelPath As String)
  Dim swDoc As ModelDoc2 = Nothing
  Dim swPart As PartDoc = Nothing
  Dim swDrawing As DrawingDoc = Nothing
  Dim swAssembly As AssemblyDoc = Nothing
  Dim boolstatus As Boolean
  Dim longstatus As Long

  '        `
  'New Document
  Dim swDoc As ModelDoc2 = Nothing
  swDoc = swApp.NewDocument(drawingTemplate, 0, 0, 0)
  swDrawing = swDoc
  boolstatus = swDrawing.Create3rdAngleViews(modelPath)
  Dim myView As View = Nothing
  myView = swDrawing.CreateDrawViewFromModelView3(modelPath, _
    "*Isometric", 0.2225, 0.1501, 0)
  Dim vAnnotations As Array = Nothing
  vAnnotations = swDrawing.InsertModelAnnotations3(0, 32776, _
    True, True, False, True)
  '
  'Zoom To Fit
```

```
swDoc.ViewZoomtofit2()
'
'Save As
Dim drawingPath As String
drawingPath = Path.ChangeExtension(modelPath, ".SLDDRW")
longstatus = swDoc.SaveAs3(drawingPath, 0, 2)
End Sub
```

14. Add a new main procedure above the DrawingFromModel procedure, within the SolidWorksMacro class.

```
Partial Class SolidWorksMacro

    Public Sub main()

    End Sub
```

Directories and Files

The main procedure will now be used to loop through all files in a folder, filtered by file extension. For each file found, we'll start by writing the paths to the debug Immediate Window. Comments have also been added to help anyone reading the code (including you) at a later date.

15. Add the following code in the main procedure to loop through all SOLIDWORKS part files in a given folder. Be sure to enter a valid folder path that contains a few SOLIDWORKS parts.

```
Public Sub main()

    'set the directory for parts
    Dim MyDir As New DirectoryInfo ("C:\Models")
    If MyDir.Exists Then
        Dim MyFile As FileInfo = Nothing
        Dim AllParts() As FileInfo
        AllParts = MyDir.GetFiles("*.sldprt")
        For Each MyFile In AllParts
            'Write the file path to the Immediate Window
            Debug.Print(MyFile.FullName)
        Next
    End If
    Stop
End Sub
```

DirectoryInfo

DirectoryInfo, from the System.IO namespace, is a great way to get information about a Windows folder (directory). DirectoryInfo is an interface or class. In many cases, to connect a variable to a new instance of a Visual Basic class, the New statement can be used to declare and connect to a specific instance. In this case, since DirectoryInfo is a folder, passing the folder path as a string connects to that specific directory.

The first code used on DirectoryInfo is its Exists property. The odd thing is that you can set DirectoryInfo to a new instance by passing the path of a folder, but that folder or directory may not yet exist. The Exists property will return False if the folder does not exist. If you needed to create a new folder, use the DirectoryInfo.Create method to a DirectoryInfo instance where Exists is False.

After a couple variables are declared, the GetFiles method of DirectoryInfo is used. GetFiles is passed a search pattern as a string value. This is a filter for which file types to get. In this example, it has been set to only get SOLIDWORKS Part files that would match the pattern `"*.sldprt"`. This method returns an array of the FileInfo type. The pattern string is not case sensitive. The declaration of `MyFiles() As FileInfo` ensures that the dynamic array `MyFiles` can handle the right type returned from the call to GetFiles. If you need to get all files in sub folders, you can pass a second argument `SearchOption.AllDirectories`. For example, the following code would find all drawings in the folder and its sub folders.

```
AllParts = mydir.GetFiles("*.slddrw", SearchOption.AllDirectories)
```

FileInfo

Since you now have an array of file paths, the For Each … Next looping method is ideal. It gets each FileInfo interface in the `MyFiles` array. Like DirectoryInfo, FileInfo is an interface to a specific file. You can check its Exists property or use its Create method to create new files. In addition, FileInfo.Name returns the name of the file, including its extension, but without its path. FileInfo.FullName is used to return the full path and file name, including extension. FileInfo.Extension gets just the extension from the file.

Debugging

Some useful debugging methods have been added to the main procedure including `Debug.Print` and the `Stop` statement.

16. Run the macro and review the following.

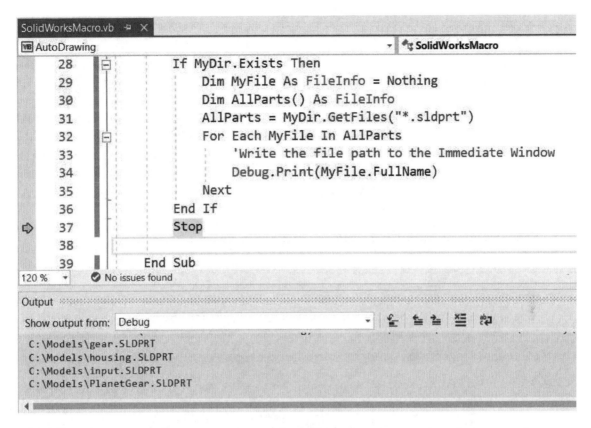

The Stop statement acts like a hard coded breakpoint. Code execution pauses and you can review debugging information. At the bottom of the VSTA interface, the Debug Output Window is displayed. If you close the window, it can be opened again by selecting Debug, Windows, Output. This window can be used to view text displayed using Debug.Print. It can also be used as an active debugging console while code execution is paused. You can enter lines of code and execute them on the fly. The Debug.Print statement wrote a list of all parts from the specified folder into the Output Window.

While code execution is paused (the Stop statement is still highlighted in yellow), hover your cursor over MyDir in its declaration line and click the ▶ to expand its properties. You can also expand its Parent and Root elements to view related objects. Try the same thing with MyFile. Notice that it doesn't display anything. MyFile is inside an If block that has already ended, so the variable is no longer available. It is "out of scope."

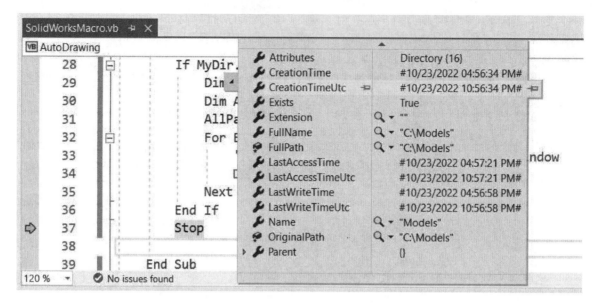

17. Open the Immediate Window by selecting Debug, Windows, Immediate or by selecting the Immediate tab from the bottom of Visual Studio. Type the following under the current text in the Immediate Window and hit Enter.

```
MyDir = Nothing
```

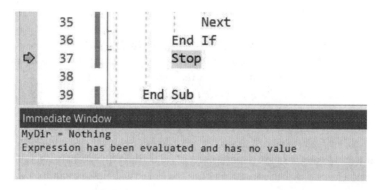

Again, hover your mouse over MyDir and notice that it has been disconnected from the folder and is now empty, shown as Nothing. The Immediate Window is helpful for running additional code while debugging and can be especially useful for testing alternatives.

18. Type the following at the bottom of the Immediate Window to re-connect MyDir to a folder. Hit Enter after typing the line. Does the folder exist?

```
MyDir = New IO.DirectoryInfo("C:\Models")
```

19. The yellow arrow next to the Stop statement indicates the next statement to be processed. Drag the arrow up to the Dim MyDir line to "rewind" your code. This does not undo any processing already done.

```
Partial Class SolidWorksMacro

    2 references
    Public Sub main()
        'set the directory for parts
        Dim MyDir As New DirectoryInfo("C:\Models")
        If MyDir.Exists Then
            Dim MyFile As FileInfo = Nothing
            Dim AllParts() As FileInfo
```

20. Step through a few lines of code, one at a time, by selecting Debug, Step Into or clicking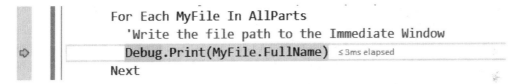
(F11). Stop when the Debug.Print line is highlighted in yellow.

```
For Each MyFile In AllParts
    'Write the file path to the Immediate Window
    Debug.Print(MyFile.FullName)   ≤3ms elapsed
Next
```

21. Again hover your mouse over MyFile to see its properties. This is the first file in the array.

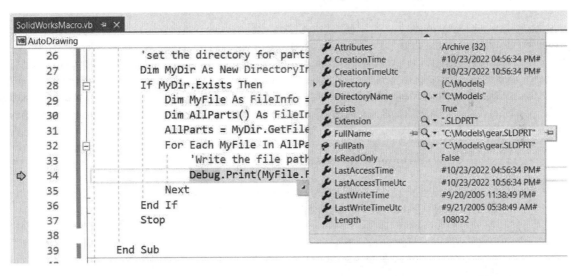

22. Continue to step through the code line at a time (F11), reviewing variable properties as you go. Click Continue ▶ (F5) when you are ready to continue running to the Stop statement.

23. Click Stop Debugging ■ (Shift+F5) to stop the macro. Notice the Immediate Window is closed while the Output Window is still visible.

24. Modify the main procedure to create a drawing for each part found in the directory using the template specified. Be sure to enter a valid template path. Comment out or delete the Stop statement.

```
Public Sub main()
```

```
'set the directory to work on
Dim MyDir As New DirectoryInfo("C:\Models")
If MyDir.Exists Then
  'enter a valid template path
  Dim templatePath As String = _
      "C:\ProgramData\SOLIDWORKS\SOLIDWORKS 2023\templates\Drawing.DRWDOT"
  Dim MyFile As FileInfo = Nothing
  Dim AllParts() As FileInfo
  AllParts = MyDir.GetFiles("*.sldprt")
  For Each MyFile In AllParts
    'Write the file path to the output window
    Debug.Print(MyFile.FullName)
    'create the drawing
    DrawingFromModel(templatePath, MyFile.FullName)
  Next
End If
'Stop
End Sub
```

25. Test the macro at this point and debug as needed.

Programming frequently involves troubleshooting. You should see your macro running, but the three projected views are not created! Don't worry yet. You are very close. There is one important step that the macro recorder did not do for you. Before most drawing view creation methods can be used, the model must be open in the SOLIDWORKS session. If you do not believe me, just open the API Help and type "CreateDrawViewFromModelView3" in the index. Scroll your way down to the bottom of the page in the Remarks section. Notice that it states "The ModelName must be an open document in the current SOLIDWORKS session."

26. Close all open documents in SOLIDWORKS.

27. In Visual Studio, within the SolidWorksMacro class, add a new Function named openModel that accepts a String argument named modelPath. The function will open any SOLIDWORKS file type.

```
Function openModel(modelPath As String) As ModelDoc2
  Dim docType As Long
  Dim errors As Long
  Dim warnings As Long
  Select Case Path.GetExtension(modelPath).ToUpper
    Case ".SLDPRT"
      docType = swDocumentTypes_e.swDocPART
    Case ".SLDASM"
      docType = swDocumentTypes_e.swDocASSEMBLY
    Case ".SLDDRW"
      docType = swDocumentTypes_e.swDocDRAWING
  End Select
  Dim swDoc as ModelDoc2
  swDoc = swApp.OpenDoc6(modelPath, docType, _
```

```
      swOpenDocOptions_e.swOpenDocOptions_Silent, "", errors, warnings)
   Return swDoc
End Function
```

Let's review the last SOLIDWORKS API call in this function first.

OpenDoc6

When this book was written, the current version of the OpenDoc method was OpenDoc7. However, OpenDoc6 will allow us to open files without introducing the more complicated DocumentSpecification interface used by OpenDoc7. Occasionally, an older method can simplify your macro code if you do not need all of the features of the latest method. We have continued to use an obsolete call to IModelDoc2.SaveAs3 in this example for the same reason. As always, check the SOLIDWORKS API Help as you write your macros to review the structure of the call you are using.

value = ISldWorks.OpenDoc6(Filename, Type, Options, Configuration, Errors, Warnings)

- **Filename** is a string representing the full path to the document.

- **Type** is a long SOLIDWORKS constant for the type of document (swDocPART for parts).

- **Configuration** should be a string for the configuration name to open or an empty or NULL string for the last saved configuration.

- **Errors** and **Warnings** finally are long variables passed. It does not matter what the values of these two variables are. The SOLIDWORKS API uses these as additional returned values. Errors and warnings might occur as a file is opened in SOLIDWORKS. Check these two variables after using the OpenDoc method to see if there were any problems opening the document. If they are zero, there were no errors or warnings. If they are not zero, the error or warning can be determined from the SOLIDWORKS constants as follows from swFileLoadWarning_e and swFileLoadError_e.

  ### Warnings
  swFileLoadWarning_IdMismatch
  swFileLoadWarning_ReadOnly
  swFileLoadWarning_SharingViolation
  swFileLoadWarning_DrawingANSIUpdate
  swFileLoadWarning_SheetScaleUpdate
  swFileLoadWarning_NeedsRegen
  swFileLoadWarning_BasePartNotLoaded
  swFileLoadWarning_AlreadyOpen
  swFileLoadWarning_DrawingsOnlyRapidDraft
  swFileLoadWarning_ViewOnlyRestrictions
  swFileLoadWarning_ViewMissingReferencedConfig
  swFileLoadWarning_DrawingSFSymbolConvert
  swFileLoadWarning_RevolveDimTolerance
  swFileLoadWarning_ModelOutOfDate
  swFileLoadWarning_DrawingSFSymbolConvert

swFileLoadWarning_MissingDesignTable
swFileLoadWarning_ComponentMissingReferencedConfig

Errors
swGenericError
swFileNotFoundError
swFileWithSameTitleAlreadyOpen
swInvalidFileTypeError
swFutureVersion
swLiquidMachineDoc
swLowResourcesError
swNoDisplayData

Since OpenDoc6 requires the document type as an argument, the rest of the function code, above OpenDoc6, determines the document type based on the file extension. A Select Case statement is ideal for checking one value against several possibilities. However, Select Case is case sensitive when comparing strings. To account for both upper and lower case file extensions, the String.ToUpper method is used. Anytime a variable or literal string is used, the ToUpper property can be appended. Now each comparison case is hard coded in upper case. Alternatively, String.ToLower could be used with comparison cases to lower case file extensions.

Function Return Statement

Since a Function returns a value when it is called, it is declared with an As clause just like a typical variable. This function has been declared as IModelDoc2 since it opens and returns a model.

Once your function has reached its goal, the Return statement sends back the right side of that statement to the code that called the function. It leaves the function when the Return statement is hit and does not run function code below the statement. The Return statement can be included anywhere in a function. Use that to your advantage when a function might return different values based on If statements or other logic.

28. Modify the main procedure to open each model before creating its drawing. Only the For loop is shown below.

```
For Each MyFile In AllParts
  'Write the file path to the output window
  Debug.Print(MyFile.FullName)
  'open the model
  Dim swDoc As ModelDoc2
  swDoc = openModel(MyFile.FullName)
  'create the drawing
  DrawingFromModel(templatePath, MyFile.FullName)
Next
```

29. Run another test of the macro and verify the completed results. All parts and drawings will remain open in SOLIDWORKS.

30. Close all open documents in SOLIDWORKS.

Adding User Input

There is still a limitation in the macro as it stands. It will only operate on a hard-coded directory. That would be fine if you always wanted to copy all your parts into one directory to make drawings. But for practical purposes that is not ideal. The next step will be to add the ability for the user to choose the directory to be used.

FolderBrowserDialog Control

One of the best things about the .NET framework is that it is easy to make use of standard Windows components without having to build them yourself. The following changes will add a standard Windows FolderBrowser control to prompt the user for a folder to process.

Since this macro has no forms or dialogs, System.Windows.Forms is not yet referenced in the solution and needs to be added manually before it can be imported.

31. Select Project, Add Reference. Click the checkbox next to System.Windows.Forms from the Assemblies category as shown, then click OK.

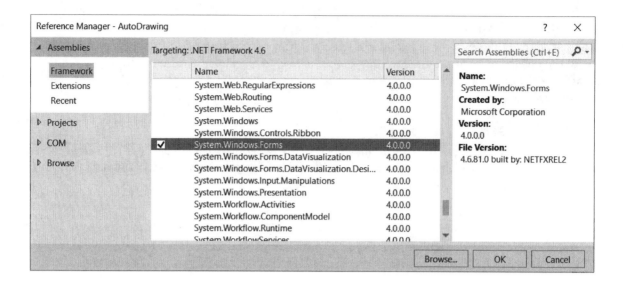

32. Make the code changes for user input as follows.

```
Imports SolidWorks.Interop.sldworks
Imports SolidWorks.Interop.swconst
Imports System.Runtime.InteropServices
Imports System
Imports System.IO
Imports System.Diagnostics
Imports System.Windows.Forms

Partial Class SolidWorksMacro

  Public Sub main()
    'browse for a folder to process
    Dim MyFolderBrowser As New FolderBrowserDialog
    Dim results As DialogResult
    MyFolderBrowser.Description = "Select a folder:"
    results = MyFolderBrowser.ShowDialog()
    Dim MyPath As String
    If results = DialogResult.OK Then
      MyPath = MyFolderBrowser.SelectedPath
    Else
      'no folder selected
      Exit Sub
    End If

    'set the directory to work on
    Dim MyDir As New DirectoryInfo(MyPath)
    If MyDir.Exists Then
      ...
```

System.Windows.Forms

The first requirement for using the Folder Browser is to import the System.Windows.Forms namespace. It contains many standard Windows form controls such as the FolderBrowserDialog, OpenFileDialog, SaveFileDialog, ColorDialog and PrintDialog. You do not need to create a form in your macro to make use of them. Simply create a new instance of them and call their ShowDialog method.

The ShowDialog method of any form returns a DialogResult value. Some common return values are OK and Cancel.

The Folder Browser has several properties that can be used to alter the way it displays to the user as well as determining what folder was selected. For example, the SelectedPath property returns a string representing the full path of the folder that was selected. The ShowNewFolderButton property sets whether or not the New Folder button will display on the form. The Description property sets the text that displays just above the folder list.

Debug

33. Run the macro as a test.

When you attempt to run the macro it will fail to build and run. Select the Error List tab in the bottom-left corner of Visual Studio to see the errors.

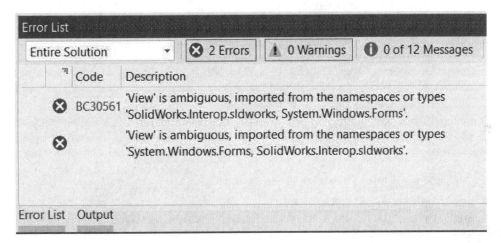

We now have a problem with the declaration of myView As View. Notice the red underlining under the type View in the DrawingFromModel procedure. As the message states, View is a member of two imported namespaces, System.Windows.Forms and SolidWorks.Iterop.sldworks. It is not uncommon to have a conflict between members of different imported namespaces. This is especially true with common terms like View, Frame, Sheet and others. To correct the conflict, you can fully qualify which View data type to use.

34. Change the declaration of myView as follows to specifically declare it as a member of SolidWorks.Interop.sldworks.

```
Dim myView As SolidWorks.Interop.sldworks.View = Nothing
```

Hint: use "Show potential fixes" to automatically change the code by hovering over the error in code.

When you run the macro it should first prompt you for the folder as shown in the image. After selecting the folder and clicking OK, the macro will build drawings for all parts in that folder.

35. Optionally close the new drawing and the part after they are created by adding the following code to the end of DrawingFromModel.

...
```
    'Save As
    Dim drawingPath As String
    drawingPath = Path.ChangeExtension(modelPath, ".SLDDRW")
    longstatus = swDoc.SaveAs3(drawingPath, 0, 2)
    swApp.CloseDoc(drawingPath)
    swApp.CloseDoc(modelPath)
  End Sub
```
...

36. Close Visual Studio.

Conclusion

Now that you know how to record and modify basic drawing operations you are ready to start expanding on the topic. There are many ways you could extend the automatic drawing application beyond what has been done in this exercise. Here are some ideas.

- Add a user form for some of the following drawing options.

- Add a dropdown list of drawing templates the user can choose from.

- Extract custom properties from the parts and output a list of missing properties to a log file to report missing information.

- Extend the tool to work with assemblies.

- If the part has sheet metal features, create a flat pattern view. Exporting flat patterns to DXF and DWG is discussed in the Data Import and Export chapter.

- Add various tables and other annotations as described in the next chapter.

- Add an additional IModelDocExtension.SaveAs method to save a PDF version of the drawing after it is complete.

C# Example

```csharp
public void main()
{
    //browse for a folder to process
    FolderBrowserDialog fbd = new FolderBrowserDialog;
    DialogResult myRes;
    fbd.Description = "Select a folder...";
    myRes = fbd.ShowDialog();
    string MyPath;
    if (myRes == DialogResult.OK)
    {
        MyPath = fbd.SelectedPath;
    }
    else
    {
        //no folder selected
        return;
    }
    //set the directory for parts
    DirectoryInfo MyDir = new DirectoryInfo(MyPath);
    if (MyDir.Exists)
    {
        //enter a valid template path
        string templatePath = "C:\\...\\Drawing.drwdot";
        FileInfo[] AllParts;
        AllParts = (FileInfo[])MyDir.GetFiles("*.sldprt");
        foreach (FileInfo MyFile in AllParts)
        {
            //Write the file path to the Immediate Window
            Debug.Print(MyFile.FullName);
            //open the model
            ModelDoc2 swDoc = openModel(MyFile.FullName);
            //create the drawing
            DrawingFromModel(templatePath, MyFile.FullName);
```

```
        }
    }
}

private ModelDoc2 openModel(string modelPath)
{
    int docType;
    int errors = 0;
    int warnings = 0;
    switch (Path.GetExtension(modelPath).ToUpper())
    {
        case ".SLDPRT":
            docType = (int)swDocumentTypes_e.swDocPART;
            break;
        case ".SLDASM":
            docType = (int)swDocumentTypes_e.swDocASSEMBLY;
            break;
        case ".SLDDRW":
            docType = (int)swDocumentTypes_e.swDocDRAWING;
            break;
        default:
            docType = (int)swDocumentTypes_e.swDocNONE;
            break;
    }
    ModelDoc2 swDoc = swApp.OpenDoc6(modelPath, docType,
        (int)swOpenDocOptions_e.swOpenDocOptions_Silent,
        "", errors, warnings);
    return swDoc;
}

public void DrawingFromModel(string drawingTemplate, string modelPath)
{
    int longstatus = 0;
    bool boolstatus;

    //
    // New Document
    ModelDoc2 swDoc = null;
    swDoc = ((ModelDoc2)(swApp.NewDocument(drawingTemplate,0,0,0)));
    DrawingDoc swDrawing = null;
    swDrawing = (DrawingDoc)swDoc;
    boolstatus = swDrawing.Create3rdAngleViews(modelPath);
    SolidWorks.Interop.sldworks.View myView = null;
    myView = swDrawing.CreateDrawViewFromModelView3(modelPath,
        "*Isometric", 0.451104, 0.2954, 0);
    Array vAnnotations = null;
    vAnnotations = ((Array)(swDrawing.InsertModelAnnotations3(0,
        32776, true, true, false, true)));
    //
    // Zoom To Fit
```

```
    swDoc.ViewZoomtofit2();
    //
    // Save As
    string drawingPath = Path.ChangeExtension(modelPath, ".SLDDRW");
    longstatus = swDoc.SaveAs3(drawingPath, 0, 0);
    swApp.CloseDoc(drawingPath);
    swApp.CloseDoc(modelPath);
    return;
}
```

Notes, Annotations and Tables

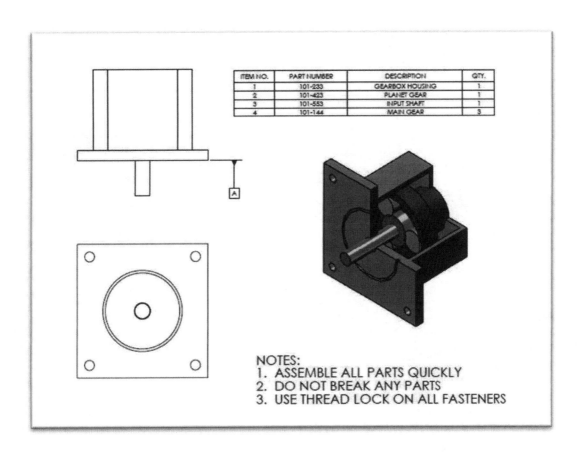

ITEM NO.	PART NUMBER	DESCRIPTION	QTY.
1	101-233	GEARBOX HOUSING	1
2	101-423	PLANET GEAR	1
3	101-553	INPUT SHAFT	1
4	101-144	MAIN GEAR	3

NOTES:
1. ASSEMBLE ALL PARTS QUICKLY
2. DO NOT BREAK ANY PARTS
3. USE THREAD LOCK ON ALL FASTENERS

- **Working with Notes**

- **Annotations**

- **Working with Tables**

Introduction

The first important point to understand about objects like tables and notes is that they are all considered annotations. Notes, dimensions, GD&T symbols, surface finish symbols, tables and balloons are all on the Annotation toolbar. The SOLIDWORKS API has a general IAnnotation interface that encapsulates all annotation types. This is the same concept as the relationship between IModelDoc2 and IPartDoc, IAssemblyDoc and IDrawingDoc. Every annotation, regardless of what type, has common settings such as layer, color, font style and position. From the IAnnotation interface, you can drill into the specific annotation type and its underlying properties.

This chapter is devoted to exploring several common annotations and their uses through the API. It will also be an exercise in building a code module with practical procedures that can be used in other macros.

Creating Notes

Creating a note on a drawing involves several API calls to both the INote interface as well as to its IAnnotation interface. A simple recorded macro will give you the foundation for adding notes to a drawing and will highlight some of the methods and properties you can use.

1. Open any drawing or create a new drawing containing part or assembly views.

2. Start recording a new macro.

3. Create a note with default font settings anywhere on the drawing sheet with some simple text.

4. Stop the macro, name it *InsertAnnotations.vbproj* and then edit the macro.

Your code should look similar to the following. I have removed all of the redundant and unnecessary lines of code to simplify this example as well as CType calls for clarity.

```
Public Sub main()

Dim swDoc As ModelDoc2 = Nothing
Dim boolstatus As Boolean = False
Dim longstatus As Integer = 0
swDoc = swApp.ActiveDoc

Dim myNote As Note = Nothing
Dim myAnnotation As Annotation = Nothing
Dim myTextFormat As TextFormat = Nothing
myNote = swDoc.InsertNote("HERE IS MY FIRST NOTE")
If (Not (myNote) Is Nothing) Then
  myNote.LockPosition = False
  myNote.Angle = 0
  boolstatus = myNote.SetBalloon(0, 0)
  myAnnotation = myNote.GetAnnotation ()
  If (Not (myAnnotation) Is Nothing) Then
```

```
      longstatus = _
      myAnnotation.SetLeader3(swLeaderStyle_e.swNO_LEADER, _
        0, True, False, False, False)
      boolstatus = myAnnotation.SetPosition(0.1681, 0.1542, 0)
      boolstatus = myAnnotation.SetTextFormat(0, True, myTextFormat)
    End If
End If
swDoc.ClearSelection2(True)
swDoc.WindowRedraw()
End Sub
```

Annotations

Inserting and positioning notes using the API is a multi-step process. The recorded code includes inserting a note and formatting the resulting note in two steps.

There are two interfaces involved in the process, Note and Annotation. As described in the chapter introduction, think of the Note interface as a child of the Annotation interface. Other children of the Annotation interface are IDisplayDimension, IWeldSymbol, ITableAnnotation and IGtol. Each annotation type records very well. So rather than hunting through the SOLIDWORKS API Help to learn how to use them, try recording annotations first.

InsertNote

Inserting a note into a part, assembly or drawing is as simple as IModelDoc2.InsertNote. Only the text for the note is required as an argument. The call returns a Note interface.

There are literally dozens of methods and properties directly related to notes, but only two are recorded. Note.Angle can be used to get or set the angle of the Note. Do not forget to use radians for setting angles.

Note.LockPosition fixes the note position on the sheet to prevent it from being moved accidentally. This property corresponds to the Lock/Unlock Note tool in the Text Format Property Manager panel.

GetAnnotation

Sometimes it is necessary to go from the INote interface back to the Annotation interface. Getting the parent Annotation interface from a note is as easy as INote.GetAnnotation. From the IAnnotation interface there are again dozens of methods and properties available. The recorded code shows the use of IAnnotation.SetLeader3, IAnnotation.SetPostion and IAnnotation.SetTextFormat. SetPosition is an obsolete method and will be replaced with the latest SetPosition2.

value = IAnnotation.SetLeader3(LeaderStyle, LeaderSide, SmartArrowHeadStyle, Perpendicular, AllAround, Dashed**)**

- **LeaderStyle** is based on swLeaderStyle_e and represent the Leader options seen in the properties of a Note. Multiple leader styles can be combined to achieve the desired results. See the API help for more detail.

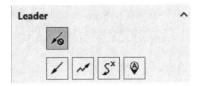

- **LeaderSide** forces the leader to the left or right, or leaves it as smart.

- **SmartArrowHeadStyle** is a Boolean. If True, the arrow style adapts to the selection. If False, you must also make a call to IAnnotation.SetArrowHeadStyleAtIndex to customize the arrow head display.

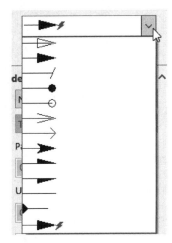

- The remaining three arguments do not apply to notes.

value = IAnnotation.SetPosition2(X, Y, Z)

- **X, Y and Z** values are doubles that locate the note. In drawings, the position is relative to the bottom-left corner of the sheet. You can ignore the Z value for drawings.

value = IAnnotation.SetTextFormat(Index, UseDoc, TextFormat)

- **Index** is for annotations with multiple text elements where each can have its own formatting. Use 0 for notes since they only have one text element.

- **UseDoc** is True to use the document default font formatting.

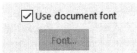

- **TextFormat** is an ITextFormat interface used to change the default font formatting. It only applies if UseDoc is False. Record an example of modifying the font formatting to see it in use.

Use some of the Note and Annotation methods and properties to change the display and position of the inserted note. Rather than static text, the note text will be linked to the file name of the part that has been inserted into the drawing. The note will be positioned relative to the bottom-right corner of the drawing sheet and rotated 90 degrees (π/2 radians).

5. Modify the code as shown to create a note linked to the file name, including its new position and rotation angle. Make sure to comment out the SetTextFormat line.

```
Public Sub main()

Dim swDoc As ModelDoc2 = Nothing
Dim boolstatus As Boolean = False
Dim longstatus As Integer = 0
swDoc = swApp.ActiveDoc

Dim myNote As Note = Nothing
Dim myAnnotation As Annotation = Nothing
Dim myTextFormat As TextFormat = Nothing
Dim myDrawing As DrawingDoc = swDoc
Dim mySheet As Sheet
Dim width As Double
Dim height As Double
mySheet = myDrawing.GetCurrentSheet
width = mySheet.GetProperties2(5)  'sheet width
height = mySheet.GetProperties2(6) 'sheet height
myDrawing.EditSheet()
myNote = swDoc.InsertNote("$PRPSHEET:""SW-File Name""")

If (Not (myNote) Is Nothing) Then
  myNote.LockPosition = False
  myNote.Angle = Math.PI / 2
  boolstatus = myNote.SetBalloon(0, 0)
  myAnnotation = myNote.GetAnnotation()
  If (Not (myAnnotation) Is Nothing) Then
    longstatus = _
    myAnnotation.SetLeader3(swLeaderStyle_e.swNO_LEADER, _
    0, True, False, False, False)
```

```
    boolstatus = myAnnotation.SetPosition2(width - 0.01, 0.015, 0)
    'boolstatus = myAnnotation.SetTextFormat(0, True, myTextFormat)
  End If
End If

swDoc.ClearSelection2(True)
swDoc.WindowRedraw()
End Sub
```

EditSheet

A new line of code was added prior to the creation of the note itself. You may have noticed that SOLIDWORKS will frequently attach the notes you create to the nearest view. This is a result of dynamic view activation. If a note is attached to a view and you move the view, the note will follow in its relative position. This is great for most notes, but is not what is wanted in this case. The call to IDrawingDoc.EditSheet causes the current drawing sheet to get the focus. Anything created by the macro after this call is associated to the sheet rather than the nearest view. It is important to clarify that this is not the same thing as editing the sheet format in SOLIDWORKS. It compares more directly to right-clicking on the sheet and selecting Lock Sheet Focus.

GetCurrentSheet and GetProperties2

You can easily access the current sheet of an IDrawingDoc interface by making a call to IDrawingDoc.GetCurrentSheet. You have captured the sheet into the variable mySheet. The Sheet interface is useful for a variety of calls including the current sheet size. You need to determine the sheet size to place notes in a relative position. If you want the note to always be 10mm from the right and 15mm from the bottom, you must at least know the width of the sheet. If you want the note to be a given distance from the top of the sheet you must know the height.

The height and width of a sheet can be accessed through Sheet.GetProperties2(5) and Sheet.GetProperties2(6) respectively. The GetProperties2 method of the Sheet returns an array of eight values including the paper size and the sheet scale numerator and denominator.

Create a Module

To make this code reusable, it will be moved into a new code module dedicated to annotations.

6. From the Project menu, select Add Module. Name it *AnnotationsMod.vb* and click Add. The new code module will be added to the project and opened.

7. Add a new procedure named AddFileNameNote as shown.

```
Module AnnotationsMod
  Sub AddFileNameNote(swDoc as ModelDoc2)

  End Sub
End Module
```

8. Switch back to *SolidWorksMacro.vb* and cut and paste everything from Dim myNote As Note = Nothing through swDoc.WindowRedraw() into the new procedure as shown.

```
Module AnnotationsMod
  Sub AddFileNameNote(ByVal swDoc As ModelDoc2)
    Dim myNote As Note = Nothing
    Dim myAnnotation As Annotation = Nothing
    Dim myTextFormat As TextFormat = Nothing
    Dim myDrawing As DrawingDoc = swDoc
...
    swDoc.ClearSelection2(True)
    swDoc.WindowRedraw()
  End Sub
End Module
```

9. Copy and paste the Imports statements from *SolidWorksMacro.vb* to the new *AnnotationsMod.vb* as shown.

```
Imports SolidWorks.Interop.sldworks
Imports SolidWorks.Interop.swconst
Imports System.Runtime.InteropServices
Imports System
```

```
Module AnnotationsMod
  Sub AddFileNameNote(ByVal swDoc As ModelDoc2)
```

10. Cut and paste the declarations of the `longstatus` and `boolstatus` variables from the `main` procedure to the top of the `AddFileNameNote` procedure as shown to eliminate the last code errors.

```
Module AnnotationsMod
  Sub AddFileNameNote(ByVal swDoc As ModelDoc2)
    Dim boolstatus As Boolean = False
    Dim longstatus As Integer = 0
    Dim myNote As Note = Nothing
    Dim myAnnotation As Annotation = Nothing
```

11. Add a call to the new procedure in the `main` procedure as shown.

```
Public Sub main()

  Dim swDoc As ModelDoc2 = Nothing
  swDoc = swApp.ActiveDoc

  AddFileNameNote(swDoc)

End Sub
```

12. Save the macro and test and debug as needed.

GD&T and Notes Attached to Models

Some annotation types require an attachment to model edges. Balloons, datums and geometric tolerances all require a connection to an edge or face. A selection must be made before making a call to insert these annotation types. This can be done with code or you can make it a requirement for the user to pre-select. Ultimately, full automation of annotations can pose a challenge since it is typically subjective based on the user's preference for location and arrangement.

13. Switch back to *AnnotationsMod.vb*. Add a new procedure to insert a datum on a pre-selected edge. The datum will use the next datum label. Notice the use of the SelectionManager interface to check that the user has made at least one pre-selection. It does not guarantee the type of selection, however.

```
Sub AddDatum(swDoc As ModelDoc2)
  'an edge must be pre-selected
  Dim selMgr As SelectionMgr = swDoc.SelectionManager
  If selMgr.GetSelectedObjectCount = 0 Then Exit Sub

  'no datum will be created otherwise
  Dim myDatumTag As DatumTag
  myDatumTag = swDoc.InsertDatumTag2
End Sub
```

InsertDatumTag2

Since a datum can be inserted in a part, assembly or drawing, the method is a member of the IModelDoc2 interface. There are no arguments, only the requirement of pre-selection for the location. If there is no selection made before calling IModelDoc2.InsertDatumTag2, the return value will be Nothing. If you need to position the newly created DatumTag interface, you would access its Annotation interface and use SetPosition in the same way you would with notes. Use IDatumTag.SetLabel if you would like to set a label other than the default.

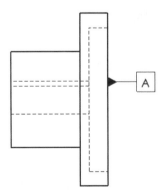

14. In the main procedure, add a call to AddDatum as shown. It must come before AddFileNameNote to maintain the user pre-selection.

```
Public Sub main()
```

```
Dim swDoc As ModelDoc2 = Nothing
swDoc = swApp.ActiveDoc

AddDatum(swDoc)
AddFileNameNote(swDoc)

End Sub
```

15. Save, then debug and test the macro. Make sure a model edge is pre-selected to create a datum. Verify that the note is created in the bottom-right corner of the drawing that links to the name of the model in the first drawing view.

Optional – Update AutoDrawing

If you would like to add this capability to the AutoDrawing macro, follow these last steps.

16. Close Visual Studio.

17. From SOLIDWORKS, select Tools, Macro, Edit to open *AutoDrawing.vbproj*.

18. From the menu, select Project, Add Existing Item. Browse to your InsertAnnotations folder, select AnnotationsMod.vb and click Add. The file is added to the Solution Explorer. *Hint: Add as Link to maintain a single code module shared between multiple macros.*

19. In *SolidWorksMacro.vb*, add the calls to insert the file name note by calling AddFileNameNote as shown within the DrawingFromModel procedure.

```
...
   Dim vAnnotations As Array = Nothing
   vAnnotations = swDrawing.InsertModelAnnotations3(0, 32776, _
     True, True, False, True)
   AddFileNameNote(swDoc)
   '
   'Zoom To Fit
...
```

20. Test the macro and debug as necessary. It will now add the file name note as it builds each drawing.

21. Close Visual Studio.

Tables

SOLIDWORKS has many different tables for drawings and models. Inserting, reading and writing to tables can all be done through the API and is relatively straight forward. Tables also have a parent Annotation interface. The table interface defines how the object displays. For example, a BOM table can be set to show parts, top level or an indented list. That setting is in the table itself. However, if you need to read or write text to a table, you need its Annotation interface.

General Tables

The following example illustrates how to insert a general two row, two column table to a drawing at a specific location. The value from one of the table cells is read and displayed in a message to the user. General tables are a good place to start since they do not require a drawing view as a reference.

22. Open the existing *InsertAnnotations.vbproj* macro by going to Tools, Macro, Edit.

23. Create a new procedure in AnnotationsMod.vb named `InsertGeneralTable`. Add the following code in the procedure.

```
Public Sub InsertGeneralTable(swDoc as ModelDoc2)
  Dim swDrawing As DrawingDoc = swDoc
  Dim myTable As TableAnnotation = Nothing
  'table definition values
  'location in meters
  Dim tableLocX As Double = 0.15
  Dim tableLocY As Double = 0.05
  'if no template is used, specify the number
  'of columns and rows
  Dim columns As Long = 2
  Dim rows As Long = 2
  Dim Anchor As Boolean = False
  'place at X,Y location
  'optionally set the full path to a table template
  Dim tableTemplate As String = ""

  'insert the table
  myTable = swDrawing.InsertTableAnnotation2 _
    (Anchor, tableLocX, tableLocY, _
    swBOMConfigurationAnchorType_e.swBOMConfigurationAnchor_TopRight, _
    tableTemplate, rows, columns)

  'add table text
  myTable.Text2(0, 0, True) = "Row 1 Column 1"
  myTable.Text2(0, 1, True) = "Row 1 Column 2"
  myTable.Text2(1, 0, True) = "Row 2 Column 1"
  myTable.Text2(1, 1, True) = "$PRP:""SW-Sheet Name(Sheet Name)"""

  'read some table text
  MsgBox(myTable.Text2(1, 1, True) & vbCrLf _
    & myTable.DisplayedText2(1, 1, True))
End Sub
```

InsertTableAnnotation2

InsertTableAnnotation2 is a method of the IDrawingDoc interface and inserts the table on the active sheet. It has seven arguments, but not all require valid values depending on your use. If you use a template, you do not need to specify the number of columns or the anchor type. The columns and anchor type of the template are used instead. If the template path is left as an empty string, you must specify the number of columns for the table as done in this example.

value = IDrawingDoc.InsertAnnotationTable2(UseAnchorPoint, X, Y, AnchorType, TableTemplate, Rows, Columns)

- **UseAnchorPoint** must be either True or False. If set to True, the newly created table is anchored to the general table anchor point defined in the drawing. If False, the X and Y coordinates are used for position.

- **X** is the x coordinate location of the table in meters.

- **Y** is the y coordinate location of the table in meters.

- **AnchorType** is a long value from swBOMConfigurationAnchorType_e. It defines where the x and y coordinates are in reference to the table itself.

⊞	swBOMConfigurationAnchor_BottomLeft
⊞	swBOMConfigurationAnchor_BottomRight
⊞	swBOMConfigurationAnchor_TopLeft
⊞	swBOMConfigurationAnchor_TopRight

- **TableTemplate** is a string representing the full path to an existing general table template.

- **Rows** is a long integer value representing the total rows in the table.

- **Columns** is a long integer value representing the total columns in the table.

- **Value** is returned as an ITableAnnotation interface. From the new interface, we can read and write text along with several other actions.

All of the required arguments are declared as variables in the macro for easier editing. The code inserts a 2 x 2 table, not attached to an anchor point, and located at 150mm from the left and 50mm from the bottom of the sheet. This location is defined at the top-right of the table based on the defined anchor point.

ITableAnnotation.Text2

General tables can include existing text and formatting by using a template. Since this table didn't use a template, it is empty to start. The Text2 method of the ITableAnnotation interface provides a way to read and write text to a table. The text added to cells can include links to properties like the sheet name in this example in row 2, column 2.

Value = ITableAnnotation.Text(Row, Column, IncludeHidden)

- **Row** is a long integer representing the row of the desired cell. Row 0 represents the first, or top row of the table.

- **Column** is a long integer representing the column of the desired cell. Column 0 represents the left-most column of the table.

- **IncludeHidden** is a Boolean value. If set to True, values will be read from hidden cells. If False, hidden cell values will not be read.

- **Value** is a string value that can be returned or can also be set. Text2 will return property link syntax rather than its evaluated value.

ITableAnnotation.DisplayedText2

If you need to get the evaluated value of cell text that is linked to a property, use DisplayedText2. Its argument format is exactly the same as Text2 but it returns a different value. The MsgBox call will display both the Text2 and DisplayedText2 values for you to compare.

24. Edit the `main` procedure in *SolidWorksMacro.vb* as shown to call the new procedure.

```
Public Sub main()

  Dim swDoc As ModelDoc2 = Nothing
  swDoc = swApp.ActiveDoc

  AddFileNameNote(swDoc)
  AddDatum(swDoc)
  InsertGeneralTable(swDoc)

End Sub
```

25. Open or create a drawing and run the macro to see the results. Notice the difference between the values in the table and MsgBox comparing Text2 and DisplayedText2.

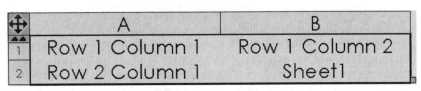

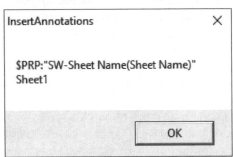

When using the Text2 property to write large amounts of data to tables, graphics refreshing can slow down the process. In the Remarks section under Text2, the API help recommends setting the table's

Visible property to False before populating the table. Then set it back to True when finished so the table display is only updated once.

BOM Tables

Bill of Materials (BOM) are the most commonly used table in SOLIDWORKS with the possible exception of Revision tables. Because of their designed automation, they provide an excellent means of collecting and reporting bill of materials information whether the table will be kept on the drawing or not.

BOM tables require a drawing view of a part or assembly. To prepare for this section of the exercise, open any drawing with at least one assembly view. Choose one with a simple assembly to start, and then try the same code on more detailed and complex designs.

26. Create a new function in *AnnotationsMod.vb* named `InsertBOMTable`. Declare the function to return a BomTableAnnotation. Add the following code to access the first drawing view.

```
Function InsertBOMTable(swDoc as ModelDoc2) As BomTableAnnotation
  Dim swDrawing As DrawingDoc = Nothing
  swDrawing = swDoc

  'get the first view from the drawing
  Dim swActiveView As View = Nothing

  Dim swSheetView As View = Nothing
  swSheetView = swDrawing.GetFirstView 'sheet
  swActiveView = swSheetView.GetNextView 'first view

End Function
```

Traversing Drawing Views

Inserting a Bill of Materials table through the SOLIDWORKS user interface requires selecting a drawing view containing a model. The view can contain a part or assembly, but assemblies are certainly the most common. To automate the process, we will assume the active drawing sheet's first drawing view contains the model we would like the BOM table to reference. Use the GetFirstView method of the IDrawingDoc interface to retrieve a View interface. It is important to note that the first View returned by this method is not what you would typically think of as a drawing view. It is actually the drawing sheet itself. Traversing the remaining drawing views requires a call to GetNextView from an existing View interface. Neither of these calls require any arguments.

The newly added function fills two variables, one for the View interface to the sheet, the other for the first actual drawing view. The second will be the view needed for the BOM table.

Note: if you need to verify the model used in any View interface, use View.ReferencedDocument. This will return the ModelDoc2 interface to the referenced model. View.ReferencedConfiguration will return the name of the model's configuration used by the same view as a string.

27. Add the bold code below to insert the BOM table and return it to the calling code. Be sure to replace the BOM template string with the full path to a BOM table template file.

```
Function InsertBOMTable(ByVal swDoc As ModelDoc2) As BomTableAnnotation
```

```
Dim swDrawing As DrawingDoc = Nothing
swDrawing = swDoc

'get the first view from the drawing
Dim swActiveView As View = Nothing

Dim swSheetView As View = Nothing
swSheetView = swDrawing.GetFirstView 'sheet
swActiveView = swSheetView.GetNextView 'first view

'BOM Table definition variables
Dim swBOMTable As BomTableAnnotation = Nothing
Dim config As String = _
    swActiveView.ReferencedConfiguration()
Dim template As String = "C:\Program Files\" _
    & "SOLIDWORKS Corp\SOLIDWORKS\lang\english\bom-standard.sldbomtbt"

'insert the table into the drawing
'based on the active view
swBOMTable = swActiveView.InsertBomTable4(False, 0.2, 0.3, _
    swBOMConfigurationAnchorType_e.swBOMConfigurationAnchor_TopRight, _
    swBomType_e.swBomType_Indented, config, template, _
    False, swNumberingType_e.swNumberingType_Detailed, False)
Return swBOMTable
End Function
```

Before making the call to insert the table, two additional variables are declared for two of the required arguments. The call to insert the BOM table will require the name of the model configuration to be referenced by the table as well as the full path to a Bill of Materials template file (*.sldbomtbt*). The configuration name can be retrieved from the View interface through its ReferencedConfiguration property.

InsertBomTable4

InsertBomTable4 is a method of the View interface. There would be nothing to report in a BOM table if there were no view containing a model. The method does not require the View to be selected, just a View interface. InsertBomTable4 has 10 arguments. You cannot leave any of the arguments out of the call, but they are not all used. Therefore, you can get away with some of them being empty strings or zero values.

value = IView.InsertBomTable4(UseAnchorPoint, X, Y, AnchorType, BomType, Configuration, TableTemplate, Hidden, IndentedNumberingType, DetailedCutList)

- **View** must be a valid drawing view containing a model.

- **UseAnchorPoint** is a Boolean value. If set to True, the newly created table will be locked to the Bill of Materials anchor point defined in the drawing. X and Y values will be ignored if this is True.

- **X** is the horizontal coordinate of the inserted table. It must be a double value in meters. Its value will be ignored if UseAnchorPoint is True.

- **Y** is the vertical coordinate of the inserted table. It will also be ignored if UseAnchorPoint is True.

- **AnchorType** is a long value represented by the same enumeration used when inserting general tables. It is the location of the table relative to the anchor or its x, y coordinate location.

- **BomType** is a long value that defines how the BOM table will be expanded from the enumeration swBomType_e. It matches the setting in the user interface for parts only, top assemblies, or a full indented BOM.

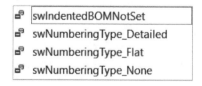

- **Configuration** is a string defining the configuration name to be used in the table.

- **TableTemplate** is a string representing the full path to the template to be used for the table. If an invalid path is given, no table will be created. If an empty string is used, a table will be created with default columns.

- **Hidden** is a Boolean value defining whether the newly created table will be visible on the drawing sheet or not. I cannot think of many times when you would want to use False here.

- **IndentedNumberingType** defines the item numbering definition of the table. This is a long value from the enumeration swNumberingType_e. Use any but the first swIndentedBOMNotSet value to correspond with the same settings from the user interface when inserting a BOM table.

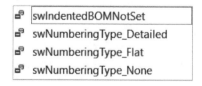

- **DetailedCutList** is a Boolean defining whether any weldment part contained in the table should have its cutlist expanded into the BOM table. Set this value to True if you would like the cutlist expanded in the BOM.

- **Value** returned by this method is the newly created IBomTableAnnotation interface.

Reading BOM Table Text

The BomTableAnnotation structure is unique compared to the general TableAnnotation. It includes methods for getting the model path names or components of any row along with methods for getting the custom properties utilized in columns. However, if you need to simply read the text from the table, you can get its underlying TableAnnotation and process that interface just as we did in the previous example.

28. Add a new procedure named ReadBOMTable to read all cells from the BOM table's underlying TableAnnotation interface.

```
Imports SolidWorks.Interop.sldworks
Imports SolidWorks.Interop.swconst
Imports System.Runtime.InteropServices
Imports System

Module AnnotationsMod
...

  Sub ReadBOMTable(swBOMTable As BomTableAnnotation)
    'read the cells from the table
    Dim genTable As TableAnnotation = swBOMTable
    Dim columns As Long = genTable.ColumnCount
    Dim rows As Long = genTable.RowCount
    For i As Integer = 0 To rows - 1
      For j As Integer = 0 To columns - 1
        Debug.Write(genTable.DisplayedText2(i, j, True) & vbTab)
      Next
      Debug.WriteLine("")
    Next

  End Sub
```

Get an Existing BOM

The BOM table in this exercise is retrieved from the InsertBOMTable function. If you needed to get an existing BOM table from a drawing, you can traverse the feature tree, looking for the feature with the correct type name. Use the traversal technique described in the Model Creation chapter. The example below will find the BOM feature, get its BOM table and then its underlying general table interface. It assumes a drawing is currently open. The BOM table can be on any sheet.

29. Add a new function named GetBOMTable that finds the BOM from the feature tree and returns its BomTableAnnotation to the calling code.

```
Function GetBOMTable(swDoc As ModelDoc2) As BomTableAnnotation
  Dim swFeat As Feature
  Dim swBOMFeat As BomFeature = Nothing

  'traverse all features
  'looking for a BomFeature
```

```
    swFeat = swDoc.FirstFeature
    Do While Not swFeat Is Nothing
      If swFeat.GetTypeName = "BomFeat" Then
        swBOMFeat = swFeat.GetSpecificFeature2
        Exit Do
      End If
      swFeat = swFeat.GetNextFeature
    Loop

    Dim myBOMTable As BomTableAnnotation
    Dim genTable As TableAnnotation
    If swBOMFeat Is Nothing Then
      MsgBox("No BOM table found.")
      Return Nothing
    Else
      If swBOMFeat.GetTableAnnotationCount = 1 Then
        myBOMTable = swBOMFeat.GetTableAnnotations(0)
        genTable = myBOMTable
        Debug.Print("First cell = " & genTable.DisplayedText2(0, 0, True))
      Else
        MsgBox("BOM is split. Only the first section will be returned.")
        Return swBOMFeat.GetTableAnnotations(0)
      End If
    End If

    Return myBOMTable
End Function
```

Reading BOM Tables

To read the table text from a BOM table, you must first get to its underlying ITableAnnotation interface. Getting the ITableAnnotation interface from an IBomTableAnnotation is the same as getting a IDrawingDoc from an IModelDoc2 interface. Simply declare a variable as TableAnnotation and set it equal to the existing BomTableAnnotation.

ITableAnnotation Columns and Rows

The ITableAnnotation ColumnCount and RowCount properties make it easy to loop through all cells in the table. Review the structure of the nested For, Next loops. The outermost loop cycles through each row. The inner loop cycles through each column. The two loops are set to count from zero to one less than the number of columns and rows since the Text property requires input starting at 0 rather than one. For example, the first cell in a table is always at the address (0, 0).

Writing Debug Text

You may ultimately need this macro to write to a text file, XML or spreadsheet application. This example uses the Debug class from System.Diagnostics to visualize the text output. The Write and WriteLine methods are another way to write text to the Output Window. WriteLine adds a return character at the end of the string while Write does not. The ReadBOMTable procedure uses Write until it gets to the last column in the row, and then uses an empty string in WriteLine to add only the return character.

30. Add the following calls to the `main` procedure to create a new BOM table and then to read its contents.

31. Comment out the first three procedures to reduce the number of annotations and tables added to drawings as you test.

```
Public Sub main()

  Dim swDoc As ModelDoc2 = Nothing
  swDoc = swApp.ActiveDoc

  'AddFileNameNote(swDoc)
  'AddDatum(swDoc)
  'InsertGeneralTable(swDoc)

  Dim BOM As BomTableAnnotation
  BOM = InsertBOMTable(swDoc)
  ReadBOMTable(BOM)

  Stop

End Sub
```

The addition of the Stop statement will cause a hard-coded breakpoint. It will help us review the macro output. As a reminder, use the Stop statement sparingly. If you leave it in your compiled macro, it will stop code execution for the user and will look like your macro has crashed.

32. Open any drawing containing at least one model view of an assembly.

33. Run and debug the macro as needed. When the code hits the Stop statement, it should pause and the Output Window should display your BOM text as shown below. If the Output Window is not visible, first make sure your code is running and paused. The Stop statement should be highlighted in yellow. If needed, show the Output Window by selecting Debug, Windows, Output.

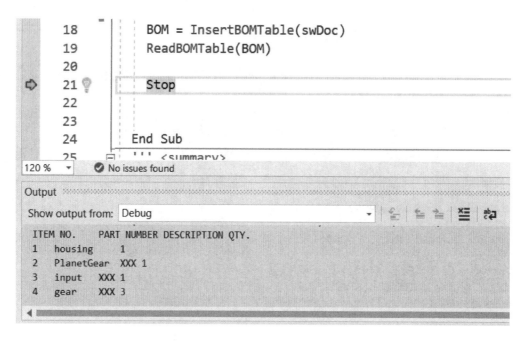

34. Delete or comment out the Stop statement before you build or run your final macro to prevent errors when running.

35. Uncomment the first three methods. Save and close Visual Studio.

Conclusion

At this point you could use the *AnnotationsMod.vb* module from this chapter to continue to build on the drawing automation macro. If an assembly is added to a drawing, you could insert a BOM table. You could also build a macro that batch processes drawings to open and read the BOM tables and output to an external format if needed.

Processing the other types of SOLIDWORKS tables is very similar. Record a brief macro to get the initial code structure. The Hole Table is one that requires a detailed selection before inserting.

Notes: The most common causes of failure in the BOM table macro are 1) no drawing views in the active drawing, 2) an invalid template path specified. If you run into any errors during testing, look for interface variables that are Nothing when they are used in methods and properties. The problem will commonly manifest itself with the following error. The solution is rarely the first troubleshooting tip. It is often the second. The object is null (Nothing) before calling on one of its methods or properties.

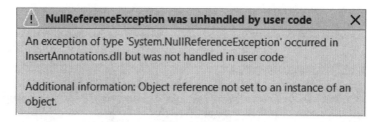

C# Example

See downloadable source files for the full C# project *Annotations.csproj.*

```csharp
public static void AddFileNameNote(ModelDoc2 swDoc)
{
    bool boolstatus = false;
    int longstatus = 0;
    //
    Note myNote = null;
    Annotation myAnnotation = null;
    TextFormat myTextFormat = null;
    DrawingDoc myDrawing = (DrawingDoc)swDoc;
    Sheet mySheet;
    double[] props;
    double width;
    double height;
    mySheet = (Sheet)myDrawing.GetCurrentSheet();
    props = (double[])mySheet.GetProperties2();
    width = props[5];   //sheet width
    height = props[6]; //sheet height
    myDrawing.EditSheet();
    myNote = ((Note)(swDoc.InsertNote("$PRPSHEET:\"SW-File Name\"")));

    if ((myNote != null))
    {
        myNote.LockPosition = false;
        myNote.Angle = Math.PI / 2;
        boolstatus = myNote.SetBalloon(0, 0);
        myAnnotation = ((Annotation)(myNote.GetAnnotation()));
        if ((myAnnotation != null))
        {
            longstatus = myAnnotation.SetLeader3((
                (int)(swLeaderStyle_e.swNO_LEADER)), 0,
                true, false, false, false);
            boolstatus = myAnnotation.SetPosition(width - 0.01, 0.015, 0);
            //boolstatus = myAnnotation.SetTextFormat(0, true, myTextFormat);
        }
    }
    swDoc.ClearSelection2(true);
    swDoc.WindowRedraw();
}
```

Building Assemblies

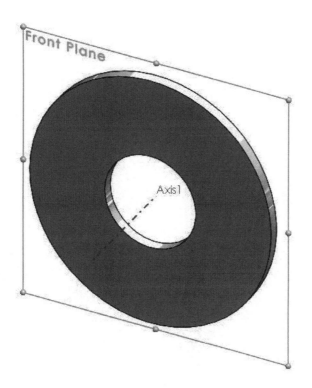

- **Topology Traversal**

- **Add Components**

- **Math Transform**

- **Add Mates**

Introduction

This chapter will focus on the task of adding components to assemblies along with a few operations that help in the process. The resulting macro will add a part and mate it to a user-selected location. There is a part named *washer.sldprt* included in the downloadable sample files that is already designed to work with the code in this example. Feel free to use your own as long as it matches the assumptions below.

In a perfect world, no matter how a user interacts with the interface, software would understand all possible solutions to every problem. Any component added to an assembly would be located perfectly by your macro. It would come up with the right solution every time. Those days may be coming – based on machine learning and big data. But until then, as a developer, you have to build in the possible solutions.

With that in mind, we will establish some rules and assumptions about the component being added to the assembly as well as requirements for user selections. These are listed below.

- The user can either select one flat face that has at least one hole in it, or multiple circular edges where the newly added part will be located.

- If the user selects circular edges, assume the edge has an adjacent flat face.

- If the user selects circular edges, parts will be added only to locations where the edge is a full circle. Partial circular edges will be ignored.

- If the user selects a face, a component will be added at the center of each full circular edge found on that face.

- The component to be added must have an axis named "Axis1." This will be mated concentric to the cylindrical face adjacent to the user-selected circular edge.

- The component to be inserted must have a plane named "Front Plane." This plane will be mated to the flat face connected to the circular edge.

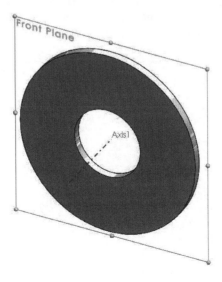

Error handling will be built to handle unexpected selections. We should also handle cases where the *Front Plane* or *Axis1* features are not found in the inserted component.

Initial Code

The basic structure of this macro consists of a form and a module. The form will collect the name of the part to be added and will give the user instructions to make their selections. The module will contain all code to process the insertion and mating of the components.

As a reminder, it is good practice to separate your processing code from the user interface whenever possible. This gives you the flexibility to create new user interfaces such as PropertyManager pages without altering any underlying process code.

1. Open the existing macro project named *AddComponents.vbproj* from the book's downloadable samples mentioned in the Introduction.

2. Review the `main` procedure in *SolidWorksMacro.vb*.

```
Public Sub main()
  Dim MyDialog As New Dialog1
  AddComponents.m_swApp = swApp
  MyDialog.Show()
End Sub
```

A new instance of `Dialog1` is created. This is a form in the macro. The module AddComponents has a variable named m_swApp, so the current swApp instance of the SOLIDWORKS application is passed to the module so it can be used in the processing code.

The form is displayed to the user by the Show method. Unlike ShowDialog, the Show method does not return anything. It simply displays the dialog and continues executing the macro code. Since there is no more code to run, we won't want the macro to exit once it hits the End Sub line, so we will need to disable the SOLIDWORKS setting "Stop VSTA debugger on macro exit."

The Show method also has the unique behavior of making the form modeless. You can still interact with the SOLIDWORKS session while the form is showing. ShowDialog keeps you from interacting with SOLIDWORKS while the form is showing.

3. Go to Tools, Options in SOLIDWORKS and turn off "Stop VSTA debugger on macro exit" from the General category.

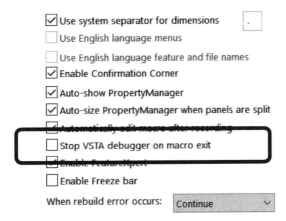

4. Review the design of *Dialog1.vb* by double-clicking on it in the Solution Explorer.

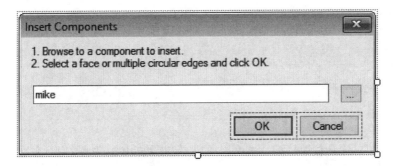

The form has a basic layout. A label provides instructions to the user. A text box and button control are used to let the user browse for a SOLIDWORKS part. The OK and Cancel buttons are the default dialog buttons that confirm or cancel the operation.

TopMost Property

Select the form itself and review its properties. The TopMost property is set to True. This is an easy way to keep a form on top of all other windows when you expect the user to interact with other applications. In fact, it keeps it in front of all applications while running, not just SOLIDWORKS. This form needs to stay visible while the user makes selections in SOLIDWORKS. If this property value was False, the form would fall behind the SOLIDWORKS interface when the user attempted to make the required selections. Since the ShowInTaskbar property is set to False, it would not show up in the Windows task bar and the user would have to move SOLIDWORKS to find the form again.

System.Windows.Forms.OpenFileDialog

The OpenFileDialog class is a standard way to let the user select a file. It only takes a few lines of code to let the user browse for a SOLIDWORKS part and then return the full path to the part to the form. The first step is to let the code know it uses the System.Windows.Forms namespace with an `Imports` statement.

5. Double-click on the BrowseButton to view the code required to present the OpenFileDialog to the user.

```
Imports System.Windows.Forms
Imports System.IO

...

Private Sub BrowseButton_Click(ByVal sender As System.Object, _
ByVal e As System.EventArgs) Handles BrowseButton.Click
  Dim OpenDia As New OpenFileDialog
  OpenDia.Filter = "SOLIDWORKS Part (*.sldprt)|*.sldprt"
  OpenDia.Title = "Browse to a part"
  Dim diaRes As DialogResult = OpenDia.ShowDialog
  If diaRes = Windows.Forms.DialogResult.OK Then
    FilePathTextBox.Text = OpenDia.FileName
  End If
End Sub
```

A new instance of OpenFileDialog is declared first. Its Filter property is used to set the File Type list and will filter the types of files displayed to the user. It is set to SOLIDWORKS part in this example. If you want to provide multiple file types in the typical drop-down list, you can repeat the same formatting separated by a pipe character. For example, if you wanted to provide a selection to show all files or SOLIDWORKS Parts, you would use the following filter text.

```
"SOLIDWORKS Part (*.sldprt)|*.sldprt|All Files (*.*)|*.*"
```

The Title property sets the text the user will see on the form header. This is the place to give the user some basic instructions.

The ShowDialog method is used to show this form to the user since we need to wait for the form to finish and determine whether the user clicked OK or Cancel. If the user clicks OK, the FileName property of the OpenFileDialog will return the full path to the file the user selected. This example passes the file name back to the text box control named FilePathTextBox.

Saving Application Settings

One of my favorite features of Visual Studio and .NET is the built-in ability to save and retrieve application settings. Previously, if you wanted to preserve settings for your macro or application, you had to read and write them to the Windows Registry, an ini file, or something similar. Using Visual Studio, you can create settings that are saved to a *.config* file. You can save previously used text values, dialog size and position and others. Settings can store form control properties and values set by the user, or can be completely controlled by the application.

To add settings to a macro, double-click on My Project in the Solution Explorer and select the Settings tab. You will notice this macro has a user setting named LastModelPath set as a String data type. Since its Scope is set to User, it can be changed by the macro while it is running and saved for the next time it is used.

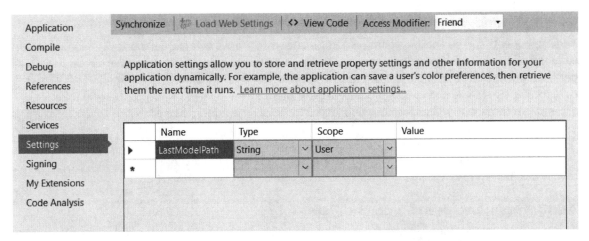

On the form, the FilePathTextBox control has its *(Application Settings), Text* value linked to an Application Setting named *LastModelPath*. You may wish to review how the FilePathTextBox control is linked. Double-click on Dialog1.vb from the Solution Explorer and select the FilePathTextBox on the form. View its properties to see the link.

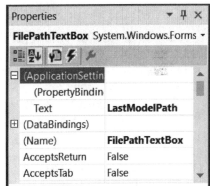

Finally, when Dialog1 is closed, the settings are saved in the following procedure. My.Settings provides access to all settings in the macro. The Save method saves those settings for use any time the macro is run.

```vb
Private Sub Dialog1_FormClosing(ByVal sender As Object, _
ByVal e As System.Windows.Forms.FormClosingEventArgs) _
Handles Me.FormClosing
  My.Settings.Save()
End Sub
```

AddComponents Module

When the user clicks the OK button, the following code launches the AddCompAndMate method from the AddComponents module. The last two lines of the OK button click event handler are the default code from the dialog to return OK and close the dialog.

```vb
Private Sub OK_Button_Click(ByVal sender As System.Object, _
ByVal e As System.EventArgs) Handles OK_Button.Click
    'process the data from the form with solidworks
    AddComponents.AddCompAndMate(FilePathTextBox.Text)

    Me.DialogResult = System.Windows.Forms.DialogResult.OK
    Me.Close()
End Sub
```

6. Review the code in *AddComponents.vb* by right-clicking on the AddComponents text in the AddComponents.AddCompAndMate line. Select Go To Definition from the menu.

Public, Friend and Private Scope

AddComponents consists of one public procedure and several private functions along with a public variable to hold a module-level reference to the SOLIDWORKS application. The functions have been declared with a Private scope, assuming they will only be used by procedures in this module. If you wish to make them available outside the module, declare them with the Friend keyword to make them accessible to code within the macro, or Public to be accessible by external applications.

The functions, IsFullCircle, GetEdgeFaces, GetAllFaceCircularEdges, GetSelectedCircularEdges, OpenPartInvisible and GetCircleCenter are operations required to process or retrieve content that is not directly available through a single API call in SOLIDWORKS. Each will be discussed as it is used.

AddComponents.AddCompAndMate

AddCompAndMate is called by Dialog1 when the user clicks OK. The procedure is declared with one argument, PartPath, as the full path to the part to be inserted.

```
Public Sub AddCompAndMate(ByVal PartPath As String)
  Dim Model As ModelDoc2 = m_swApp.ActiveDoc
  Dim MyView As ModelView = Model.ActiveView
  Dim errors As Long
  If Model.GetType <> swDocumentTypes_e.swDocASSEMBLY Then
    MsgBox("For use with assemblies only.", MsgBoxStyle.Exclamation)
    Exit Sub
  End If
  Dim MyAssy As AssemblyDoc = Model
  'continue adding processing code here

End Sub
```

The existing code gets the active document's IModelDoc2 interface, then the ModelDoc's IModelView interface and determines whether the active document is an assembly. If it is not, a message is presented to the user and the procedure exits. Otherwise, a specific reference to the IAssemblyDoc interface is retrieved from the ModelDoc.

Processing Selections

Before the part is added to the assembly, some validation needs to be done to make sure the user has made the right selections, or any selection at all for that matter.

Selection Count and Selection Type

Based on our assumptions, we need to check to see if the user has selected a face or multiple edges.

> 7. Add the following code to the AddCompAndMate procedure.

```
Public Sub AddCompAndMate(ByVal swApp As SldWorks, ByVal PartPath As String)
  Dim Model As ModelDoc2 = swApp.ActiveDoc
  Dim MyView As ModelView = Model.ActiveView
  If Model.GetType <> swDocumentTypes_e.swDocASSEMBLY Then
    MsgBox("For use with assemblies only.", MsgBoxStyle.Exclamation)
    Exit Sub
```

```
End If
Dim MyAssy As AssemblyDoc = Model

'get the current selection
'user should have selected the edge of a hole
'or an entire face for parts to be inserted on
Dim Edges As New Collection
Dim selMgr As SelectionMgr = Model.SelectionManager
Dim SelectedObject As Object = Nothing
Dim SelCount As Integer = selMgr.GetSelectedObjectCount2(-1)
If SelCount > 0 Then
  'make sure they have selected a face
  SelectedObject = selMgr.GetSelectedObject6(1, -1)
  Dim SelType As Integer = selMgr.GetSelectedObjectType3(1, -1)
  If SelType = swSelectType_e.swSelFACES And SelCount = 1 Then
    'get all circular edges on the face

  ElseIf SelType = swSelectType_e.swSelEDGES Then
    'get all circular edges selected

  End If
Else
  MsgBox("Please select a face or hole edges.", MsgBoxStyle.Exclamation)
  Exit Sub
End If

End Sub
```

The added code makes use of the SOLIDWORKS ISelectionManager interface. For instance, the number of selected items is retrieved and stored as SelCount using the GetSelectedObjectCount2 method. Pass a value of -1 to get all selections, regardless of their Mark.

If the user has made at least one selection, the ISelectionManager interface's GetSelectedObjectType3 method is used to determine if the selection is a face. The first argument is the selection index and the second is the Mark. An index of 1 returns the type of the first selected item. Selection types can be derived from the swSelectType_e enumeration. There are many selection types available in SOLIDWORKS. Refer to the API Help for a complete listing.

The If statement combines the logic of two tests. If the selected item is a face and there is only one selection, then we will need to get all of the circular edges on that face. Alternatively, the ElseIf part of the statement will run if the selection type is an edge. It does not matter how many edges the user selects, so no additional logical check is used.

If the user has not selected anything, a message is given and the procedure exits.

Traversing Topology

We now have two distinct paths our procedure can take. From our assumptions, if the user selected a face, all full circular edges on that face must be collected. If they select edges, all full circular edges must be collected. All other edges can be ignored.

Collections

Collections are a special type of array and are especially helpful when adding elements one at a time. They behave much more like a List form control. Collections have methods for adding and removing members instead of an array's index-based method. They are heavier than arrays since they do not require an explicit data type definition. Each element of a collection can hold any data type. After a collection of elements is built, it can be traversed using a For Each Next loop. Notice the declaration of Edges as a new collection. You have to use the New keyword to create an instance of a collection before adding elements to it.

Get all Edges from a Face

8. Review the function named GetAllFaceCircularEdges. Its job is to gather all of the circular edges from the selected face, adding them to the Edges collection only if they are a full circle. It expects a single argument, a reference to an IFace2, and returns a collection of edges.

```
Private Function GetAllFaceCircularEdges _
(SelectedFace As Face2) As Collection
  Dim Edges As New Collection
  Dim FaceEdges() As Object = SelectedFace.GetEdges
  For Each MyEdge As Edge In FaceEdges
    If IsFullCircle(MyEdge) Then
      Edges.Add(MyEdge)
    End If
  Next
  Return Edges
End Function
```

From an IFace2 interface, you can call GetEdges to return an Object array of IEdge interfaces. Some SOLIDWORKS API calls do not directly return an array of the type of interface it contains. In these cases, you should declare your array as an Object type since it can hold any data type.

IsFullCircle Function

The IsFullCircle function in AddComponents uses the edge's underlying curve definition to determine if the user has selected a full circular edge. The function uses several properties and methods of the edge and its curve to determine if the selected edge is a full circle. IsFullCircle simply returns a Boolean value of True if the edge is a full circle, False if not.

The topology of a SOLIDWORKS model is deeper than you might expect. A part might first include multiple IBody interfaces. These could include solids, graphics or surfaces. Each IBody is composed of Face2, Vertex and Edge interfaces. These are the user-selectable items in SOLIDWORKS that allow you to add mates, start sketches, convert entities, etc. However, you cannot determine if an edge is a circle without getting its underlying ICurve interface. In the same respect, you cannot determine if a face is flat or cylindrical without its underlying ISurface interface. It takes some digging to determine if the user selected edges are actually full circles. We take much of this for granted in the user interface, but your code has to be explicit when you use the API to interrogate geometry and topology.

```
  Private Function IsFullCircle(EdgeToCheck As Edge) As Boolean
    Dim MyCurve As Curve = EdgeToCheck.GetCurve
    If MyCurve.IsCircle Then
```

```
      'you have a circular edge
      'is it a complete circle?
      If EdgeToCheck.GetStartVertex() Is Nothing Then
        'full circle
        Return True
      End If
    End If
    Return False
  End Function
```

The function first retrieves the ICurve interface by calling the GetCurve method of the edge. From the curve, you can call IsCircle. This returns True if the curve from the edge is circular. However, it returns True even if it is a partial circle.

An additional check is used to determine if the edge is a full circle. The IEdge interface has a method called GetStartVertex. A full circle has no vertices. So if this method returns Nothing, you can expect the edge is a complete circle.

If you want the user to be able to select any partial circular edge, not just full circles, you could remove the GetStartVertex check.

9. Add the call to GetAllFaceCircularEdges in the AddCompAndMate procedure as shown.

```
...
If SelCount > 0 Then
  'make sure they have selected a face
  SelectedObject = selMgr.GetSelectedObject6(1, -1)
  Dim SelType As Integer = selMgr.GetSelectedObjectType3(1, -1)
  If SelType = swSelectType_e.swSelFACES And SelCount = 1 Then
    'get all circular edges on the face
    Edges = GetAllFaceCircularEdges(SelectedObject)

  ElseIf SelType = swSelectType_e.swSelEDGES Then
    'get all circular edges selected

  End If
Else
...
```

Processing Selected Edges

We now need to handle the case where the user may have selected any number of edges. Since we do not know how many edges the user may have selected, a For loop is used to work through the selections. The GetSelectedCircularEdges function will be used to handle edge selections.

10. Review the function named GetSelectedCircularEdges that filters out anything that isn't a full circular edge, and then return the collection.

```
Private Function GetSelectedCircularEdges _
(selMgr As SelectionMgr) As Collection
```

```
   Dim SelectedObject As Object
   Dim Edges As New Collection
   For i As Integer = 1 To selMgr.GetSelectedObjectCount2(-1)
     SelectedObject = selMgr.GetSelectedObject6(i, -1)
     Dim SelType As Integer = selMgr.GetSelectedObjectType3(i, -1)
     'make sure each selection is an edge
     If SelType = swSelectType_e.swSelEDGES Then
       Dim MyEdge As Edge = SelectedObject
       If IsFullCircle(MyEdge) Then
         Edges.Add(MyEdge)
       End If
     End If
   Next
   Return Edges
End Function
```

At this point we only know that the first user selected item was an edge. A quick check of each selected item's type will prevent non-edges from being evaluated further. Next, simply send the edge through the IsFullCircle function. If it is a full circle, it is added to the collection of edges. The resulting collection is then returned.

    11. Add the call to GetSelectedCircularEdges in the AddCompAndMate procedure as shown.

```
...
If SelCount > 0 Then
   'make sure they have selected a face
   SelectedObject = selMgr.GetSelectedObject6(1, -1)
   Dim SelType As Integer = selMgr.GetSelectedObjectType3(1, -1)
   If SelType = swSelectType_e.swSelFACES And SelCount = 1 Then
     'get all circular edges on the face
     Edges = GetAllFaceCircularEdges(SelectedObject)

   ElseIf SelType = swSelectType_e.swSelEDGES Then
     'get all circular edges selected
     Edges = GetSelectedCircularEdges(selMgr)

   End If
Else
...
```

Adding a Part to an Assembly

To add a part to an assembly, the part must be opened in memory, just like adding drawing views. We have already covered the OpenDoc6 method of the ISldWorks interface so we will not go into detail on that call again. For fast performance, we are going to disable the visibility of the newly opened part. This allows us to load the part into memory, but not display it in a window to the user.

    12. Review the existing function named OpenPartInvisible to open the selected part invisibly. Include one argument for the part's file path.

```
Private Function OpenPartInvisible(PartPath As String) As PartDoc
  Dim errors As Integer
  Dim warnings As Integer
  'open the file invisibly
  m_swApp.DocumentVisible(False, swDocumentTypes_e.swDocPART)
  Dim Part As PartDoc = m_swApp.OpenDoc6 _
    (PartPath, swDocumentTypes_e.swDocPART, _
    swOpenDocOptions_e.swOpenDocOptions_Silent, "", errors, warnings)
  m_swApp.DocumentVisible(True, swDocumentTypes_e.swDocPART)

  If Part Is Nothing Then
    MsgBox("Unable to open " & PartPath, MsgBoxStyle.Exclamation)
    Return Nothing
  End If
  Return Part
End Function
```

The DocumentVisible method of the ISldWorks interface controls whether newly opened models of a specific type will be invisible or not. The first argument is whether it is visible, the second is the document type from the swDocumentTypes_e enumeration. After opening the part, the DocumentVisible setting is returned to its default of opening parts visibly. If you left it at False, the user could not open parts visibly for the rest of the SOLIDWORKS session.

The OpenDoc6 method opens the part, returning its IModelDoc2 interface. If it is Nothing after the OpenDoc6 call, there was a problem opening the part.

13. Add the following to AddCompAndMate to call OpenPartInvisible, get its result, and then verify we have a part and a collection of edges. The procedure will exit if there is either no part or edges. There will be some additional clean up work to do later.

```
...
If SelCount > 0 Then
  'make sure they have selected a face
  SelectedObject = selMgr.GetSelectedObject6(1, -1)
  Dim SelType As Integer = selMgr.GetSelectedObjectType3(1, -1)
  If SelType = swSelectType_e.swSelFACES And SelCount = 1 Then
    'get all circular edges on the face
    Edges = GetAllFaceCircularEdges(SelectedObject)

  ElseIf SelType = swSelectType_e.swSelEDGES Then
    'get all circular edges selected
    Edges = GetSelectedCircularEdges(selMgr)

  End If
Else

  'open the part now
  Dim Part As PartDoc = OpenPartInvisible(PartPath)
  If Part Is Nothing Or Edges Is Nothing Then
    'Clean up
```

```
    Exit Sub
  End If
. . .
```

14. Add the following code to insert an instance of the part to the assembly for every full circular edge that is in the collection. The part will be added at the center of each circular edge.

```
. . .
'open the part now
Dim Part As PartDoc = OpenPartInvisible(PartPath)
If Part Is Nothing Or Edges Is Nothing Then
  'Clean up
  Exit Sub
End If

'get the component that was selected
Dim SelectedComp As Component2 = Nothing
SelectedComp = selMgr.GetSelectedObjectsComponent2(1)

'turn off assembly graphics update for speed
'add the new part to each circular edge
MyView.EnableGraphicsUpdate = False
For Each CircEdge As Edge In Edges
  'get the center of the circular edge
  Dim Center() As Double
  Center = GetCircleCenter(CircEdge.GetCurve, SelectedComp)
  'insert the part at the circular edge's center
  Dim MyComp As Component2 = MyAssy.AddComponent4(PartPath, _
    "", Center(0), Center(1), Center(2))

Next

  End Sub
```

The first bit of code gets the IComponent2 interface of the component that was selected. The one that has the flat face or circular edges. This interface is used later to add the part to the assembly at the center of the circular edge. The ISelectionManager interface gives you direct access to the selected component by calling GetSelectedObjectsComponent2. The argument is the selection index. This code gets the first selected component. In this example, we are using an old method for simplicity. GetSelectedObjectsComponent4 is the latest call, but requires another argument that is not needed for this example.

For performance reasons, it is more efficient to add multiple components after turning off graphical updating as discussed in the Model Creation chapter. Use the EnableGraphicsUpdate method of IModelView to turn it off. Just do not forget to add the code to turn it back on at any point your macro exits. If your code exits before it is turned back on, the user must re-open the model to reset the graphics.

GetCircleCenter Function

The next part of the code uses another function from AddComponents named GetCircleCenter. It returns an array of double values representing the x, y, z center of the circular edge. This again uses several API methods in combination to get a useful result.

```
'return an array of doubles for the x, y, z circle center
'relative to the assembly
Private Function GetCircleCenter(ByVal MyCurve As Curve, _
ByVal Comp As Component2) As Double()
    Dim MyCenter(2) As Double
    Dim returnValues As Object = MyCurve.CircleParams
    MyCenter(0) = returnValues(0)
    MyCenter(1) = returnValues(1)
    MyCenter(2) = returnValues(2)
    Dim Radius As Double = returnValues(6)

    Dim MathUtil As MathUtility = m_swApp.GetMathUtility
    Dim mPoint As MathPoint = Nothing
    mPoint = MathUtil.CreatePoint(MyCenter)

    Dim CompTransform As MathTransform = Comp.Transform2
    mPoint = mPoint.MultiplyTransform(CompTransform)
    'return the x,y,z location in assembly space
    Return mPoint.ArrayData
End Function
```

IMathUtility and Transforms

The function uses the IMathUtility, IMathPoint and an IMathTransform interfaces. SOLIDWORKS has many built-in functions for dealing with the complexities of transforming positions, locations and rotations from one coordinate space to another. For example, a part has geometry created in its own coordinate space. However, when the part is in an assembly, it is in the assembly's coordinate space and its position and orientation is relative to the assembly. If you want to determine the position of a hole in a part at the assembly level, you start with the hole's position in part coordinates. You then transform that location into the assembly's coordinates. Most engineers take courses on matrix math in college to learn how to do these transformations. However, not many remember how to do it. If you do, you can write your own matrix math transformation tools if you prefer. As for me, I'll stick with the math utilities SOLIDWORKS provides so I do not have to break out my old math books.

CircleParams

The first part of the function gets the x, y, z center of the circular edge in the component's local coordinates. The CircleParams method of the ICurve interface returns an array of data related to the curve. For circles, the first three array elements are the center x, y and z positions respectively. Element 6 of the array is the curve's radius. The radius is not returned by the function in this example, but if you wanted to do some auto-sizing of the inserted part, you could make use of the radius to determine which configuration or part to add.

IMathPoint

Before we can transform the x, y, z location from component coordinates into assembly coordinates we must turn the array into an IMathPoint interface. The IMathUtility interface makes that easy with a call to CreatePoint. Pass in an array of three doubles representing the point's x, y and z location.

IMathTransform

The new IMathPoint must then be transformed to assembly coordinates by first getting the selected component's IMathTransform interface. This represents its location and rotation in assembly space. The Transform2 method of IComponent2 gets the MathTransform.

The MultiplyTransform method of the IMathPoint is then used to get a new IMathPoint transformed to the root component – being the assembly. Finally, the IMathPoint's array of x, y and z are extracted into an array of doubles using the ArrayData method.

IAssemblyDoc.AddComponent4

Now that we have the insertion location relative to the assembly, the part can be added using the AddComponent4 method of IAssemblyDoc. The first argument is the name of the part or assembly to add. The second is the configuration name. Passing an empty string uses the active configuration. The third, fourth and fifth arguments are the x, y and z insertion location of the component. Be aware that this does not mean that the part's origin will be at that location. It is more like the center of the volume of the part.

This is another example where we have used an older API call for simplicity. SOLIDWORKS 2010 introduced AddComponent5. However, it has several new arguments that do not add value to our current process. They provide options for inserting a sub assembly and configuring it in the same process. Since we are only inserting parts, there is no need for the complexity of AddComponent5.

Adding Mates

Now that the component is in the assembly, mates should be added to constrain its position. Reviewing our assumptions, we need to add a mate between the plane named "Front Plane" of the added part to the flat face adjacent to each edge. We then need to add a concentric mate between "Axis1" in the added part and the cylindrical face adjacent to the edge.

15. Add the following before the Next statement of the For loop to add the necessary mates in AddComponentAndMate.

```
...
'turn off assembly graphics update
'add it to each circular edge
MyView.EnableGraphicsUpdate = False
For Each CircEdge As Edge In Edges
  'get the center of the circular edge
  Dim Center() As Double
  Center = GetCircleCenter(CircEdge.GetCurve, SelectedComp, swApp)
  'insert the part at the circular edge's center
  Dim MyComp As Component2 = MyAssy.AddComponent4(PartPath, _
    "", Center(0), Center(1), Center(2))
```

```
    'get the two faces from the edge
    'set the first face to the cylinder
    'the second to the flat face
    Dim MyFaces() As Face2 = GetEdgeFaces(CircEdge)
    'Add Mates
    'add a coincident mate bewteen the flat face
    'and the Front Plane of the added component
    Dim MyPlane As Feature = MyComp.FeatureByName("Front Plane")
    Dim MyMate As Mate2
    If Not MyPlane Is Nothing Then
      MyPlane.Select2(False, -1)
      MyFaces(1).Select(True)

      MyMate = MyAssy.AddMate5 _
        (swMateType_e.swMateCOINCIDENT, _
        swMateAlign_e.swMateAlignALIGNED, False, _
        0, 0, 0, 0, 0, 0, 0, 0, False, False, 0, errors)
    End If

    'mate the cylinder concentric to "Axis1"
    Dim MyAxis As Feature = MyComp.FeatureByName("Axis1")
    If Not MyAxis Is Nothing Then
      MyAxis.Select2(False, -1)
      'select the cylindrical face
      MyFaces(0).Select(True)

      MyMate = MyAssy.AddMate5(swMateType_e.swMateCONCENTRIC, _
        swMateAlign_e.swMateAlignCLOSEST, False, _
        0, 0, 0, 0, 1, 0, 0, 0, False, False, 0, errors)
    End If
  Next

End Sub
```

GetEdgeFaces Function

The first part of the newly added code gets an array of IFace2 interfaces from the IEdge interface by using the custom GetEdgeFaces function. The first element of the array is the cylindrical face adjacent to the edge, the second is the flat face. The function's code is explained here.

```
'function to get and return the
'two adjacent faces of the edge
Private Function GetEdgeFaces(ByVal MyEdge As Edge) As Face2()
  Dim tmpFaces(1) As Face2
  Dim tmpFace0 As Face2 = MyEdge.GetTwoAdjacentFaces2(0)
  Dim tmpSurf0 As Surface = tmpFace0.GetSurface
  'check if the surface is a cylinder
  If tmpSurf0.IsCylinder Then
    tmpFaces(0) = MyEdge.GetTwoAdjacentFaces2(0)
    tmpFaces(1) = MyEdge.GetTwoAdjacentFaces2(1)
```

```
    Else
      tmpFaces(0) = MyEdge.GetTwoAdjacentFaces2(1)
      tmpFaces(1) = MyEdge.GetTwoAdjacentFaces2(0)
    End If
    'the zero element should be a cylinder
    Return tmpFaces
  End Function
```

The GetTwoAdjacentFaces2 method of the IEdge interface returns an array of the two faces that meet to form the edge. The GetEdgeFaces function is only necessary to determine which of the two faces is the cylinder. Just like with edges and curves, you must get to the underlying ISurface interface of a face to determine whether it is cylindrical or not. The IsCylinder property of ISurface returns True if the surface is a cylinder.

IComponent2.FeatureByName

Now that we have two of the faces necessary for mates, we need to get the two named features from the added part. The FeatureByName method of IComponent2 is an easy way to get a feature if you know its name. As we determined in the assumptions, we are assuming the user is selecting a component that meets these requirements.

Select and Select2

Once you have an IFeature interface from the FeatureByName method, it can be selected using the Select2 method of IFeature. The first argument is False if you want to start a new selection or True to add to the existing selection. The second argument is the selection Mark. Again, a value of -1 indicates no mark.

Faces must be selected using the Select method of IFace. Be careful which select method you use. Each interface has its own select method. The only argument for the Select method is a True if you want to add to the selection and False if you want to start a new selection.

AddMate5

The AddMate5 method of IAssemblyDoc adds a mate between the previously selected entities. This is another obsolete method used here because the new CreateMate method is significantly more complex than necessary for most mate conditions. AddMate5 has fifteen arguments, most of which are not applicable to the coincident and concentric mates needed here. The critical arguments are described below. Reference the API help if you need to explore additional mate types and their required arguments. AddMate5 returns a Mate2 interface, collected here in the variable MyMate.

value = instance.AddMate5(MateTypeFromEnum, AlignFromEnum, Flip, Distance, DistanceAbsUpperLimit, DistanceAbsLowerLimit, GearRatioNumerator, GearRatioDenominator, Angle, AngleAbsUpperLimit, AngleAbsLowerLimit, ForPositioningOnly, LockRotation, WidthMateOption, ErrorStatus)

- **MateTypeFromEnum** controls the mate type with a long data type from the swMateType_e enumeration.

- **AlignFromEnum** sets the alignment condition based on the swMateAlign_e enumeration. It can be either closest, aligned or anti-aligned.

- **ErrorStatus** a long variable is passed to this argument so the method can use it as an additional output. If there are errors in the mate operation they will show up in this error variable after the method is run.

The process of selection and mate is repeated twice, once for the flat face and plane and then again for the axis and cylindrical face. Always use the first, and sometimes second, mate to set the overall alignment. Subsequent mates can use the swMateAlignCLOSEST condition. Use the Plane Display settings under Document Properties to set unique front and back plane colors to make it easier to understand mate alignment to planes.

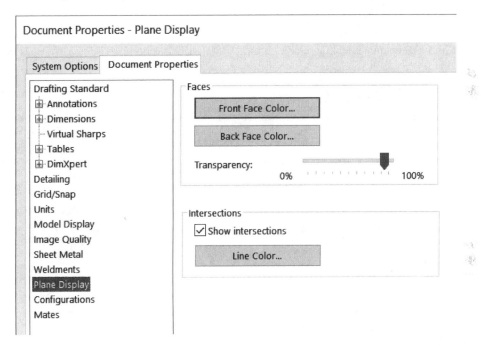

Finish Up

Since we turned off graphical updates to the assembly's ModelView earlier in the macro, it should be reset to keep everything working properly. Adding another procedure that can be called at any exit point will simplify the code.

16. Create a new procedure named CleanUp as shown.

```
Private Sub CleanUp(ByVal MyView As ModelView, ByVal Model As ModelDoc2)
    'turn graphics updating back on
    MyView.EnableGraphicsUpdate = True
    Model.ClearSelection2(True)
End Sub
```

17. Finish the macro by calling the CleanUp procedure from the two possible exit points.

```
...
    'open the part now
    Dim Part As PartDoc = OpenPartInvisible(PartPath)
    If Part Is Nothing Or Edges Is Nothing Then
      CleanUp(MyView, Model)
      Exit Sub
    End If
...
      MyMate = MyAssy.AddMate5(swMateType_e.swMateCONCENTRIC, _
        swMateAlign_e.swMateAlignCLOSEST, False, 0, 0, 0, _
        0, 1, 0, 0, 0, False, False, 0, errors)
    Next
    CleanUp(MyView, Model)
  End Sub
```

18. Test and debug your new macro as needed. Watch for common problems with mate alignment and failure to select the expected geometry prior to calling AddMate5.

If you would like to see the detailed steps while they happen, comment out the `MyView.EnableGraphicsUpdate` line in AddComponentAndMate.

```
    ...
    'turn off assembly graphics update
    'add it to each circular edge
    'MyView.EnableGraphicsUpdate = False
    For Each CircEdge As Edge In Edges
      'get the center of the circular edge
    ...
```

Also, comment out the DocumentVisible line from OpenPartInvisible.

```
Private Function OpenPartInvisible(PartPath As String) As PartDoc
  Dim errors As Integer
  Dim warnings As Integer
  'open the file invisibly
  'm_swApp.DocumentVisible(False, swDocumentTypes_e.swDocPART)
  Dim Part As PartDoc = m_swApp.OpenDoc6 _
    (PartPath, swDocumentTypes_e.swDocPART, _
    swOpenDocOptions_e.swOpenDocOptions_Silent, "", errors, warnings)
  ...
```

Then run another test of your macro. This time you will see every selection, component opened, and mate solved as it happens. You should also notice a significant difference in speed. The graphics updates represent a large part of the entire process and are simplified by only updating once.

Conclusion

You now have a functional macro to insert multiple parts at one time. This is somewhat similar to the SOLIDWORKS SmartFastener tool without the automatic size and length capabilities. Spend some time testing and trying different parts.

C# Example

See the complete *AddComponents.csproj* in the downloadable example files for C# source.

Working with File References

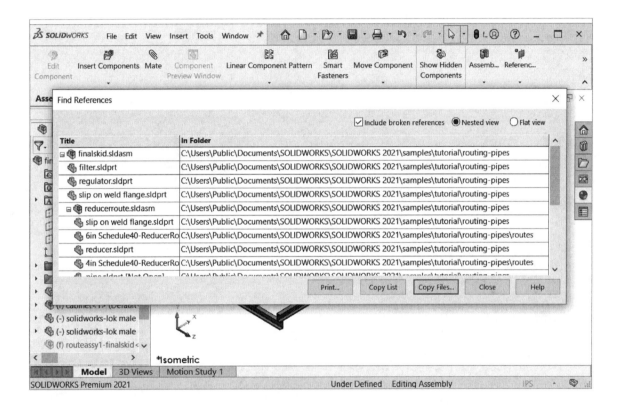

- **Finding File References**

- **Using DataTables**

- **Saving the References List**

Introduction

One of the strengths of SOLIDWORKS is the file associativity – you change the part and it changes the assembly and drawing where it is used. In turn, it requires careful management of SOLIDWORKS documents to maintain this associativity. Product data management or PDM is the common solution because of this very relationship.

Understanding SOLIDWORKS file references is the basis of developing file manipulation and reporting tools. Once you master topics in this exercise, you could build a utility that tracks file references, provides where used searches and automates revision history control.

This exercise will begin with a macro that allows you to view and save a list of file references. It will then add the ability to track whether a file has been changed in the current session.

All of this capability already exists in SOLIDWORKS in Find References and Pack and Go. This chapter is intended to introduce the concepts rather than create something new.

Finding File References

1. Create a new macro with the name *ListReferences.vbproj*.

2. Add a new dialog to the project by selecting Project, Add Windows Form. Select the Dialog template and name it *Dialog1.vb*.

3. Add a DataGridView control as shown. DataGridView controls are the perfect container to display tabulated data.

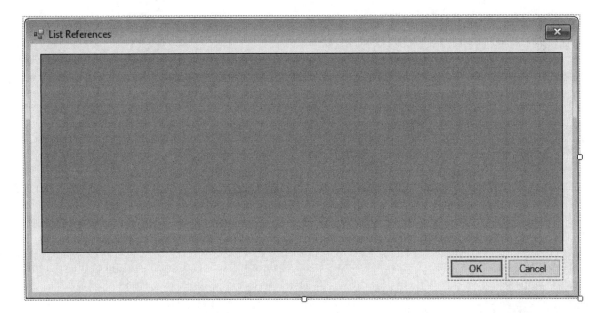

4. Set the form and DataGridView properties as shown below.

Form Properties:
Text = List References
TopMost = True
FormBorderStyle = Sizeable

DataGridView Properties:
Name = FilesGridView
Anchor = Top, Bottom, Left, Right
AllowUserToAddRows = False
AllowUserToDeleteRows = False
AutoSizeColumnsMode = Fill
ReadOnly = True
RowHeadersVisible = False

5. Add a new module to the macro by selecting Project, Add Module. Name the new module *ReferencesMod.vb*.

6. Add the following code to the new module, including the SOLIDWORKS Interop libraries, a public variable for a module level reference to SOLIDWORKS, and a function named ReadRefs.

```
Imports SolidWorks.Interop.sldworks
Imports SolidWorks.Interop.swconst
Imports System.Data

Module ReferencesMod
  Public m_swApp As SldWorks

  Function ReadRefs() As DataTable

  End Sub
End Module
```

System.Data

The Visual Basic.NET Data namespace includes many tools for working with tables of data without having to use complicated arrays or other containers like Excel spreadsheets. The DataTable class from the Data namespace is the basic container for rows and columns of data. The ReadRefs function has been declared to return a DataTable interface that will be populated with the file references from the active SOLIDWORKS assembly.

The DataTable class was chosen for another important reason. A DataGridView control can display a DataTable by simply passing a DataTable to the DataGridView's DataSource property. All of the columns and column headers you would want to see in the DataGridView will come right from the DataTable.

7. Add the following code to the ReadRefs function to get the active document from the SOLIDWORKS session, get its file path and its DirtyFlag property.

```
Function ReadRefs() As DataTable
  Dim swDoc As ModelDoc2
```

```
Dim PathName As String
Dim UpdateStamp As Long

swDoc = m_swApp.ActiveDoc
PathName = swDoc.GetPathName
UpdateStamp = swDoc.GetUpdateStamp
```

```
Exit Function
```

GetPathName

After the required declarations and connecting to the active document, the macro sets a variable PathName to the full path of the top-level document using the IModelDoc2.GetPathName method. For example, this call might return a string like "C:\Automating SolidWorks\File References\Gearbox.SLDDRW" if *Gearbox.SLDDRW* was the active document.

GetUpdateStamp

IModelDoc2.GetUpdateStamp is a great way to check if a model has been changed during the SOLIDWORKS session. This method returns a long integer value. Its value is essentially meaningless unless it is compared to the same model's update stamp at another point in time. The update stamp of a model only changes when the model is changed in a way that causes or requires a rebuild. Editing features, editing sketches and suppressing features would cause it to change. Changing the name of a feature would not.

8. Add the following to get all file references from the active document.

```
Function ReadRefs() As DataTable
  Dim swDoc As ModelDoc2
  Dim PathName As String
  Dim UpdateStamp As Long
  Dim fileRefs() As String

  swDoc = m_swApp.ActiveDoc
  PathName = swDoc.GetPathName
  UpdateStamp = swDoc.GetUpdateStamp

  'get an array of file references from the active document
  fileRefs = swDoc.Extension.GetDependencies(True, True, _
    False, True, False)


Exit Function
```

GetDependencies

A full list of file references, or dependencies, can be collected with one simple method from the IModelDocExtension interface. These include parts or sub-assemblies used in assemblies, parts or assemblies shown in drawings and externally referenced parts or assemblies used in top down design. It also includes part references to any type of derived part such as the parent of a mirrored part.

value = IModelDocExtension.GetDependencies(TraverseFlag, SearchFlag, AddReadOnlyInfo, ListBrokenRefs, AppendImportedPaths)

- **TraverseFlag** should be set to True if you want all dependencies listed. Setting it to False only returns top level dependencies. In an assembly, for example, if traverseFlag is False then you will get a list of only top level parts and sub-assemblies.

- **SearchFlag** is set to True if you want to use the SOLIDWORKS search rules to find the referenced documents. Setting it to False will search only the last saved location of the documents. If you have lightweight documents, setting the value to True can result in a bad value because the current search rules will be applied to the document. *(Hint: resolving lightweight components or explicitly setting the SOLIDWORKS search paths while the macro runs can be effective ways to prevent returning an invalid reference if you use lightweight assemblies or drawings.)*

- **AddReadOnlyInfo** should be set to True if you want to know if the document is opened read-only. The array will contain either two values per document if addReadOnlyInfo is set to False or three values per document if addReadOnlyInfo is set to True. The value will be a string value of TRUE if the file is opened ReadOnly or FALSE if not.

- **ListBrokenRefs** can display references that have been broken by the user. If set to False, broken references will be ignored like this example. Locked references will always be returned.

- **AppendImportedPaths** is designed for use with 3D Interconnect. If a part or assembly has a reference to another format file such as Creo or Solid Edge, and if this is set to True, the SOLIDWORKS file path along with the original format file path will be listed together with a separator character. Set it to False if you do not use 3D Interconnect.

The method returns an array of strings for the document name and the full path location of the document. In this exercise, `fileRefs` is populated with the array. Notice that it was declared as an Object to match the return value documented in the API help.

9. Set a breakpoint at Exit Function to prepare for a Debug test.

```
swDoc = m_swApp.ActiveDoc
PathName = swDoc.GetPathName
UpdateStamp = swDoc.GetUpdateStamp

'get an array of file references from the active document
fileRefs = swDoc.Extension.GetDependencies(True, True, _
    False, False, False)

End Function
```

10. Open an assembly in SOLIDWORKS and start running the macro.

11. When the code gets to the breakpoint at End Function, right-click on the `fileRefs` variable and select Add Watch. Review the Watch tab at the bottom of Visual Studio and explore the array values.

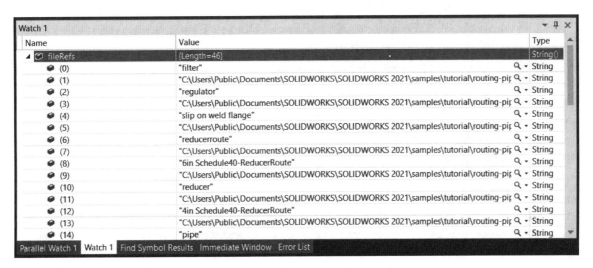

12. Stop debugging by clicking 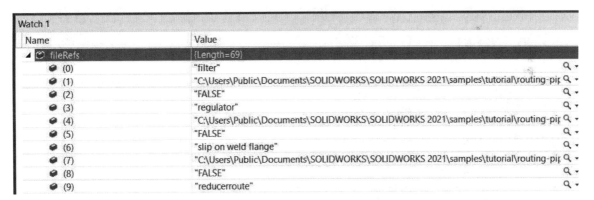.

Experiment with other settings for GetDependencies. Setting the AddReadOnly, or the third argument, to True would result in something like the following, displaying three array values per file reference.

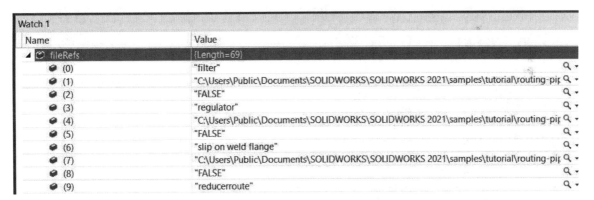

DataTable

The function is ready to start collecting values and adding them to the DataTable. But the first step is to define the columns for the table. Since the table and its column names will be used in more than just this function, they will be declared at the module level.

13. Add the following declarations at the top of the ReferencesMod module.

```
Module ReferencesMod
    Public m_swApp As SldWorks
    Dim filesTable As New DataTable
    Dim filePathColumnName As String = "File path"
    Dim dirtyFlagColumnName As String = "Dirty flag"
```

The New keyword is used to immediately create a new instance of a DataTable and the column name variables are initialized with the desired values when the macro is first run.

14. Add the following code to the ReadRefs function to create two columns in the DataTable (named `filesTable`).

```
...
  'get an array of file references from the active document
  fileRefs = swDoc.Extension.GetDependencies(True, True, _
    True, False, False)

  'set up the table columns
  filesTable.Columns.Add(filePathColumnName)
  filesTable.Columns.Add(dirtyFlagColumnName)
```

From any DataTable, the Columns property simply gives access to the columns of the table. Add columns by calling the Add method. The DataColumns Add method has several Overloads. An overloaded method enables the method to accept a variety of arguments. For example, you can call the Add method without any arguments and a column will be added at the next position. You can pass a single string value as the column name and a new column will be created with that specific name. There are several additional overloads for the Add method for you to explore. The column order is defined by the order they are created.

15. Add the following code, including declarations for two new variables, adding a row to the table, a For loop structure, and a Return statement to send back the DataTable to the calling code.

```
Function ReadRefs() As DataTable
  Dim fileRefs() As String
  Dim PathName As String
  Dim UpdateStamp As Long
  Dim swDoc As ModelDoc2
  Dim RowData(1) As String
  Dim i As Integer

  swDoc = m_swApp.ActiveDoc
  PathName = swDoc.GetPathName
  UpdateStamp = swDoc.GetUpdateStamp

  'get an array of file references from the active document
  fileRefs = swDoc.Extension.GetDependencies(True, True, _
    True, False, False)

  'set up the table columns
  filesTable.Columns.Add(filePathColumnName)
  filesTable.Columns.Add(dirtyFlagColumnName)

  'add the top file name and UpdateStamp
  RowData(0) = PathName
  RowData(1) = UpdateStamp.ToString
```

```
filesTable.Rows.Add(RowData)

For i = 1 To UBound(fileRefs) Step 2

Next

Return filesTable
```

End Function

The simplest way to add a row to a DataTable is to pass an array of values. The size of the array must be no larger than the number of columns in the table. The table row will populate each column with values from the array starting with the first element. The RowData array is declared with two elements to match the two columns of filesTable DataTable. The first element of the array is populated with the PathName variable and the second with the UpdateStamp variable converted to a string value.

Add rows to any DataTable through its Rows collection, just like adding a column. The Rows.Add method has two overloads, but the simplest is the array method.

For Loops with Step

The For loop must now parse the fileRefs array for file paths. As discussed earlier, fileRefs has two elements per file. The first file path is not element 0, but rather element 1. The next file path is at element 3. Unless you set a True value to the AddReadOnlyInfo argument of GetDependencies, in which case, the next file path would be element 4. The Step statement in a For loop forces the loop counter to increment in a step size other than 1. Two in this case. This is especially useful when you need to traverse elements of an array or collection at a given interval.

Now it is time to get the desired values out of the fileRefs array, getting each ModelDoc2 interface so the UpdateStamp value can also be added to the table.

16. Add the following code to the For loop, collecting the file paths into the first RowData array element, getting the ModelDoc2 interface for its UpdateStamp value, and adding those values to the filesTable DataTable.

```
For i = 1 To UBound(fileRefs) Step 2
  RowData(0) = fileRefs(i)
  'activate the document to get its modeldoc
  swDoc = m_swApp.GetOpenDocumentByName(fileRefs(i))
  If swDoc Is Nothing Then
    RowData(1) = "Not loaded"
  Else
    UpdateStamp = swDoc.GetUpdateStamp
    RowData(1) = UpdateStamp.ToString
  End If

  filesTable.Rows.Add(RowData)
Next
```

Because the assembly is open, each part and sub assembly is also open in memory. So, instead of having to open each file reference to get its ModelDoc2 interface, you can use the GetOpenDocumentByName method of the SOLIDWORKS application. It is a simple call that requires only the name of the open document as an argument. Passing the full file path works too.

There is one caveat. If a part is lightweight or suppressed, GetOpenDocumentByName returns Nothing. An If statement checks for a valid ModelDoc2 reference before calling the GetUpdateStamp method. Otherwise, a substitute string of "Not loaded" is set in its place.

This is a perfect place to test and debug. DataTables can be viewed in debug mode to verify their results.

17. Switch back to *SolidWorksMacro.vb* and add the following code to the main procedure.

```
Public Sub main()

    Dim refDataTable As DataTable
    ReferencesMod.m_swApp = swApp
    refDataTable = ReferencesMod.ReadRefs()

End Sub
```

The macro is now ready for a test. A variable is declared to store the returned DataTable. A reference to the SOLIDWORKS application is sent to the module level variable m_swApp. The new ReadRefs function is then called and its return value is sent to `refDataTable`.

18. Add a breakpoint to the End Sub line of the main procedure. As a shortcut, click on the End Sub line and hit F9.
19. Make sure you have an assembly open and start debugging the macro.
20. With the breakpoint highlighted, hover your cursor over `refDataTable` to show its quick info bar.

21. Click the magnifying glass icon 🔍 to view the DataTable in the DataSet Visualizer.

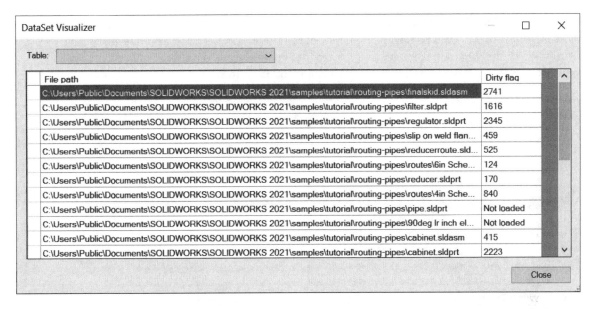

You should see two columns with the column names defined in the ReferencesMod module. The first should be the full file paths of each part and sub assembly and the second will display the current update stamp. If a file reference is suppressed or lightweight, the second column will display "Not loaded."

Verify the table format at this point. Correct any problems with the For loop step increment or other errors that might have occurred.

22. Stop debugging by clicking 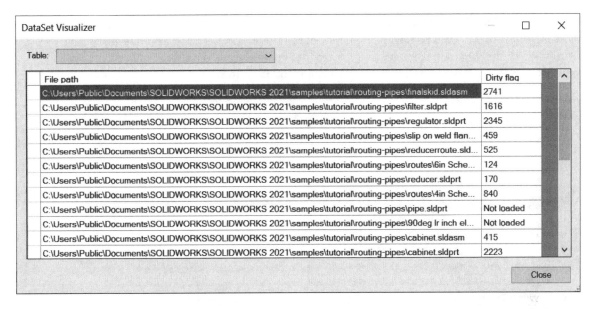.

Sending Data to the Form

After collecting a list of all references in the DataTable, it is time to display the content to the user. We also want to show them the current update stamp value. The DataGridView in the form Dialog1 simply needs to be populated with the DataTable and shown to the user.

23. Add the following to the main procedure to pass the DataTable to the DataSource property of the DataGridView named FilesGridView on the form refDia. Then show the form using ShowDialog.

```
Public Sub main()

  Dim refDataTable As DataTable
  ReferencesMod.m_swApp = swApp
  refDataTable = ReferencesMod.ReadRefs()

  Dim refDia As New Dialog1
  refDia.FilesGridView.DataSource = refDataTable
  refDia.ShowDialog()

End Sub
```

It is as easy as that. The DataSource of a DataGridView can be set directly to the DataTable instance. Now the DataGridView will display the matching data. This is another good example of keeping processing code separate from the user interface, or Separation of Concerns. If this were needed as part of a larger application, the module code is ready for any interface.

24. Run the macro to display the form with the columns from the DataTable. Adjust column widths and the form size to see more column detail.

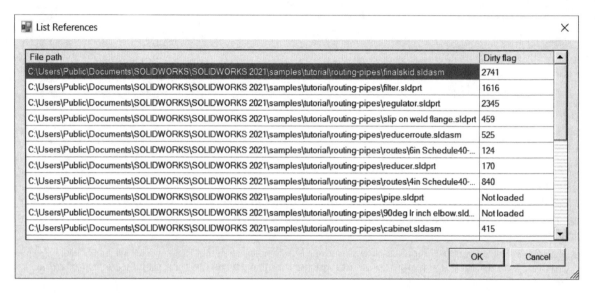

25. Stop debugging after reviewing the results.

Functionality Additions

After building the list, it would be helpful to be able to activate any document from that list. It may also prove useful to save the references list to a text file, similar to what can be done from Find References.

Code Changes

26. Switch to the *ReferencesMod.vb* tab and add a new procedure named ShowModel. Add the following code to the procedure.

```
Sub ShowModel(filename As String)
  'activate the currently selected item in the list
  Dim swDoc As ModelDoc2
  swDoc = m_swApp.ActivateDoc3(filename, True, 0, Nothing)
End Sub
```

ISldWorks.ActivateDoc3

You can make a document visible that is already loaded into memory by calling the ActivateDoc3 method of the SOLIDWORKS application. The method also returns the ModelDoc2 interface of the activated document.

value = ISldWorks.ActivateDoc3(Name, UseUserPreferences, Option, Errors)

- **Name** is the file to open. This can be the full file path or just the file name.
- **UseUserPreferences** should be True to use the assembly rebuild on activate settings from the Performance section of the System Options. Pass False if you would rather force or ignore a rebuild.
- **Options** will only apply if the previous argument is False. Use any option from the enumeration swRebuildOnActivation_e.
- **Errors** is a returned integer value from swActivateDocError_e. If you'd like to check for errors when activating a model, pass an integer variable and test its value after the call. Nothing is passed in this example to simplify the call and to ignore any returned error code.

As an example of interacting with a DataGridView, we will call this new procedure when the user double-clicks a filename cell. We will make use of a DataGridView event handler named CellContentDoubleClick.

27. From the *Dialog1.vb* code window select FilesGridView from the Class Name list and CellContentDoubleClick from the Method Name list. A new procedure outline is built.

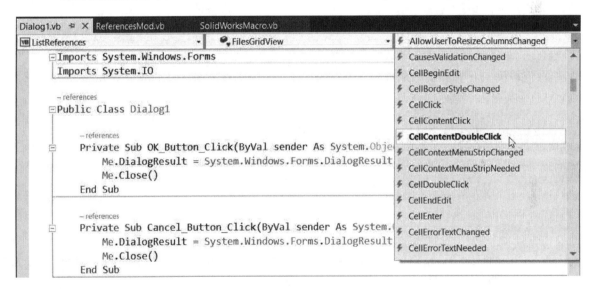

28. Add the following code to the new procedure.

```vb
Private Sub FilesGridView_CellContentDoubleClick _
  (ByVal sender As Object, _
  ByVal e As System.Windows.Forms.DataGridViewCellEventArgs) _
  Handles FilesGridView.CellContentDoubleClick
    If FilesGridView.CurrentCell.ColumnIndex = 0 Then
      Dim filename As String
      filename = FilesGridView.CurrentCell.Value
      ReferencesMod.ShowModel(filename)
    End If
End Sub
```

The macro now has the ability to activate any document in the list. The CurrentCell property of the DataGridView control returns the clicked cell. To make sure the user has double-clicked on a file name and not a cell from the wrong column, the cell's ColumnIndex property is checked. A column index starts at zero. The call to show the model will only be made if the current cell is in the first column. The Value property of a cell returns the text entered in that cell.

Saving the References List

It may prove useful to save the list of references to a text file. You learned how to write data to a text file in the Custom Properties exercise. The first step will be to convert the DataTable to a formatted string that can be written to a file.

29. Add the following function to *ReferencesMod.vb*.

```
Function FilesTableToString() As String
  Dim output As String = ""
  For Each row As DataRow In filesTable.Rows
    output = output & row.Item(filePathColumnName) & vbCrLf
  Next
  Return output
End Function
```

DataRows

The new function reads each row of the filesTable DataTable using a For Each loop described in the Data Import and Export chapter. The row variable is defined as a DataRow type since the DataTable.Rows property returns a collection of DataRows. The DataRow class has an Items collection representing each cell in the row. You can access the Items by index with the first cell at index 0, or by column name. By defining the column names as module level variables we simplify the process of getting the file path value regardless of the order of columns in the DataTable. Each row is concatenated with the previous row, including a return character, into the string variable named output. Finally, the function returns the output string to the calling code.

The user interface now needs a button to hold the code to save a file.

30. Switch back to Dialog1.vb [Design]. Add a save button to your form as shown, using the following properties.

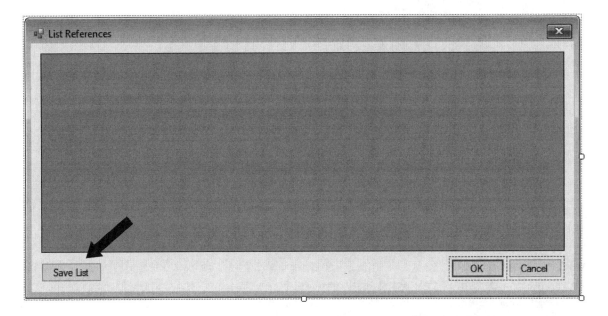

Button properties:
Name = SaveButton
Anchor = Bottom, Left
Text = Save List

31. Double-click the new button to build the button click event handler.

```
Private Sub SaveButton_Click(ByVal sender As System.Object, _
  ByVal e As System.EventArgs) Handles SaveButton.Click

End Sub
```

32. Add an imports statement to the top of the *Dialog1.vb* code window to make it easier to utilize the System.IO namespace as shown.

```
Imports System.Windows.Forms
Imports System.IO
```

33. Add the following to the SaveButton's click event handler to save the results of the FilesTableToString function to a text file.

```
Private Sub SaveButton_Click(ByVal sender As System.Object, _
  ByVal e As System.EventArgs) Handles SaveButton.Click

  Dim DefaultFilePath As String = FilesGridView.Rows(0).Cells(0).Value _
    & ".REFS.txt"
  Dim saveDia As New SaveFileDialog
  saveDia.FileName = Path.GetFileName(DefaultFilePath)
  saveDia.InitialDirectory = Path.GetDirectoryName(DefaultFilePath)
  saveDia.Filter = "Text file (*.txt)|*.txt"
  If saveDia.ShowDialog = Windows.Forms.DialogResult.OK Then
```

```
    Try
        Dim output As String = ReferencesMod.FilesTableToString()
        My.Computer.FileSystem.WriteAllText(saveDia.FileName, output, False)
    Catch ex As Exception
        MsgBox(ex.Message, MsgBoxStyle.Exclamation)
        Exit Sub
    End Try

    If MsgBox("Open file now?", MsgBoxStyle.YesNo _
        + MsgBoxStyle.Question) = MsgBoxResult.Yes Then
        Shell("notepad " & saveDia.FileName, AppWinStyle.NormalFocus)
    End If
  End If
End Sub
```

Code Description

A string variable DefaultFilePath is first created by combining the first cell in the first row of the DataGridView and the string ".REFS.txt". A new variable saveDia was declared as a New SaveFileDialog class from the System.Windows.Forms namespace.

SaveFileDialog

SaveFileDialog is very similar to the OpenFileDialog introduced in the Drawing Automation chapter. Its FileName property sets the default new file name. InitialDirectory sets the default save folder. These two properties were filled by using the System.IO.Path class methods of GetFileName and GetDirectoryName. The Filter property is set to define the file type selections in the dialog. If the user clicks OK, the remaining code is run. The SaveFileDialog's FileName property will return the full path to the new file to be created, regardless of the destination folder or user-defined file name.

The output variable in this routine is populated by the FilesTableToString function. The resulting string is written to a file using My.Computer.FileSystem.WriteAllText as was done in previous chapters. Remember that this is easily accessible as a code snippet.

Try, Catch Blocks

Some methods and functions return data to the .NET Exception handler. This is a built-in error handling system. A Try, Catch block is a quick way to watch for these exceptions because if you don't, your macro will stop running with an unhandled error. You may have noticed that when you type Try and hit enter, the rest of the Try, Catch block structure is written for you by Visual Studio. If any code under the Try statement fails, the code within the Catch block begins. The variable ex is declared as the Exception data type from the System namespace. The Exception.Message property returns the pre-built error message describing the problem. The macro would then display the problem message to the user through a message box. To avoid additional errors, the SaveButton_Click procedure exits if an exception occurs.

Use Try, Catch blocks liberally in your macros anywhere a call may result in an error. Sometimes it is difficult to avoid all possible code errors.

A final message box is used to ask the user if he would like to open the newly created text file. The combined options MsgBoxStyle.YesNo and MsgBoxStyle.Question are used to get a MessageBox to look like the one shown.

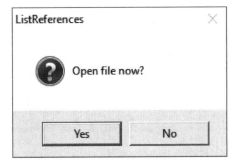

Shell

The Visual Basic Shell function is used to open the newly created text file in Notepad if the user answers yes to the request. The Shell function is another way to run any executable file. It does not attach to the application like CreateObject or GetObject. The Shell technique is especially useful when the application does not have an available API but allows command line arguments.

You can test this same code using the run command in Windows. Select Start, Run and type in the text "notepad c:\test.txt" where *test.txt* is a text file in the root directory on your computer. Shell is doing essentially the same thing.

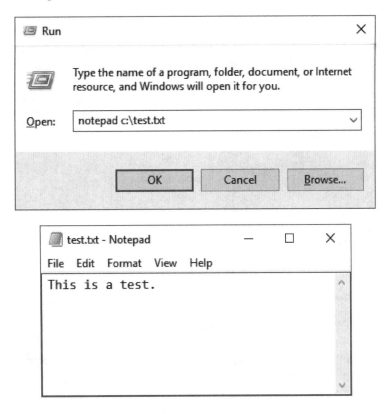

Debug and Test

34. Run the macro with an assembly open in SOLIDWORKS. Correct any errors you find.

You should be presented with a list of references and the corresponding update stamp values. Try activating a model from the list by double-clicking on the file path. Double-click on the top assembly file path to make it active. Focus your attention on one part and make note of its update stamp value. Close the dialog. Make a change to that model and switch back to the assembly or drawing. Run the macro again and you should see a new update stamp value for that part.

35. Click the Save List button and then click Yes to open the text file in Notepad.

36. Close Visual Studio.

Conclusion

IModelDoc2.GetUpdateStamp and IModelDocExtension.GetDependencies are two simple SOLIDWORKS API methods that allow you to track where files are used and which ones have been changed in the current session. Consider how you might move forward from this exercise. Here are a few ideas to enhance the tool.

- Store current file locations, dependencies lists and update stamps in text files, Excel or a database as part of a check-out process.

- Compare file locations, dependency lists and update stamps to their current values to process a check-in operation.

- Change custom properties such as revision in the modified documents and save them with a new name. *(Hint: use the SaveAs4 method discussed in the Drawing Automation exercise, setting saveAsCopy to False.)*

C# Example

See the full project *ListReferences.csproj* in the downloaded example files for C# code.

```
public static DataTable ReadRefs()
{
    string[] fileRefs;
    string PathName;
    int UpdateStamp;
    ModelDoc2 swDoc;
    var RowData = new string[2];
    int i;
    swDoc = (ModelDoc2)m_swApp.ActiveDoc;
    PathName = swDoc.GetPathName();
    UpdateStamp = swDoc.GetUpdateStamp();

    // get an array of file references from the active document
    fileRefs = (string[])swDoc.Extension.GetDependencies(true, true,
        false, false, false);
```

```
    // set up the table columns
    filesTable.Columns.Add(filePathColumnName);
    filesTable.Columns.Add(dirtyFlagColumnName);

    // add the top file name and UpdateStamp
    RowData[0] = PathName;
    RowData[1] = UpdateStamp.ToString();
    filesTable.Rows.Add(RowData);
    int loopTo = fileRefs.Length;
    for (i = 1; i <= loopTo; i += 2)
    {
        RowData[0] = fileRefs[i];
        // activate the document to get its modeldoc
        swDoc = (ModelDoc2)m_swApp.GetOpenDocumentByName(fileRefs[i]);
        if (swDoc == null)
        {
            RowData[1] = "Not loaded";
        }
        else
        {
            UpdateStamp = swDoc.GetUpdateStamp();
            RowData[1] = UpdateStamp.ToString();
        }

        filesTable.Rows.Add(RowData);
    }

    return filesTable;
}
```

Document Manager API

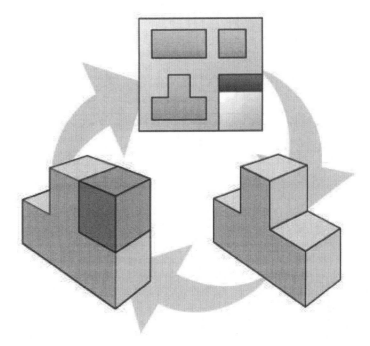

- **Custom Properties**
- **Tables**
- **References**

Introduction

Up to this point, everything we have done has relied on files open in SOLIDWORKS. This chapter will introduce a way to work with SOLIDWORKS file data, including properties and references, without using the SOLIDWORKS interface. The SOLIDWORKS Document Manager enables many of the capabilities you see in SOLIDWORKS File Utilities as well as SOLIDWORKS PDM. It is a powerful tool, but can also damage files if not used carefully. It provides the greatest benefit when batch processing since the Document Manager can access file information much faster than fully loading and resolving them in SOLIDWORKS.

Unlike the general API in SOLIDWORKS, the Document Manager is enabled using a license code. This license code is specific to your company or to you as an individual, depending on how your SOLIDWORKS license is registered.

The Document Manager can also be used for development on a computer without an installed SOLIDWORKS license. You will find the Document Manager SDK in the SOLIDWORKS installation media under the *swdocmgr* folder. The installer is named *SOLIDWORKS Document Manager API.msi*.

Requesting a Document Manager License

To request a license key, log into your SOLIDWORKS Customer Portal account with a valid serial number. Select API Support under the My Support category in Self-Service. You will find a link to the Document Manager Key Request page there. Select the link to Request New Document Manager Key and fill in the form. Since this license is specific to you, it should not be shared in readable code outside your company. You can also access the key request page here.

http://www.solidworks.com/sw/support/api-support.htm

Referencing the Document Manager

SOLIDWORKS macros do not have a reference to the Document Manager class by default, so it will need to be added manually to your macro project.

1. Start a new macro in SOLIDWORKS and save it as *SWDMProperties.vbproj*.

2. Add a reference to the Document Manager by selecting Project, Add Reference. Select the Browse tab and browse to the SOLIDWORKS installation directory, then api\redist folder. Select SolidWorks.Interop.swdocumentmgr.dll and click Add. *Hint: this folder contains all SOLIDWORKS libraries needed for .NET development.*

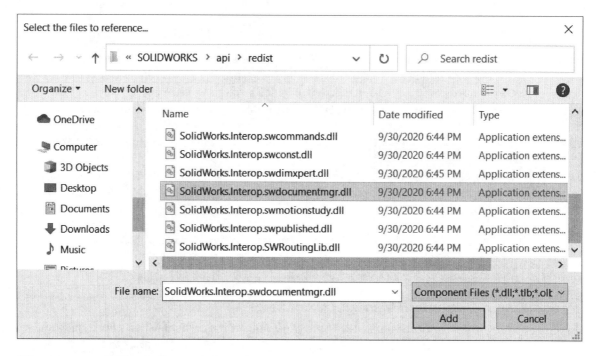

Classes and Class Properties

The Document Manager includes methods that can be useful in many situations. Connecting to the Document Manager and applying your license string is needed every time it is used. Rather than adding the same code to the beginning of each macro project, make use of a separate code file and add it to each project. This example will make use of a new Class. The concept of a VisualBasic.NET Property will also be introduced as a way to share data between your code files without having to declare variables globally.

3. Add a new code file to your project by selecting Project, Add Class. Enter the name *MyDocMan.vb* and select Add.

4. Reference the Document Manager and System.Diagnostics namespaces by adding the following Imports statements.

```
Imports SolidWorks.Interop.swdocumentmgr
Imports System.Diagnostics
Class MyDocMan

End Class
```

Visual Basic.NET Property

A Visual Basic.NET Property is a data element, similar to a variable, but comes with greater flexibility. Consider the previous examples that have used a module variable to maintain an instance of the SOLIDWORKS application class named m_swApp. As long as the macro populates this variable with an instance of SOLIDWORKS prior to calling any operational code, it works. But what would happen if a module level procedure or function was called and the SOLIDWORKS instance had not yet been set? The code would fail with the typical "Object reference not set to an instance of an object" error.

A Property has Get and Set procedures enabling you to process the request with code. In this example, a Property will be used to return the Document Manager Application to the calling code. If there is no instance of the Document Manager Application, a new one will be initialized. That creates a failsafe preventing code from failing due to a missing or invalid object reference.

Within the class, make use of a Visual Basic.NET ReadOnly Property to initialize the Document Manager Class Factory and return the Document Manager Application. The property will be Public in scope to allow access from anywhere.

5. Add the following code to create a Property.

```
Class MyDocMan
    Public ReadOnly Property DocManApp() As SwDMApplication

End Class
```

Hitting Enter after typing the Property declaration will automatically generate the additional required structure shown below. A ReadOnly Property has only a Get process. It is like a Function without any arguments. If the ReadOnly term, or specifier, were left off, the Property would be created with both Get and Set processes.

```
Public ReadOnly Property DocManApp() As SwDMApplication
    Get

    End Get
End Property
```

Connecting to Document Manager

The Document Manager Class Factory runs on its own, outside of the SOLIDWORKS process. So rather than connecting to the SOLIDWORKS application as done in the previous projects, use CreateObject to connect.

6. Add the following code within the Property between Get and End Get. Replace "YOUR LICENSE STRING HERE" with the license string that was emailed to you.

```
Imports SwDocumentMgr
Class MyDocMan
  Dim swDMApp As SwDMApplication

  Public ReadOnly Property DocManApp() As SwDMApplication
    Get
      If swDMApp Is Nothing Then
        Dim swDMClassFact As SwDMClassFactory
        swDMClassFact = _
          CreateObject("SwDocumentMgr.SwDMClassFactory")
        swDMApp = swDMClassFact.GetApplication( _
        "YOUR LICENSE STRING HERE")
      End If
      Return swDMApp
```

```
      End Get
   End Property
End Class
```

First, let's examine the use of a ReadOnly Property. The first time this property is requested, the variable swDMApp will be Nothing. The contents of the If statement will be run and swDMApp will be set. The Get process acts like a Function and returns the value of swDMApp. The return value must match the data type declared at the property level. After the first use of the property, the local class variable will have a valid reference to the Document Manager and the property will simply return the instance.

Connecting to the Document Manager takes two steps. The first is to connect to the class by using CreateObject. This is the SwDMClassFactory interface, declared in the code as swDMClassFact. The second is to connect to the SwDMApplication interface declared as swDMApp. This variable has been declared at the class level. The only way to get to it will be through the Property. Think of the Class Factory as the lock to the Document Manager Application. Its only method is GetApplication and you must pass your license key as a string to return an instance of SwDMApplication, which can be considered the Document Manager itself.

7. Switch back to *SolidWorksMacro.vb* and add the Imports statement for the Document Manager. Edit the main procedure as shown to use the new class.

```
Imports SolidWorks.Interop.sldworks
Imports SolidWorks.Interop.swconst
Imports System.Runtime.InteropServices
Imports System
Imports SolidWorks.Interop.swdocumentmgr

Partial Class SolidWorksMacro

    Public Sub main()
        Dim MyDM As New MyDocMan


    End Sub
```

You now have the base macro and code for using the Document Manager.

Common Interfaces

There are well over 100 interfaces available in the Document Manager Namespace. Many are related to DimXpert which will not be covered by this book. Others are much more common such as documents, components, configurations, sheets, views and tables. The following table shows the most common interfaces from the Document Manager and their accessor.

Interface	Accessor
ISwDMDocument	ISwDMApplication.GetDocument
ISwDMConfigurationMgr	ISwDocument.ConfigurationManager
ISwDMConfiguration	ISwDMConfigurationMgr.GetConfigurationByName
ISwDMComponent	ISwDMConfiguration2.GetComponents
ISwDMSheet	ISwDMDocument10.GetSheets
ISwDMView	ISwDMDocument10.GetViews
ISwDMTable	ISwDMDocument10.GetTable

Inheritance

At this point it is worth introducing a programming technique called Inheritance, introduced with the .NET language. The Document Manager API makes heavy use of this concept. Inheritance allows developers to expand the capability of a class without having to maintain duplicate code in multiple interfaces. Since future class development does not risk damaging existing properties, methods or procedures, it is a safer development technique. Class interfaces can be related to each other with all parent data available through children. To illustrate the point, previous chapters have made heavy use of IModelDoc2 and its methods such as EditRebuild and ClearSelection. However, the SOLIDWORKS API still has older versions of these calls maintained in its API to support older code. The IModelDoc interface is still there with its own EditRebuild and ClearSelection methods. This is not inheritance, but versioning.

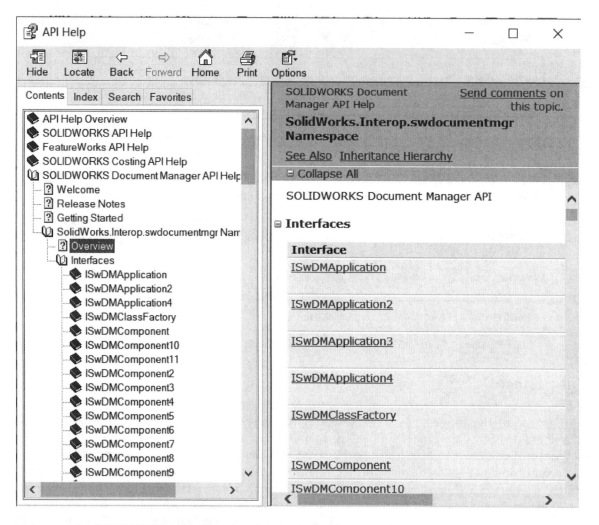

Browse to the SOLIDWORKS API help, select the Contents tab. Expand the SOLIDWORKS Document Manager API Help chapter, then SolidWorks.Interop.swdocumentmgr Namespace, then Interfaces. You will notice a numbering convention on the interfaces that looks similar to the numbering of SOLIDWORKS API calls. In a similar way, the numbered interfaces were created when new capabilities were added to the software or exposed to the API over time. However, in the Document Manager, these numbers do not represent current and obsolete interfaces. Each successive interface inherits the methods and properties of the previous.

Review ISwDMApplication as an example. Browse to its members and you will see the following.

Name	Description
CopyDocument	Copies a single document and optionally updates references to it.
GetDocument	Gets the specified document.
GetLatestSupportedFileVersion	Gets the latest supported SOLIDWORKS version number that this utility can read.
GetSearchOptionObject	Gets an ISwDMSearchOption object.
MoveDocument	Moves a single document and optionally updates references to it.

Now look at the methods of the newer ISwDMApplication2.

Name	Description
GetDocument2	Gets the specified document from an IStream or IStorage storage.

Notice that ISwDMApplication2 does not have the CopyDocument or MoveDocument methods. But because ISwDMApplication2 inherits from ISwDMApplication, if you declare your application variable as ISwDMApplication2, or even ISwDMApplication4, it will still have these methods available. Each successive interface builds on its predecessors rather than maintaining separate, similar methods and properties.

When you find a method or property in the Document Manager API that you would like to use, make sure to declare your parent interface to the child version that supports that method or higher. If you do not, the call will fail and indicate that the property or method does not exist. For example, if you want to use GetDocument2 from ISwDMApplication, you must declare your variable as either ISwDMApplication2 or ISwDMApplication4.

As an example, to create a new class that inherits from the existing MyDocMan class, add the Inherits statement shown at the top of the new MyDocMan2 class.

```
Public Class MyDocMan
  Dim swDMApp As SwDMApplication4

  Public ReadOnly Property DocManApp() As SwDMApplication4
    Get ...

  End Property
End Class

Public Class MyDocMan2
  Inherits MyDocMan

End Class
```

The new class can utilize the DocManApp property from its parent. If you were to change the declaration in the main procedure to the MyDocMan2 class, you could still access the DocManApp property.

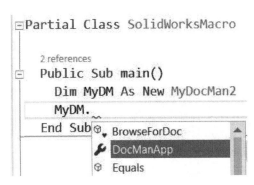

Opening and Closing Documents

The remainder of the macro will focus on opening a SOLIDWORKS file, then reading file and configuration specific properties. The new capabilities will be created in the MyDocMan class so they can be used in many other macros. The process will be split into multiple functions to give the greatest flexibility.

8. Add a new function named BrowseForDoc to *MyDocMan.vb* to allow the user to browse for a file and then return the file path as a string. Add the function after the DocManApp property block but before the end of the class.

```
Friend Function BrowseForDoc() As String
  'open the document for read/write
  Dim FilePath As String = ""
  Dim ofd As New Windows.Forms.OpenFileDialog
  ofd.Filter = "SOLIDWORKS Part (*.sldprt)|*.sldprt"
  If ofd.ShowDialog = Windows.Forms.DialogResult.OK Then
    FilePath = ofd.FileName
  End If
  If FilePath = "" Then Return Nothing

  Return FilePath
End Function
```

The function begins by making use of the OpenFileDialog from the Windows.Forms Namespace. This was described previously in the chapter on adding assembly components. If the user selects a file and clicks OK, the full path to the file is passed to the variable FilePath. If FilePath is an empty string, the macro exits with a return value of Nothing. Otherwise, the function returns the FilePath value.

9. Add another function named OpenDoc. It will take a filepath as an argument and open the file with the Document Manager. It will then return the resulting SwDMDocument interface.

```
Friend Function OpenDoc(ByVal FilePath As String, _
ByVal isReadOnly As Boolean) As SwDMDocument
  Dim openerrors As Integer
  Dim dmDoc As SwDMDocument
  dmDoc = DocManApp.GetDocument(FilePath, _
    SwDmDocumentType.swDmDocumentPart, isReadOnly, openerrors)
  Return dmDoc
End Function
```

Using Class Properties

This function makes use of our previously created property DocManApp. Though it is used just like a variable, its Get procedure is run whenever it is referenced.

SwDMApplication.GetDocument

The GetDocument method of SwDMApplication, referencing the classes property DocManApp, will open the SOLIDWORKS file for use with the Document Manager and returns a SwDMDocument interface. The method has the following structure.

value = SwDMApplication.GetDocument(FullPathName, docType, allowReadOnly, result)

- **FullPathName** is a string representing the full file path including the name and extension.

- **docType** is an Integer from the SwDmDocumentType enumeration.

- **allowReadOnly** is a Boolean value that dictates if the file will be open for read-only access. Use True for read-only and False if you intend to make changes.

- **result** is a returned integer value reporting any errors defined in the SwDmDocumentOpenError enumeration.

The example code above expects a part file to be selected since the docType argument is set to SwDmDocumentType. swDmDocumentPart. The OpenFileDialog Filter property was set to make sure the user selects only parts in the `BrowseForDoc` function. The part will be opened with full read/write access to enable reading and writing of custom properties if the `isReadOnly` argument is passed False. Otherwise the file will be opened read-only. The open result is passed to the variable `openerrors` and could be evaluated as needed for additional error handling.

10. Switch back to SolidWorksMacro.vb and add the following code to call the two newly created functions from the MyDocMan class.

```
Public Sub main()
  Dim MyDM As New MyDocMan

  Dim dmDoc As SwDMDocument
  Dim FilePath As String
  FilePath = MyDM.BrowseForDoc()
  dmDoc = MyDM.OpenDoc(FilePath, False)

  dmDoc.CloseDoc()

End Sub
```

SwDMDocument.CloseDoc

The first SwDMDocment method used is CloseDoc. Just like working with files in the SOLIDWORKS application, you must close them in the Document Manager to make them available for other users, especially if you open them for write access as is done here.

Getting and Setting Custom Properties

Now that we have the open and close structure in place and a reference to a document, a new function to read properties can be added to the MyDocMan class and then called from the main procedure. The properties will be written to the Immediate Window in this example for simplicity.

> 11. Add the following function in *MyDocMan.vb* to loop through all file properties in the open document. The document is passed to the function as an argument. The property names and values will be merged into a formatted string and returned.

```
Friend Function ReadAllProps(ByVal dmDoc As SwDMDocument) As String
  'read all file custom properties
  Dim output As String
  output = dmDoc.FullName & " Properties:"
  Dim propNames As Object = dmDoc.GetCustomPropertyNames()
  If Not propNames Is Nothing Then
    For Each propName As String In propNames
      Dim propVal As String
      propVal = dmDoc.GetCustomProperty(propName, Nothing)
      output = output & vbCrLf & propName & vbTab & propVal
    Next
  End If
  Return output
End Function
```

SwDMDocument.GetCustomPropertyNames

The GetCustomPropertyNames method of the SwDMDocument interface will return an array of strings containing all of the custom property names in the file. The return is an Object type even though the Object contains an array of Strings, so make sure to declare the variable that gets the result as Object. The array does not include configuration specific properties which will be evaluated later.

If there are no custom properties in the file, the array will be empty and will be Nothing. The If statement makes sure there are properties before looping through them. If your code attempted to use the For Each, Next loop on an empty variable it would result in the code stopping with an error.

With an array of strings, the macro then makes use of a loop to iterate through the elements of the returned array. Since these are the property names, the GetCustomProperty method of SwDMDocument gets the value of the property.

SwDMDocument.GetCustomProperty

GetCustomProperty requires a string as the first argument and defines the specific property by name. The second argument is a returned Integer that identifies the value type, whether it is text, date, yes/no, number or unknown. Since we expect text values back in this example, we are not getting the result and simply passing Nothing to the argument. GetCustomProperty returns the custom property value as a string.

The property value and name are then concatenated into the output variable along with a return character between lines and a tab character between the property name and value.

12. Add the following code to the main procedure in SolidWorksMacro.vb to call the new function and then display its return value in the Immediate Window.

```
Public Sub main()
  Dim MyDM As New MyDocMan

  Dim dmDoc As SwDMDocument
  Dim FilePath As String
  FilePath = MyDM.BrowseForDoc()
  dmDoc = MyDM.OpenDoc(FilePath, False)

  Dim props As String
  props = MyDM.ReadAllProps(dmDoc)
  Debug.Print(props)

  dmDoc.CloseDoc()

  Stop

End Sub
```

Stop

The Stop statement is added before End Sub so that results from the Debug statements can be reviewed. Remember that the Stop statement acts much like a breakpoint, but persists. If you compile and run a macro with a Stop statement, the macro will always stop at that point while breakpoints are only functional while debugging. Be careful to remove all Stop statements before compiling final versions of your macros.

13. Run the macro. When the Open dialog appears, browse to any SOLIDWORKS part and select Open. When the code pauses at the Stop statement, review the results in the Immediate Window.

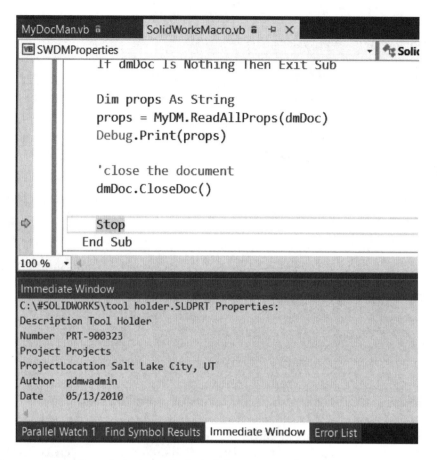

The image above assumes the selected part included these properties: Description, Number, Author, Date and others. If there were no properties in the file, the Immediate window would be empty under the file name header string.

14. Stop the macro.

Document Configurations

The next step will be to access all configuration specific properties. This is valid for parts and assemblies, but not drawings.

15. Add another new function to MyDocMan named ReadConfigProps.

```
Friend Function ReadConfigProps(dmDoc As SwDMDocument) As String
  'read all configurations
  Dim configMan As SwDMConfigurationMgr
  configMan = dmDoc.ConfigurationManager()
  Dim configNames As Object
  configNames = configMan.GetConfigurationNames
  Dim output As String = "Configurations:"
  For Each config As String In configNames
    output = output & vbCrLf & "Configuration: " & config
    'Get configuration specific properties
```

```
      Dim dmConfig As SwDMConfiguration
      dmConfig = configMan.GetConfigurationByName(config)

    Next
    Return output
End Function
```

SwDMDocument.ConfigurationManager

Before reading through the configurations in the document, we need access to the configuration manager or SwDMConfigurationMgr interface. Accessing the configuration manager is straightforward. Declare a variable with the SwDMConfigurationMgr type and use SwDMDocument.ConfigurationManager to get the interface.

The GetConfigurationNames method of the configuration manager will return an array of strings representing the configuration names in the document. This is similar to GetCustomPropertyNames. It returns an Object data type that contains the array. In the case of GetConfigurationNames, you will always get an array of at least one element since every part and assembly has at least one configuration.

The next loop should look familiar. Since we have another array of strings, a For Each, Next loop is used to iterate through all of the configuration name strings.

SwDMConfigurationMgr.GetConfigurationByName

The function gets each configuration, or SwDMConfiguration interface, through the GetConfigurationByName method from the configuration manager interface. The reason for getting the configuration is that you can get a list of property names from a SwDMDocument or SwDMConfiguration interface. The document returns the file properties and the configuration returns the configuration specific properties.

16. Add the following code to the function to append all of the configuration specific properties to the variable output.

```
Friend Function ReadConfigProps(dmDoc As SwDMDocument) As String
  'read all configurations
  Dim configMan As SwDMConfigurationMgr
  configMan = dmDoc.ConfigurationManager()
  Dim configNames As Object
  configNames = configMan.GetConfigurationNames
  Dim output As String = "Configurations:"
  For Each config As String In configNames
    output = output & vbCrLf & "Configuration: " & config
    'Get configuration specific properties
    Dim dmConfig As SwDMConfiguration
    dmConfig = configMan.GetConfigurationByName(config)
    Dim propNames As Object = dmConfig.GetCustomPropertyNames()
    If Not propNames Is Nothing Then
      For Each propName As String In propNames
        Dim propVal As String
        propVal = dmConfig.GetCustomProperty(propName, Nothing)
        output = output & vbCrLf & propName & vbTab & propVal
```

```
        Next
    End If
  Next
  Return output
End Function
```

The added code could be copied and edited from the ReadAllProps function since it performs the same property loop. The only difference is that it gets the property names from the configuration interface and gets the property value from the same configuration.

17. Modify the main procedure in *SolidWorksMacro.vb* and add the call to the new ReadConfigProps function.

```
Public Sub main()
  Dim MyDM As New MyDocMan

  Dim dmDoc As SwDMDocument
  Dim FilePath As String
  FilePath = MyDM.BrowseForDoc()
  dmDoc = MyDM.OpenDoc(FilePath, False)

  Dim props As String
  props = MyDM.ReadAllProps(dmDoc)
  Debug.Print(props)

  props = MyDM.ReadConfigProps(dmDoc)
  Debug.Print(props)
  'close the document
  dmDoc.CloseDoc()

  Stop
End Sub
```

18. Run the macro, browse to a SOLIDWORKS part that has properties and configurations and review the Immediate Window for results. Debug as needed, stopping the macro when done.

Adding and Changing Properties

The next step in the exercise will be to write a new property or change a property, then save the changes to the file. Since we have full access to file and configuration specific properties, you can use the following technique on either.

19. Add a new procedure named WriteProp to MyDocMan to add or update a file property.

```
Friend Sub WriteProp(dmDoc As SwDMDocument, _
ByVal pName As String, ByVal pVal As String)
  'write a property
  Dim addRes As Boolean
  addRes = dmDoc.AddCustomProperty(pName, _
    SwDmCustomInfoType.swDmCustomInfoText, pVal)
```

```
  If Not addRes Then
    dmDoc.SetCustomProperty(pName, pVal)
  End If
End Sub
```

The new procedure takes three arguments: a document, a property name and its value. If you need a procedure to add configuration-specific properties, it would have the identical structure. Just replace the SwDMDocument with SwDMConfiguration.

SwDMDocument.AddCustomProperty

The AddCustomProperty method is available to both document and configuration interfaces. It requires three arguments. The first is a string representing the property name. The second is the custom property type. It could be any of the following, with swDmCustomInfoText as the most common, all coming from the SwDmCustomInfoType enumeration.

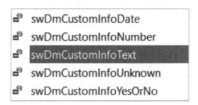

If a property of the same name already exists in the document, AddCustomProperty will return False. The If statement checks for a True result to process the code inside the statement. So using the Not keyword, we are able to run the code inside the statement if the result variable is False.

SwDMDocument.SetCustomProperty

Following the previous logic, if AddCustomProperty fails because it already exists, SetCustomProperty will be used instead. This method is only successful if the property exists. It requires only two arguments since the property type would have already been established. The first argument is the property name and the second is the value. Both arguments are string values. Using variables declared for both the property name and its value allows the code to be more portable. If you choose to change the property name or value, you only need to do so in one place.

20. Add the following code to the main procedure to add or change the Description file property before getting property values.

```
Public Sub main()
  Dim MyDM As New MyDocMan

  Dim dmDoc As SwDMDocument
  Dim FilePath As String
  FilePath = MyDM.BrowseForDoc()
  dmDoc = MyDM.OpenDoc(FilePath, False)

  'add/change Description
  MyDM.WriteProp(dmDoc, "Description", "MY DESCRIPTION")

  Dim props As String
```

```
props = MyDM.ReadAllProps(dmDoc)
Debug.Print(props)

props = MyDM.ReadConfigProps(dmDoc)
Debug.Print(props)
'close the document
dmDoc.CloseDoc()

  Stop
End Sub
```

21. Run the macro at this point and review the Immediate Window results. Look for the added or changed Description property. Stop the macro when finished.

Saving Your Changes

After running the macro on a part, open the part in SOLIDWORKS and look at its custom properties. Look specifically for the changed or added Description. It isn't there. It shouldn't be, after all, because the changes were never saved.

22. Add the following code in the `main` procedure to save the document. Make sure to add it before the `CloseDoc` method.

```
...
props = MyDM.ReadConfigProps(dmDoc)
Debug.Print(props)
'save the document
dmDoc.Save()
'close the document
dmDoc.CloseDoc()

  Stop
End Sub
```

SwDMDocument.Save

The Save method is simple and does not require any arguments. It returns a Long value representing any errors during save, though this macro example does not set the result to a variable. A well written tool should include error checking on the results of the Save operation and logging if necessary.

23. Run and debug the macro as needed.

Access Multiple Document Types

The current macro is hard-coded to only accept SOLIDWORKS parts. The GetDocument method would fail if you browsed for a drawing or assembly. GetDocument expects the second argument to define the document type. But the document type is also in the selected file extension. Rather than writing the macro explicitly for one file type, a few adjustments and additions will make the code functional for a broad range of applications.

24. Modify the BrowseForDoc function in *MyDocMan.vb* to return a SwDMDocument interface based on any valid SOLIDWORKS file path string. Add the error handling for failures as well.

```
Friend Function BrowseForDoc() As String
  'open the document for read/write
  Dim FilePath As String = ""
  Dim ofd As New Windows.Forms.OpenFileDialog
  ofd.Filter = "SOLIDWORKS Files (*.sldprt, " _
  & "*.sldasm, *.slddrw)" _
  & "|*.sldprt;*.sldasm;*.slddrw|Parts " _
  & "(*.sldprt)|*.sldprt" _
  & "|Assemblies (*.sldasm)|*.sldasm|Drawings " _
  & "(*.slddrw)" _
  & "|*.slddrw|All files (*.*)|*.*"
  If ofd.ShowDialog = Windows.Forms.DialogResult.OK Then
    FilePath = ofd.FileName
  End If
  If FilePath = "" Then Return Nothing

  Return FilePath
End Function
```

Notice the use of the extended OpenFileDialog Filter property to allow a variety of file selections.

25. Modify the OpenDoc function as follows. This will allow multiple document types to be opened as well as add error handling if there is a failure.

```
Friend Function OpenDoc(FilePath As String, _
isReadOnly As Boolean) As SwDMDocument
  Dim ext As String = System.IO.Path.GetExtension(FilePath).ToUpper
  Dim docType As Integer
  Select Case ext
    Case "SLDPRT"
      docType = SwDmDocumentType.swDmDocumentPart
    Case "SLDASM"
      docType = SwDmDocumentType.swDmDocumentAssembly
    Case "SLDDRW"
      docType = SwDmDocumentType.swDmDocumentDrawing
    Case Else
      docType = SwDmDocumentType.swDmDocumentUnknown
  End Select

  Dim openerrors As Integer
  Dim dmDoc As SwDMDocument
  dmDoc = DocManApp.GetDocument(FilePath, _
    docType, isReadOnly, openerrors)

  'error handling
  If dmDoc Is Nothing Then
    Dim errMessage As String = "Error accessing " & FilePath
```

```
      Select Case openerrors
        Case SwDmDocumentOpenError.swDmDocumentOpenErrorNonSW
          errMessage = errMessage & vbCrLf & "Non SOLIDWORKS file."
        Case SwDmDocumentOpenError.swDmDocumentOpenErrorNoLicense
          errMessage = errMessage & vbCrLf & "Invalid Document " _
            & "Manager license."
        Case SwDmDocumentOpenError.swDmDocumentOpenErrorFutureVersion
          errMessage = errMessage & vbCrLf & "Unsupported file version."
        Case SwDmDocumentOpenError.swDmDocumentOpenErrorFileNotFound
          errMessage = errMessage & vbCrLf & "File not found."
        Case SwDmDocumentOpenError.swDmDocumentOpenErrorFail
          errMessage = errMessage & vbCrLf & "Failed to access the document."
        Case SwDmDocumentOpenError.swDmDocumentOpenErrorFileReadOnly
          errMessage = errMessage & vbCrLf & "File is Read Only or is " _
            & "open in another application."
      End Select
      MsgBox(errMessage, MsgBoxStyle.Exclamation)
      Return Nothing
    Else
      Return dmDoc
    End If
End Function
```

The file extension is evaluated to determine the document type. The Visual Basic ToUpper method is used to ensure that the Select Case comparison will work regardless of upper or lower case extensions. Select Case compares strings literally. For example, "SLDPRT" and "sldprt" would not be the same.

The final addition is for error handling. If there is a problem accessing the file for any reason, a message will be presented to the user and Nothing, or a NULL value, will be returned by the function. Otherwise, if successful, the document will be returned. The SwDmDocumentOpenError enumeration includes all of the reasons a file may fail to open. A Select Case comparison is used to determine the error and report a more meaningful message to the user.

Notice that the code returning a successful SwDMDocument was also moved into the If block under the Else condition.

Additional Error Handling

Now that BrowseForDoc may not return a document, the main procedure needs a slight modification to exit.

26. Modify the main procedure and add a check of the dmDoc variable after the call to BrowseForDoc.

```
Public Sub main()
  Dim MyDM As New MyDocMan

  Dim dmDoc As SwDMDocument
  Dim FilePath As String
```

```
FilePath = MyDM.BrowseForDoc()
dmDoc = MyDM.OpenDoc(FilePath, False)
If dmDoc Is Nothing Then Exit Sub

'add/change Description
...
```

Drawings do not have configurations and that could cause another problem. If the Document Manager attempts to make a call to GetConfigurationNames on a drawing, it will crash with an error. A Try, Catch block can be added to ignore that condition.

27. Modify the ReadConfigProps function as shown below to ignore configurations for drawings.

```
Friend Function ReadConfigProps(ByVal dmDoc As SwDMDocument) As String
  'read all configurations
  Dim configMan As SwDMConfigurationMgr
  configMan = dmDoc.ConfigurationManager()
  Dim configNames As Object
  Try
    configNames = configMan.GetConfigurationNames
    Dim output As String = "Configurations:"
    For Each config As String In configNames
      output = output & vbCrLf & "Configuration: " & config
      'Get configuration specific properties
      Dim dmConfig As SwDMConfiguration
      dmConfig = configMan.GetConfigurationByName(config)
      Dim propNames As Object = dmConfig.GetCustomPropertyNames()
      If Not propNames Is Nothing Then
        For Each propName As String In propNames
          Dim propVal As String
          propVal = dmConfig.GetCustomProperty(propName, Nothing)
          output = output & vbCrLf & propName & vbTab & propVal
        Next
      End If
    Next
    Return output
  Catch ex As Exception
    'ignore the error for drawings
    Return ""
  End Try
End Function
```

28. Run the modified macro and test different file types. Debug as needed.

29. Close Visual Studio.

Bill of Materials Tables

Another common use of the Document Manager is to access the Bill of Materials (BOM) table data from SOLIDWORKS documents. These can exist in any document, but this section will focus on drawings. The text from BOM tables can be accessed cell-by-cell in a similar way to reading an Excel spreadsheet. Or, the entire table text can be read into a single array. Table size, based on row and column count, can also be determined. The example macro will use the second method along with the BOM size to extract a tab-delimited text file containing all BOM text.

> 30. Create a New macro and name it *SWDMTableExport.vbproj*.

> 31. Add a reference to SolidWorks.Interop.swdocumentmgr.dll from your SOLIDWORKS installation directory, api\redist. Select Project, Add Reference and find it on either the Recent or Browse.

Linking to Existing Code Files

The new macro will make use of the MyDocMan class created in the previous section. It will add new procedures and functions as well to expand its capability. However, if you simply add an existing code file to a new project, it is copied by default. The existing file remains as-is while any edits are only applied to the copy in the active macro project.

In some instances, it is beneficial to have several macros link to a shared code file. This is one of those cases. We will continue to use and expand MyDocMan. Instead of just adding a copy to the project, the existing code file will be linked.

> 32. Add the existing code module named *MyDocMan.vb* from the previous macro. Select Project, Add Existing Item…, browse into the *SwMacro* folder in *SWDMProperties* and select *MyDocMan.vb*. Instead of clicking Add, click the dropdown icon on the Add button and select Add As Link.

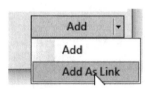

The resulting code file will show the shortcut icon shown below in the Solution Explorer. That is your indication that the file is linked rather than copied.

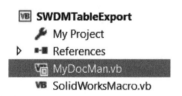

> 33. Add the same base code to *SolidWorksMacro.vb* that was used in the previous macro as shown below. Include the Imports statements as shown. Copy and paste to make it easier. The commented code has been removed for clarity.

```
Imports SolidWorks.Interop.sldworks
Imports SolidWorks.Interop.swconst
Imports System.Runtime.InteropServices
Imports System
Imports SolidWorks.Interop.swdocumentmgr
```

```
Imports System.Diagnostics

Partial Class SolidWorksMacro
Public Sub main()
  Dim MyDM As New MyDocMan

  'open the document for read/write
  Dim dmDoc As SwDMDocument
  Dim FilePath As String
  FilePath = MyDM.BrowseForDoc()
  dmDoc = MyDM.OpenDoc(FilePath, False)
  If dmDoc Is Nothing Then Exit Sub

  'close the document
  dmDoc.CloseDoc()

End Sub

''' <summary>
''' The SldWorks swApp variable is pre-assigned
''' </summary>
Public swApp As SldWorks


End Class
```

Accessing Tables

Accessing tables in a document is similar to accessing configurations. There can be several tables of different types. Both BOM and revision tables are directly accessible using the same methods. The first step to read tables is to get all table names of a specific type from the document. An individual table can then be retrieved by name.

34. Add a new function to MyDocMan by opening *MyDocMan.vb* from the Solution Explorer. Name the new function GetBOMTableText. It will access each BOM table by name in a document.

```
Friend Function GetBOMTableText(dmDoc As SwDMDocument10) As String
  Dim tableText As String
  Dim tableNames As Object
  tableNames = dmDoc.GetTableNames(SwDmTableType.swDmTableTypeBOM)
  For Each table As String In tableNames
    Dim dmTable As SwDMTable5
    dmTable = dmDoc.GetTable(table)
    Debug.Print(dmTable.Name)
  Next
End Function
```

This function is the first example of the use of Inheritance. A variable named dmDoc is declared as SwDMDocument10 type as an argument to the function. SwDMDocument10 inherits from SwDMDocument, so it has all of the same properties and methods as its parents. In fact, it also includes those from SwDMDocument3, SwDMDocument4 and so on. SwDMDocument10 is necessary because it includes the GetTableNames method.

Let's review how you might find this relationship without having to look through every SwDMDocument type. The function needs to retrieve the SwDMTable interface. Use the SOLIDWORKS API Help to browse the Index for ISwDMTable Interface. Double-click on the topic to show the overview. Look at the Accessors section and notice that the only accessor is ISwDMDocument10::GetTable.

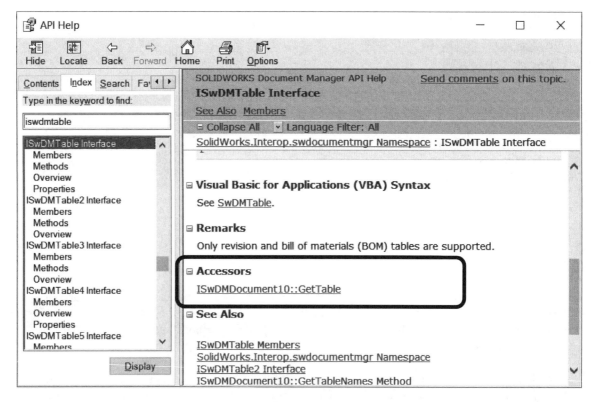

Once the parent interface is known, you must then declare your variable as the same type to get to the specific method.

GetTableNames

GetTableNames will return an Object type. But the returned object contains an array of strings, just like SwDMConfigurationManager.GetConfigurationNames. It requires one argument and IntelliSense helps make it easy. The values come from the SwDMTableType enumeration. This macro will use SwDmTableType.swDmTableTypeBOM.

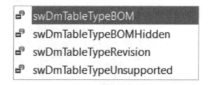

GetTable

Within the For, Each, Next loop, GetTable is used to retrieve a SwDMTable5 interface by passing the table name as a string. As mentioned, GetTable is a member of SwDMDocument10. The loop will enable the macro to retrieve every table in the document. The current code will simply print the name of each BOM table to the Immediate Window using the System.Diagnostics.Debug class.

35. Add the following code to call the new GetBOMTable function within the `main` procedure in *SolidWorksMacro.vb*.

```
Public Sub main()
  Dim MyDM As New MyDocMan

  'open the document for read/write
  Dim dmDoc As SwDMDocument
  Dim FilePath As String
  FilePath = MyDM.BrowseForDoc()
  dmDoc = MyDM.OpenDoc(FilePath, False)
  If dmDoc Is Nothing Then Exit Sub

  Dim tableText As String
  tableText = MyDM.GetBOMTableText(dmDoc)

  'close the document
  dmDoc.CloseDoc()

End Sub
```

36. Switch back to the *MyDocMan.vb* tab. Place a breakpoint at the Next line in GetBOMTableText and run the macro. Browse to a SOLIDWORKS drawing that has at least one BOM table and review the Immediate Window for table names.

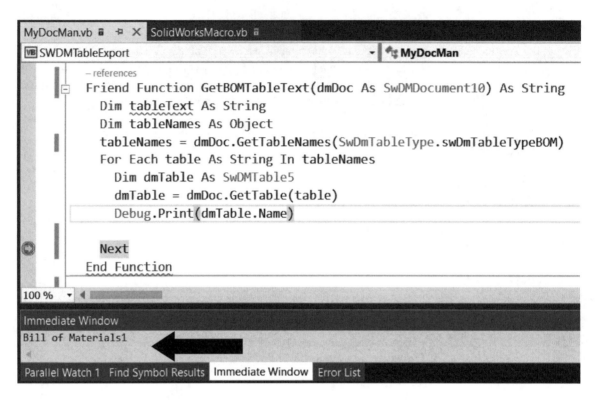

37. Run the macro to completion until the code no longer breaks at the Next line. Stop the macro if necessary.

GetTableCellText and GetCellText

Now that we have the tables, we can get the text from each as part of the GetBOMTableText function. As discussed earlier, there are two ways to get table text. SwDMTable.GetCellText will return a string value from a specific cell. This method requires you to pass the cell's row and column index as integers. It also assumes you will call GetColumnCount and GetRowCount from the table interface to index through the entire table. SwDMTable3.GetTableCellText returns an array of strings containing the text of the table in its entirety, including headers. It also returns the number of rows and columns in the one call. Each method has its advantages, but this example will use the latter for easy access to the full table text.

38. Add the following to GetBOMTableText to get an array of all text in each BOM table. Only the For, Each, Next loop is shown for brevity.

```
...
For Each table As String In tableNames
  Dim dmTable As SwDMTable5
  dmTable = dmDoc.GetTable (table)
  Debug.Print(dmTable.Name)
  Dim rows As Integer
  Dim columns As Integer
  Dim ttext As Object
  ttext = dmTable.GetTableCellText(Nothing, rows, columns)
Next
```

. . .

GetTableCellText is a member of SwDMTable3. However, the table variable was declared as SwDMTable5. When you have interfaces that inherit from each other, it is not a bad idea to use the most current interface. It will give you access to all of the available methods and properties. You only need to ensure it is compatible with the SOLIDWORKS software version the users will have installed.

GetTableCellText uses three arguments, all of which are ByRef variables which means they will each be populated with return values when the call is made.

value = SwDMTable.GetTableCellText(Error, RowCount, ColCount)

- **Value** is an Object type containing an array of strings populated with all of the table text. It is ordered by the display style of the BOM table.

- **Error** is returned as an integer value from SwDmTableError. A value of 0, or SwDmTableErrorNone means success.

- **RowCount** is an integer representing the number of rows in the table, including headers.

- **ColCount** is also an integer representing the number of columns.

GetCellText has a slightly different structure and use than GetTableCellText. It returns an integer representing success or failure. Two of the three arguments are the column and row index of the desired cell. The third is a ByRef variable where the specified cell text is returned.

value = SwDMTable.GetCellText(RowIndex, ColumnIndex, CellTextOut)

- **Value** is an integer from SwDmTableError where a value of 0, or SwDmTableErrorNone again represents success.

- **RowIndex** is an integer specifying the row number of the cell.

- **ColumnIndex** is an integer for the column of the cell.

- **CellTextOut** will be populated with the text found in the cell.

GetTableCellText is used in this example because it gives us the results of the entire table. The next step is to arrange the array of strings in a useable format. The following code will loop through each row, and each column within each row. Each cell string in a row will be separated with Tab characters, and each row will be separated with a return character. The resulting string is returned by the function for use in the main procedure where it will be saved to a file.

Formatting and Saving the BOM

39. Add the following code to the function to build a formatted string and return it.

. . .

```
Dim ttext As Object
ttext = dmTable.GetTableCellText (Nothing, rows, columns)
```

```
      tableText = "Table: " & dmTable.Name & vbCrLf
      'format the cell text array into a tab
      'delimited text file
      Dim cells As Integer = 0
      For i As Integer = 1 To rows
        Dim rowText As String = ""
        For j As Integer = 1 To columns
          rowText = rowText & ttext(cells) & vbTab
          cells += 1
        Next
        tableText = tableText & rowText & vbCrLf
      Next
    Next
    Return tableText
End Function
```

The variable `ttext` contains the array of strings. The variables `rows` and `columns` were used to collect the table size. Rather than using a For, Each, Next loop, a simple For loop is used based on the row and column sizes. A variable named `cells` is used to increment through the elements in the array. The shortcut `cells += 1` is the equivalent of `cells = cells + 1`.

40. Add the following to the `main` procedure in *SolidWorksMacro.vb* to save the resulting BOM text to a file just after closing the document.

```
Public Sub main()
  Dim MyDM As New MyDocMan

  'open the document for read/write
  Dim dmDoc As SwDMDocument
  Dim FilePath As String
  FilePath = MyDM.BrowseForDoc()
  dmDoc = MyDM.OpenDoc(FilePath, False)
  If dmDoc Is Nothing Then Exit Sub

  Dim tableText As String
  tableText = MyDM.GetBOMTableText(dmDoc)

  'close the document
  dmDoc.CloseDoc()

  Stop 'look at the tabletext in the watch text visualizer
  Dim bomFile As String = IO.Path.ChangeExtension( _
    dmDoc.FullName, ".BOM.txt")
  My.Computer.FileSystem.WriteAllText(bomFile, tableText, True)

End Sub
```

The ChangeExtension method of IO.Path is used to convert the file path string, retrieved from dmDoc using the FullName property, from a SOLIDWORKS file extension to a text file. As a shortcut, the extension ".BOM.txt" was used to make the file name meaningful. And finally, WriteAllText was used to save the formatted text string to the new file. WriteAllText can be found in the Visual Basic.NET Code snippets under the fundamentals\filesystem category named "Write text to a file."

Debug and Review Results

The code should be ready to go. The Stop statement will give you an opportunity to review the resulting formatted text string and compare it to your BOM table before removing it to finish the macro.

41. Run the finished code and browse to a SOLIDWORKS drawing containing a Bill of Materials table. When the Stop statement is hit, right-click on the tableText variable and select QuickWatch.

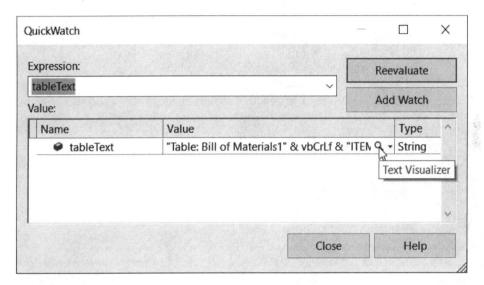

From the QuickWatch dialog, select the Text Visualizer icon $\mathcal{Q}$ ▾ at the right side of the Value column to view the full formatted string. The results should look something like the following. If you have multiple BOM tables in the drawing, they will all be shown, each with a table name header.

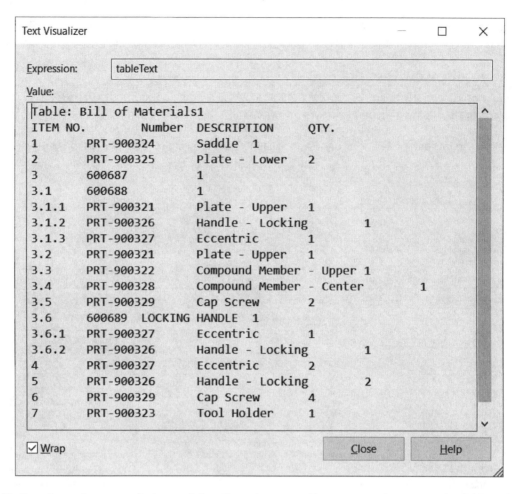

42. Run the code to completion and then Stop the macro if necessary. Browse to the folder where the drawing was selected and review the resulting text file.

```
tool vise.BOM.txt - Notepad                        —    □    ×

File  Edit  Format  View  Help
Table: Bill of Materials1
ITEM NO.         Number   DESCRIPTION      QTY.
1          PRT-900324    Saddle  1
2          PRT-900325    Plate - Lower    2
3          600687        1
3.1        600688        1
3.1.1      PRT-900321    Plate - Upper    1
3.1.2      PRT-900326    Handle - Locking        1
3.1.3      PRT-900327    Eccentric        1
3.2        PRT-900321    Plate - Upper    1
3.3        PRT-900322    Compound Member - Upper 1
3.4        PRT-900328    Compound Member - Center      1
3.5        PRT-900329    Cap Screw        2
3.6        600689  LOCKING HANDLE  1
3.6.1      PRT-900327    Eccentric        1
3.6.2      PRT-900326    Handle - Locking        1
4          PRT-900327    Eccentric        2
5          PRT-900326    Handle - Locking        2
6          PRT-900329    Cap Screw        4
7          PRT-900323    Tool Holder      1
```

43. Remove or comment out the Stop statement, then save and close Visual Studio.

Working with File References

Have you ever wished for a tool that would report back all of the missing files in your drawings and assemblies? You can use Pack and Go as a reporting tool. But it only lets you report one file at a time. This section will explore the calls available in the Document Manager to report references and replace them when needed. This gives a similar result to the File References chapter, but assumes the file is not open in a SOLIDWORKS session. This technique could be used in a stand-alone application as well as in a macro.

44. Create a New macro and name it *SWDMReferences.vbproj*.

45. Add a reference to SolidWorks.Interop.swdocumentmgr.dll. Select Project, Add Reference and find it on either Recent or Browse.

46. Add the existing code module named *MyDocMan.vb* from the SWDMProperties macro. Select Project, Add Existing Item..., browse to the *SWDMProperties* folder and select *MyDocMan.vb*. Add it as a link to allow changes to continue to update the original class file.

47. Add the following base code to *SolidWorksMacro.vb*, similar to the code used in the previous macro. Include the Imports statements as shown.

```
Imports SolidWorks.Interop.sldworks
Imports SolidWorks.Interop.swconst
Imports System.Runtime.InteropServices
```

```
Imports System
Imports SolidWorks.Interop.swdocumentmgr
Imports System.Diagnostics

Partial Class SolidWorksMacro
Public Sub main()
  Dim MyDM As New MyDocMan

  'Browse for a file
  Dim FilePath As String
  FilePath = MyDM.BrowseForDoc()

End Sub

''' <summary>
''' The SldWorks swApp variable is pre-assigned for you.
''' </summary>
Public swApp As SldWorks

End Class
```

48. Open *MyDocMan.vb* from the Solution Explorer and create a new procedure to report all missing references from a given file. All missing references will be printed to the Immediate Window. The optional replace argument will be used to enable replacement of missing references.

```
'Recursive routine to report all missing file references
Public Sub GetRefs(FilePath As String, _
Optional replace As Boolean = False)
  Debug.Print("=== " & FilePath & " REFERENCES ===")
  'get references
  Dim searchOpt As SwDMSearchOption
  Dim brokenRefs As Object = Nothing
  Dim isVirtual As Object = Nothing
  Dim timeStamp As Object = Nothing
  Dim refs As Object
  searchOpt = DocManApp.GetSearchOptionObject()

  searchOpt.SearchFilters = _
    SwDmSearchFilters.SwDmSearchExternalReference

  Dim dmDoc As SwDMDocument21
  dmDoc = OpenDoc(FilePath, False)
  If dmDoc Is Nothing Then Exit Sub

  refs = dmdoc.GetAllExternalReferences5( _
    searchOpt, brokenRefs, isVirtual, timeStamp, Nothing)

  Dim ref As String
```

```
  For i As Integer = 0 To refs.Length - 1
    ref = refs(i)
    If Not isVirtual(i) Then
      If Not IO.File.Exists(ref) Then
        Debug.Print("MISSING: " & ref)
      Else
        Debug.Print(ref)
        If IO.Path.GetExtension(ref).ToUpper = ".SLDASM" Then
          'recursively search the assembly
          'for its references
          getRefs(ref, replace)
        End If
      End If
    End If
  Next
  'save if replacing references
  If replace Then dmdoc.Save()
  dmdoc.CloseDoc()
End Sub
```

GetAllExternalReferences5

The API Help describes GetAllExternalReferences5 as a tool to be used for drawings. However, it can still be an effective reporting tool for assemblies as well. It is a method of SwDMDocument21 and has the following structure.

value = SwDMDocument21.GetAllExternalReferences5(pSrcOption, brokenRefVar, IsVirtual, TimeStamp, ImportedPaths)

- **Value** is an Object data type but, like other SOLIDWORKS API calls, contains an array of strings, each representing a fully qualified file path to each child document in the parent. If you select an assembly, it will report all top-level references.

- **pSrcOption** is of the type SwDMSearchOption. This is a specific interface where you can set the search routine options for how the method will resolve the reference documents. This will be explored further below.

- **brokenRefVar** is a ByRef Object. It will be populated with an array of broken external references. Do not confuse this output list with missing file references. These are feature and part references such as in-context features and mirrored parts.

- **IsVirtual** is another ByRef Object. As a result, it will contain an array of Boolean values indicating if the associated reference is virtual and does not have an external file.

- **TimeStamp** is a ByRef Object. It will contain an array of the times each reference was associated to the parent. The times are in time_t format, meaning the number of seconds elapsed since midnight of January 1, 1970. This is only a useful format if you compare to others or convert to a real date or time.

- **ImportedPaths** is another ByRef Object. It will contain array of strings for external file references coming from 3DInterconnect. We are ignoring these references by passing Nothing rather than an Object variable.

Each of these arguments and outputs is declared as the appropriate data type in the macro code. Before calling GetExternalReferences5, the SwDMSearchOption variable named `searchOpt` has to be pre-populated.

GetSearchOptionObject is a method of SwDMApplication and returns the default search routine settings used by the Document Manager and SOLIDWORKS. There are only minor search routine differences that are hardly worth discussing. See the API Help for more detail if desired.

After getting the default SwDMSearchOption interface, you can set search filters, or rules, by setting its SearchFilters property. Search filters are available in the enumeration SwDmSearchFilters as shown here.

SwDmSearchExternalReference
SwDmSearchForAssembly
SwDmSearchForDrawing
SwDmSearchForPart
SwDmSearchInContextReference
SwDmSearchRootAssemblyFolder
SwDmSearchSubfolders

In this example, the search filters are set to search basic references, including parts and assemblies. Multiple options can be added together to get a combined behavior. You can also use the AddSearchPath method of the SwDMSearchOption interface to set additional search folders if you need to search beyond the scope of the selected file.

It is important to finally note that GetAllExternalReferences5 does not directly report missing file references. It uses the search routine to locate the referenced files. If they are found, their known file path is reported. If they are not found, their last-saved file path is reported. Your code must now determine if the files actually exist on disc.

Traversing the Reference Array

The For loop first checks the current `isVirtual` array element, a Boolean value, to ignore virtual parts and assemblies. They will not have external file paths. A counter is used in the loop to access the associated elements in any of the output arrays.

IO.File.Exists

Next, the macro uses the Exists method of the IO.File interface. This method returns a Boolean, True if the file exists on disc, False if not. If the file does not exist, the file path is printed to the Immediate Window. It is a missing file reference.

The last check is to see if the file is an assembly. If it is, the same procedure needs to be run to report the sub assembly's children. The procedure is defined in a way that it can recursively call itself with only a file path.

The final line of code in the procedure is to close the open document currently being accessed. Since we don't need to return anything back to the calling code, this procedure will close the document.

49. Switch back to *SolidWorksMacro.vb* and add the following code to the `main` procedure to call the new GetRefs method.

```
Public Sub main()
  Dim MyDM As New MyDocMan

  'Browse for a file
  Dim FilePath As String
  FilePath = MyDM.BrowseForDoc()

  MyDM.GetRefs(FilePath, False)
End Sub
```

The selected file could be an assembly or drawing and it will report all references along with any missing files to the Immediate Window. The macro is now ready for testing.

50. Set a breakpoint at End Sub in the `main` procedure.

51. Run the macro and browse to an assembly or drawing. When the breakpoint is hit, review the results in the Immediate Window. Any missing references will be reported with the preceding text MISSING. The reported path will be the last-known location of the missing file.

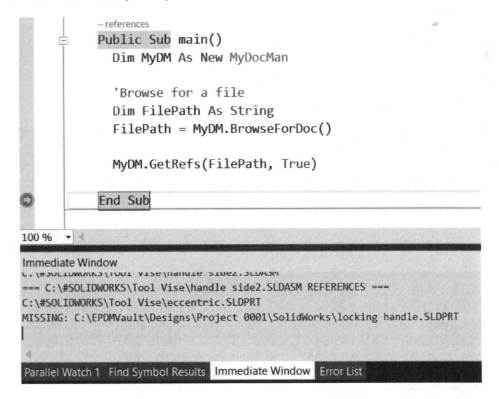

52. Stop the macro and remove the breakpoint.

Replacing References

It's time to make this macro do something more helpful. If a reference is missing, present a dialog to the user allowing them to browse for a replacement in a similar way to the SOLIDWORKS user interface.

When a missing file is found, the macro should let the user know and then make it easy to find a replacement. The OpenFileDialog has the ability to pre-set a file path that we can make use of.

53. Modify the BrowseForDoc function in *MyDocMan.vb* and add an optional argument for FileName.

```
Friend Function BrowseForDoc(Optional FileName As String = "") As String
  'open the document for read/write
  Dim FilePath As String = ""
  Dim ofd As New Windows.Forms.OpenFileDialog
  ofd.Filter = "SOLIDWORKS Files (*.sldprt, " _
  & "*.sldasm, *.slddrw)" _
  & "|*.sldprt;*.sldasm;*.slddrw|Parts " _
  & "(*.sldprt)|*.sldprt" _
  & "|Assemblies (*.sldasm)|*.sldasm|Drawings " _
  & "(*.slddrw)" _
  & "|*.slddrw|All files (*.*)|*.*"
  ofd.FileName = FileName
  If ofd.ShowDialog = Windows.Forms.DialogResult.OK Then
    FilePath = ofd.FileName
  End If
  If FilePath = "" Then Return Nothing

  Return FilePath
End Function
```

When the Optional keyword is added before an argument, you must include a default value. In this example, if no argument is passed to the function, the variable `FileName` will be an empty string. Optional variables are a convenient way to expand functions and procedures while keeping their use as simple as possible. All existing calls to BrowseForFile do not need to be changed since they will use the default empty string value.

The FileName property of the OpenFileDialog can be used before ShowDialog is called to pre-set the dialog with a file name. In our example, it will be populated with a missing reference so the user knows which file they should replace.

54. Add the following code to GetRefs immediately following the MISSING message. When a missing file is found, use a Message Box control to ask the user if they would like to find the missing file.

```
'Recursive routine to report all missing file references
Public Sub GetRefs(ByVal FilePath As String, _
Optional ByVal replace As Boolean = False)
```

```
Debug.Print("=== " & FilePath & " REFERENCES ===")
'get references
Dim searchOpt As SwDMSearchOption
Dim brokenRefs As Object = Nothing
Dim isVirtual As Object = Nothing
Dim timeStamp As Object = Nothing
Dim refs As Object
searchOpt = DocManApp.GetSearchOptionObject()

searchOpt.SearchFilters = _
  SwDmSearchFilters.SwDmSearchExternalReference

Dim dmDoc As SwDMDocument13
dmDoc = OpenDoc(FilePath, False)
If dmDoc Is Nothing Then Exit Sub

refs = dmdoc.GetAllExternalReferences4( _
  searchOpt, brokenRefs, isVirtual, timeStamp)

Dim ref As String
For i As Integer = 0 To refs.Length - 1
  ref = refs(i)
  If Not isVirtual(i) Then
    If Not IO.File.Exists(ref) Then
      Debug.Print("MISSING: " & ref)
      If replace Then
        Dim msgRes As MsgBoxResult
        msgRes = MsgBox("Missing: " & vbCrLf & ref & vbCrLf _
          & "Would you like to replace it?", _
          MsgBoxStyle.YesNo + MsgBoxStyle.Question)
        If msgRes = MsgBoxResult.Yes Then
          Dim newFilePath As String
          newFilePath = BrowseForDoc(IO.Path.GetFileName(ref))
          If newFilePath <> "" Then
            dmdoc.ReplaceReference(ref, newFilePath)
          End If
        End If
      End If
    Else
      Debug.Print(ref)
      If IO.Path.GetExtension(ref).ToUpper = ".SLDASM" Then
        'recursively search the assembly
        'for its references
        getRefs(ref, replace)
      End If
    End If
  End If
Next
'save if replacing references
If replace Then dmdoc.Save()
```

```
    dmdoc.CloseDoc()
End Sub
```

The BrowseForDoc procedure is used again, but this time, a file name is passed rather than using the default (optional) empty string. This should help the user remember the file they are trying to find. The GetFileName method of IO.Path is used to get only the file's name from its full path.

ReplaceReferences

A method of SwDMDocument, ReplaceReferences requires only two arguments and updates all references in the document to the new file. The first argument is the full path to the document to replace. In our example, that is the missing reference. GetAllExternalReferences5 has returned that string from its results. The second argument is the full path to the replacement file. If the user selects a file from the BrowseForDoc method, the resulting path is passed as a replacement. If the user cancels the file browse operation, ReplaceReferences is skipped.

To make the changed file path stick, the document has to be saved. The Save method of SwDMDocument is used in the code to make that happen. But it is only called if the procedure is called with an explicit True value for the replace argument.

> 55. Modify the main procedure in *SolidWorksMacro.vb*. Enter True for the replace argument when getRefs is called.

```
Public Sub main()
  Dim MyDM As New MyDocMan

  'browse for a file
  Dim FilePath As String
  FilePath = MyDM.BrowseForDoc()

  MyDM.GetRefs(FilePath, True)
End Sub
```

Test and Debug

The code is now ready for some additional testing. Find an assembly with a broken reference, or manually rename a part in an existing assembly without updating the assembly reference. Then run the macro and browse to that file.

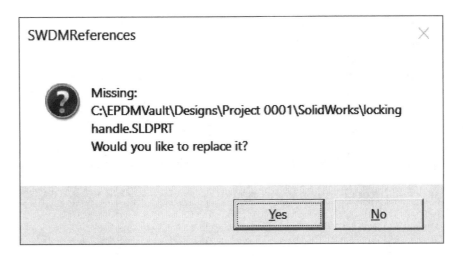

When a missing reference is found, the above message will display. Select Yes to browse to a new file reference. Run the macro a second time to verify the fixed file reference. And it all happens without opening the files in a SOLIDWORKS session!

There are some common problems that cause the macro to fail. If the file or folder is Read Only, or if the file is already opened in SOLIDWORKS or eDrawings, the Document Manager will not be able to open the file with write access. The error handling in the GetMyDocument function will catch the problem and report the error in a message.

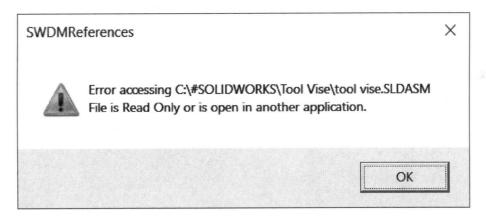

Conclusion

You now have a class, MyDocMan, that is flexible and can be re-used. Go back and edit the original macro from this chapter and look over the changes that have been made to the class since it was linked to the other macros. Linked code modules are great for commonly used operations. But as with SOLIDWORKS models, when you reference a file in multiple places, make sure the changes you make in one macro don't damage others that use the same module.

The Document Manager can be a powerful tool for modification and extraction of data from SOLIDWORKS files. It is much faster than opening files in SOLIDWORKS and is incredibly capable, far beyond the scope of this chapter. Any time you are considering a batch process, think of the Document Manager API as a possible solution.

Keep in mind that the Document Manager can also alter SOLIDWORKS files incorrectly. The Replace References section of this chapter only considers the child references. It doesn't do any validation if the user selects the wrong replacement and does not check to see if the part might have an in-context reference to another parent assembly that may need to change. As a good practice, back-up all files before testing or running any batch process macros.

C# Example

For a full C# example using the Document Manager, see the full projects (*SWDMProperties.csproj, SWDMTableExport.csproj, SWDMReferences.csproj*) included in the downloadable examples.

```csharp
Using SolidWorks.Interop.swdocumentmgr;

class MyDocMan
{
    SwDMApplication swDMApp;
    public SwDMApplication DocManApp
    {
        get {
            if (swDMApp == null)
            {
                SwDMClassFactory swDMClassFact = new SwDMClassFactory();
                swDMApp = (SwDMApplication)swDMClassFact.GetApplication
                    ("ENTER LICENSE KEY HERE");
            }
            return swDMApp; }
    }
}
```

Creating Add-Ins

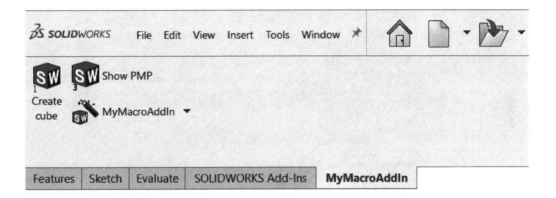

- **SOLIDWORKS API SDK**

- **PropertyManager Pages**

Introduction

At some point after creating and using macros regularly, you may begin to realize that deploying and sharing them with many users is complicated. You can share them across the network, or through your SOLIDWORKS PDM, but how do you make sure all users get the same layout for new buttons in the user interface?

Building SOLIDWORKS add-ins can make deployment simpler across your teams. The great news is that you don't have to start from scratch. You can reuse your existing code modules when you use the Separation of Concerns development technique. SOLIDWORKS also provides a comprehensive add-in template through the SOLIDWORKS API SDK. This can make the whole process simple.

Create Your Add-In

To create SOLIDWORKS Add-ins, both Visual Studio and the SOLIDWORKS API SDK are now needed.

SOLIDWORKS API SDK

The SOLIDWORKS API SDK contains add-in templates for Visual Studio in C++, C# and Visual Basic languages. You can create an add-in from scratch, but using a template makes it much simpler. The SDK installer is included in a SOLIDWORKS installation media kit under the *apisdk* folder. If you don't have a full SOLIDWORKS media download available, you can download just the SDK by following the instructions in the API Help. Find the instructions by selecting the Contents tab, then browsing to SOLIDWORKS API Help > Getting Started > Overview.

1. Make sure Visual Studio is not running. Then install the SOLIDWORKS API SDK by running *SOLIDWORKS API SDK.msi* from your installation media or download. Keep all of the default settings through the installation.

Visual Studio Projects

Now that the SDK is installed, you are ready to create your first Visual Studio SOLIDWORKS add-in. Rather than starting Visual Studio from the SOLIDWORKS Macros menu, it will be run stand-alone. It will need to run as administrator to register the project output.

2. Launch Visual Studio as administrator by right-clicking on Visual Studio from the Windows Start Menu. Select More, Run as administrator.

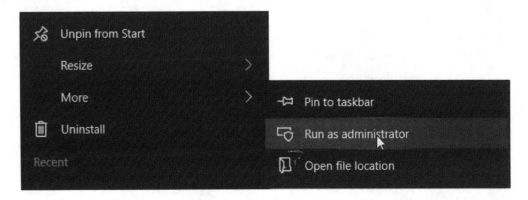

3.   From the Start Page, select Create a new project.

4.   Using the search bar, search for "sw" to find the SwVBAddIn template. Select the template and tap Next.

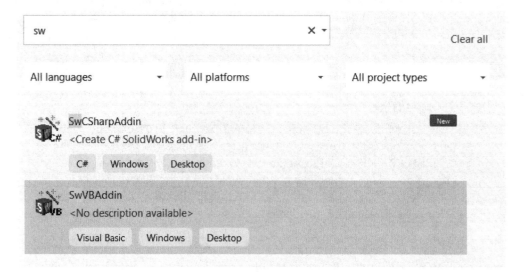

5.   Name the project *MyMacroAddIn*, verify the location folder, and tap Create.

There are cases where the SDK does not copy the project templates properly in Visual Studio. If that is the case on your system, use the example add-in project in the downloaded example files referenced in the introduction. Copy the folder and edit the project in Visual Studio for each new add-in. You can also visit My.SolidWorks.com and search for "visual studio project template" for up-to-date discussions on the topic.

Note: the SDK also includes a C# add-in template.

Add-in Structure

The add-in template creates several code files and icon image files in the new project. It also adds the common Interops needed to work with SOLIDWORKS. Examine the Solution Explorer to see the content.

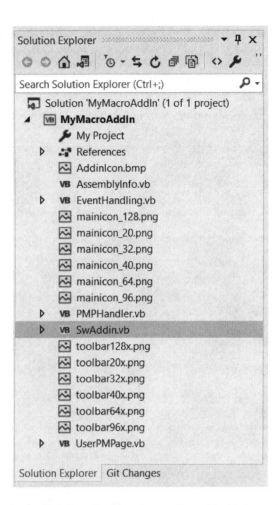

Expand the References tree in the Solution Explorer to see the added Interop libraries.

Hint: you can add references directly from the Solution Explorer by right-clicking on the References node.

You should recognize SolidWorks.Interop.sldworks and SolidWorks.Interop.swconst from the previous macros. SolidWorks.Interop.swpublished is used to create PropertyManager pages and will be discussed later in the chapter. SolidWorks.Interop.SolidWorksTools is a new namespace used only for add-ins. It includes tools for adding attributes to the add-in like COM-Visible, its title, description and its GUID.

The primary code file is *SwAddin.vb*.

   6.   Open SwAddin.vb from the Solution Explorer and review the various sections of the add-in.

Each section is pre-built with example code. We will use much of what has been created. The code regions are shown below, but not yet expanded.

```
<Guid("0d5999f2-f34b-40c6-8cb8-28a88dc65afb")> _
<ComVisible(True)> _
<SwAddin( _
       Description:="MyMacroAddIn description", _
       Title:="MyMacroAddIn", _
       LoadAtStartup:=True _
       )> _
Public Class SwAddin
    Implements SolidWorks.Interop.swpublished.SwAddin

#Region "Local Variables"

#Region "SolidWorks Registration"

#Region "ISwAddin Implementation"

#Region "UI Methods"

#Region "Event Methods"

#Region "Event Handlers"

#Region "UI Callbacks"

End Class
```

Class Attributes

Above the Class declaration of SwAddin are a few class attributes. Attributes allow you to add metadata to your project that can be referenced in other areas of the project, or even outside the project. The first attribute defines a unique identifier of the Class called a GUID. Windows uses this unique ID to distinguish one application from another. The add-in template in Visual Studio will generate a random GUID each time you use it. Your GUID should not match the example above. If you ever copy an add-in project, make sure to create a unique GUID for the new project. There are many GUID creators available on the internet.

The other attributes define the add-in description, title and whether it is loaded when SOLIDWORKS starts up. Edit these attributes to customize your add-in name and description shown in the SOLIDWORKS Add-in manager.

Implements SwAddin

Inside the SwAddin class, the first line marks the class as implementing the swpublished.SwAddin class. The Implements statement indicates the class will use pre-defined procedures or events from the SwAddin class.

7. Expand the region "ISwAddin Implementation" to review the two required functions.

The ConnectToSW function runs each time the add-in is turned on. If it is set to start with SOLIDWORKS, it will be run during startup. The function takes two arguments, an instance of ISldWorks application passed from SOLIDWORKS to the add-in, and a unique identifying integer value. The function returns True to the SOLIDWORKS application if it successfully loads.

The template code has all of the operations needed for a typical add-in. In this example, we only need to set up the user interface to run existing macros. This will include adding a CommandManager tab and menu items. The pre-defined AddCommandMgr procedure has the structure needed to add them.

You will also see sections to set up event handlers and create a sample PropertyManager page. These will be discussed later in the chapter.

DisconnectFromSW runs each time the add-in is unloaded or when SOLIDWORKS shuts down. This function performs any last cleanup, removing user interface customizations and event handlers.

User Interface Elements

Understanding the architecture of the SOLIDWORKS user interface will help you customize your add-in.

The interface is composed of CommandGroups (1) containing command buttons and/or menu items, and CommandTabs (2) also containing command buttons as well as Flyout Toolbars.

A CommandGroup is the most basic and is a grouping of commands that can be in menus, toolbars or on the CommandManager in a tab. A CommandGroup will also appear in the Customize dialog under Commands. A CommandGroup may also include Flyout Toolbar groups.

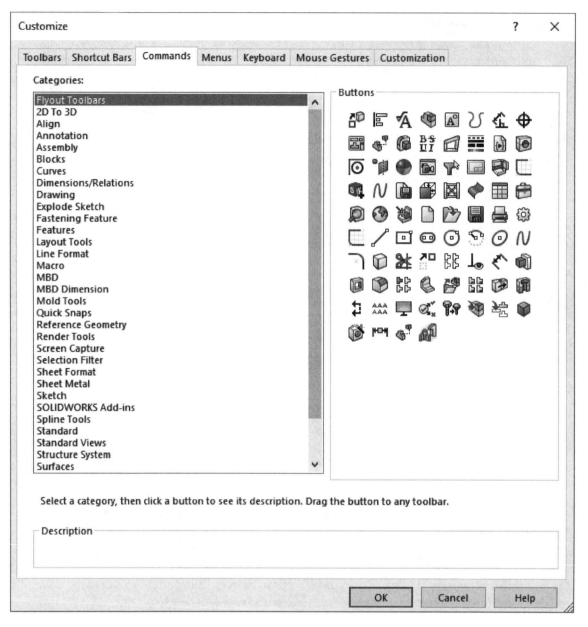

A CommandTab interface represents a tab on the CommandManager. CommandTabs are document-specific. Drawings, parts and assemblies each need unique commands and they are created by document type. CommandTabs include CommandBox interfaces that group commands together on the tab, with vertical separator bars between CommandBoxes.

8. Expand the "UI Methods" section of SwAddin class to review the procedures to build a CommandGroup, CommandTab and CommandBox as well as commands.

The first and largest procedure is AddCommandMgr, called by ConnectToSW. The top of the procedure contains required variable declarations and functions that ensure a consistent UI down to creation of a

CommandGroup. The CommandManager interface variable is named `iCmdMgr` and was connected to the SOLIDWORKS CommandManager in the ConnectToSW function.

9. Scroll down to the following code that creates the CommandGroup.

```
...
cmdGroup = iCmdMgr.CreateCommandGroup2(mainCmdGroupID, Title, _
  ToolTip, "", -1, ignorePrevious, cmdGroupErr)
If cmdGroup Is Nothing Or thisAssembly Is Nothing Then
  Throw New NullReferenceException()
End If
...
```

ICommandManager.CreateCommandGroup2

The CreateCommandGroup2 method of ICommandManager creates a new tab on the SOLIDWORKS CommandManager. The two arguments of interest are Title and ToolTip. These define the name of the CommandManager tab. The template has generic names of "VB Addin" for both. The method returns a CommandGroup interface.

The If statement following the CreateCommandGroup2 call verifies the CommandGroup was created successfully and that the add-in (`thisAssembly`) is still valid. The line inside the If block will trigger an exception, or error if one of the two conditions is True.

10. Scroll to the top of the procedure and edit the variables `Title` and `ToolTip` for your add-in CommandManager tab.

```
...
Dim Title As String = "MyMacroAddin"
Dim ToolTip As String = "MyMacroAddin"
...
```

Now that a command group has been created, each command needs to be defined. Before doing so, the command group needs icons for its commands. The template creates several example icons from PNG files and adds them to icon arrays. You can continue to use those icons or create your own to personalize your add-in.

```
...
' Add bitmaps to your project and set them as embedded resources or
' provide a direct path to the bitmaps
Dim mainIcons(6) As String
Dim icons(6) As String
icons(0) = iBmp.CreateFileFromResourceBitmap("TestAddIn.toolbar20x.png", _
    thisAssembly)
icons(1) = iBmp.CreateFileFromResourceBitmap("TestAddIn.toolbar32x.png", _
    thisAssembly)
icons(2) = iBmp.CreateFileFromResourceBitmap("TestAddIn.toolbar40x.png", _
    thisAssembly)
icons(3) = iBmp.CreateFileFromResourceBitmap("TestAddIn.toolbar64x.png", _
    thisAssembly)
```

```
icons(4) = iBmp.CreateFileFromResourceBitmap("TestAddIn.toolbar96x.png", _
    thisAssembly)
icons(5) = iBmp.CreateFileFromResourceBitmap("TestAddIn.toolbar128x.png", _
    thisAssembly)

mainIcons(0) = iBmp.CreateFileFromResourceBitmap("TestAddIn.mainicon_20.png",
    thisAssembly)
mainIcons(1) = iBmp.CreateFileFromResourceBitmap("TestAddIn.mainicon_32.png",
    thisAssembly)
mainIcons(2) = iBmp.CreateFileFromResourceBitmap("TestAddIn.mainicon_40.png",
    thisAssembly)
mainIcons(3) = iBmp.CreateFileFromResourceBitmap("TestAddIn.mainicon_64.png",
    thisAssembly)
mainIcons(4) = iBmp.CreateFileFromResourceBitmap("TestAddIn.mainicon_96.png",
    thisAssembly)
mainIcons(5) =
    iBmp.CreateFileFromResourceBitmap("TestAddIn.mainicon_128.png",
    thisAssembly)

cmdGroup.IconList = icons
cmdGroup.MainIconList = mainIcons
...
```

Icons can be several supported sizes and must be included in the arrays IconList and MainIconList. The arrays of strings should be paths pointing to BMP or PNG files. You can customize the icons by simply editing them in any image editing software that supports transparent backgrounds. Each icon image file can contain multiple icons for your application. For example, toolbar128x.png has three icons. This is how it looks in Visual Studio.

When you need to use icon 1 from the image, it will be in position 0 like a zero-based array. Icon 3 is in position 2. The SOLIDWORKS user interface scaling takes care of which element from the IconList and MainIconList arrays to use. Your application will specify which icon position is used from the image file.

The IconList and MainIconList arrays can have up to five elements, always starting with the smallest icon size in the first element. The IconList and MainIconList arrays must be matching in the number of array elements and icon sizes. Supported icon sizes are as follows.

Supported Icon Sizes

20 x 20 pixels
32 x 32 pixels
40 x 40 pixels
64 x 64 pixels
96 x 96 pixels
128 x128 pixels

ICommandGroup.AddCommandItem2

After defining the command group icons, two specific commands are added to the command group using the AddCommandItem2 method. AddCommandItem2 can make the commands available on toolbars and the CommandManager as well as in menus.

value = ICommandGroup.AddCommandItem2(Name, Position, HintString, ToolTip, ImageListIndex, CallbackFunction, EnableMethod, UserID, MenuTBOption)

- **Name** is a string representing the name of the command or menu item. The name is displayed if the command is put in a menu.

- **Position** is an integer that defines where the command is added. 0 will add it to the left side of the command group, -1 will add it after the last, or on the right.

- **HintString** is a string that displays in the status bar.

- **ToolTip** is a string that displays next to the command on the CommandManager when you display next to or below the button. This is typically the same as Name.

- **ImageListIndex** is the integer location of the desired icon image from the command group IconList array. The array is zero-based, so the first icon in the image file, the left-most, will be 0.

- **CallbackFunction** is a string that assigns the procedure to be called when the command is clicked. This is similar to event handlers but does not take the same code formatting. Make sure your syntax matches your callback procedure name exactly.

- **EnableMethod** is a string that assigns a procedure that can enable or disable a command. The procedure is optional. Leave an empty string if no enable procedure is needed. As an example, if you need to disable a command that is only for assemblies whenever you are in a drawing or part, a custom procedure is needed to do the filtering.

- **UserID** is an integer. It should be unique for each command.

- **MenuTBOption** are available from swCommandItemType_e. This assigns whether the command will be added to the CommandManager and/or the menus.

- **Value** is a returned integer specifying the index, or location, of the command in the command group.

Review the two commands added to the command group. Variables have been declared and pre-assigned for the UserID and MenuTBOption arguments as well as to receive the returned value.

```
cmdIndex0 = cmdGroup.AddCommandItem2("CreateCube", -1,
  "Create a cube", "Create cube", 0, "CreateCube", "",
  mainItemID1, menuToolbarOption)
cmdIndex1 = cmdGroup.AddCommandItem2("Show PMP", -1,
  "Display sample property manager", "Show PMP", 2,
  "ShowPMP", "PMPEnable", mainItemID2, menuToolbarOption)
```

The first, assigned to the variable `cmdIndex0`, is a command named CreateCube. It has been assigned to the first icon in the bitmap at index 0. It defines a callback function named `CreateCube` and has no enable procedure. Its ID was pre-defined in the variable `mainItemID1` and is placed on the CommandManager and in the menus through the variable `menuToolbarOption`.

The second, assigned to `cmdIndex1`, is Show PMP. It uses the third bitmap icon at index 2. It expects a callback function named ShowPMP and an enable procedure named PMPEnable. Its ID comes from the variable `mainItemID2` and uses the same placement options as the previous command.

11. Expand the "UI Callbacks" code region and review the menu callbacks and enable function of the two commands, `CreateCube`, `ShowPMP` and `PMPEnable`. The methods, procedures and API calls have already been explored in previous chapters with the exception of PropertyManager pages.

The template code includes an example of a FlyoutGroup to create Flyout Toolbars. For simplicity, this example will not use the FlyoutGroup if you choose.

12. Optionally, return to the AddCommandMgr procedure and comment out or delete the references to the FlyoutGroup as shown.

```
...
cmdGroup.HasToolbar = True
cmdGroup.HasMenu = True
cmdGroup.Activate()

'Dim flyGroup As FlyoutGroup
'flyGroup = iCmdMgr.CreateFlyoutGroup2(flyoutGroupID, ...

'flyGroup.AddCommandItem("FlyoutCommand 1", ...

'flyGroup.FlyoutType = swCommandFlyoutStyle_e.swCommandFlyoutStyle_Simple
...
```

The next section of code creates a CommandTab interface for each document type and adds the previously created commands to CommandBoxes on the CommandTab. All commands are added to all document types in the template code. If you would like to filter commands per document type, add an If statement within the For loop that iterates over document types.

13. Optionally, modify the code as shown below to comment out the addition of the FlyoutGroup based on document type.

```
...
For Each docType As Integer In docTypes
  Dim cmdTab As ICommandTab = iCmdMgr.GetCommandTab(docType, Title)
  Dim bResult As Boolean

  If Not cmdTab Is Nothing And Not getDataResult Or ignorePrevious Then
     'if tab exists, but we have ignored the registry info, re-create the tab.
     'Otherwise the ids won't matchup and the tab will be blank
     Dim res As Boolean = iCmdMgr.RemoveCommandTab(cmdTab)
     cmdTab = Nothing
  End If

  If cmdTab Is Nothing Then
    cmdTab = iCmdMgr.AddCommandTab(docType, Title)

    Dim cmdBox As CommandTabBox = cmdTab.AddCommandTabBox

    Dim cmdIDs(3) As Integer
    Dim TextType(3) As Integer

    cmdIDs(0) = cmdGroup.CommandID(cmdIndex0)
    TextType(0) = _
      swCommandTabButtonTextDisplay_e.swCommandTabButton_TextHorizontal

    cmdIDs(1) = cmdGroup.CommandID(cmdIndex1)
    TextType(1) = _
      swCommandTabButtonTextDisplay_e.swCommandTabButton_TextHorizontal

    cmdIDs(2) = cmdGroup.ToolbarId
    TextType(2) = _
      swCommandTabButtonTextDisplay_e.swCommandTabButton_TextHorizontal


    bResult = cmdBox.AddCommands(cmdIDs, TextType)

    'Dim cmdBox1 As CommandTabBox = cmdTab.AddCommandTabBox()
    'ReDim cmdIDs(1)
    'ReDim TextType(1)

    'cmdIDs(0) = flyGroup.CmdID
    'TextType(0) =
    'swCommandTabButtonTextDisplay_e.swCommandTabButton_TextBelow

    'bResult = cmdBox1.AddCommands(cmdIDs, TextType)
```

```
    'cmdTab.AddSeparator(cmdBox1, cmdIDs(0))

  End If
Next
...
```

CommandTab.AddCommandTabBox

If the CommandTab doesn't exist, the commands are added to the CommandTab through a new CommandBox interface. The AddCommandTabBox method of the CommandTab creates a new grouping that can hold commands. There are no arguments to the method, but it returns a new CommandBox interface.

CommandBox.AddCommands

The AddCommands method of the CommandBox adds commands through two arrays. The first argument is an array of commands based on their ID. The second is the layout style from swCommandTabButtonTextDisplay_e. The options are shown below and relate to the user customization options on the CommandManager.

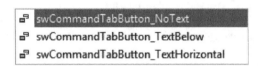

14. Modify one of the commands with a different layout style as shown below.

```
...
cmdIDs(0) = cmdGroup.CommandID(cmdIndex0)
TextType(0) = swCommandTabButtonTextDisplay_e.swCommandTabButton_TextBelow

cmdIDs(1) = cmdGroup.CommandID(cmdIndex1)
TextType(1) =
    swCommandTabButtonTextDisplay_e.swCommandTabButton_TextHorizontal
...
```

Debugging Add-ins

Add-ins are compiled dlls like macros. They cannot run alone. To debug, they have to run in-context with SOLIDWORKS. Luckily, the add-in template is already pre-configured to do this.

15. Select Project, MyMacroAddin Properties. Select the Debug tab. Review the Start Action settings that are set to start SOLIDWORKS while debugging.

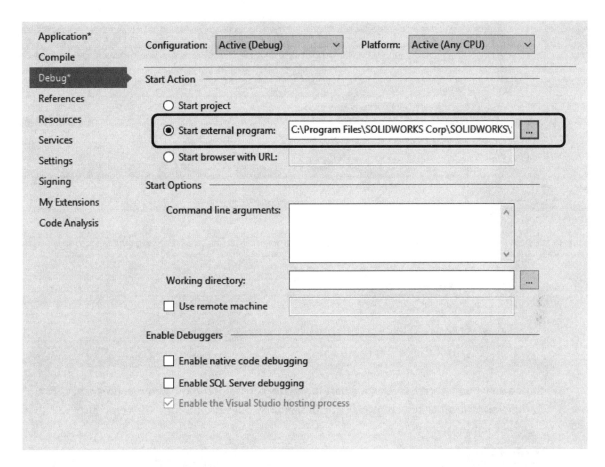

Before debugging, verify the path to *sldworks.exe*. Browse to the correct executable if necessary.

If you get an error like the following, you may have a code error or may need to run Visual Studio as administrator. Save your project and close Visual Studio. Then right-click on the Visual Studio icon from the Windows Start menu, choose More, and Run as administrator. Then re-open your project.

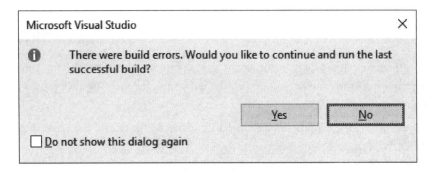

Running Visual Studio as administrator can also correct add-in registration problems. If your add-in does not display in the list of add-ins while debugging, it may need to be manually registered.

16. Return to *SwAddin.vb*. Add a breakpoint as shown in the CreateCube procedure in the UI Callbacks region and start debugging by selecting Debug, Start or click ▶ Start.

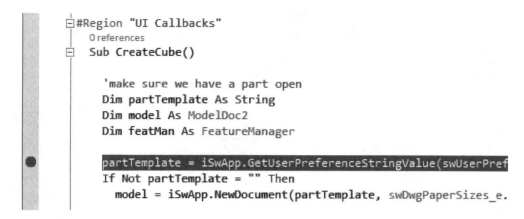

```
#Region "UI Callbacks"
    0 references
    Sub CreateCube()

        'make sure we have a part open
        Dim partTemplate As String
        Dim model As ModelDoc2
        Dim featMan As FeatureManager

        partTemplate = iSwApp.GetUserPreferenceStringValue(swUserPref
        If Not partTemplate = "" Then
            model = iSwApp.NewDocument(partTemplate, swDwgPaperSizes_e.
```

A new session of SOLIDWORKS will launch with your add-in enabled in the Other Add-ins section.

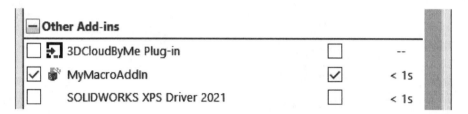

17. Start a new part or open an existing part to see the commands in the CommandManager. The CommandManager layout created by the procedure AddCommandMgr. The commands will also be visible in the Tools menu.

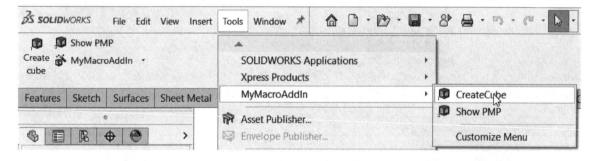

18. Click Create cube. The code will stop at the breakpoint in Visual Studio. You can now step through the procedure one line at a time to review the process. The method creates a new part, then builds a sketch and extrude.

19. Select Debug, Stop Debugging or click Stop Debugging ■ to stop the code and automatically close SOLIDWORKS.

20. Turn off the breakpoint by selecting it again or selecting Debug, Delete All Breakpoints.

Referencing Macro Code

Now that the add-in is functional and we have the interface structure for two commands, you are ready to add your macro code. Start by adding the code from AddComponents. Recall that the macro was built with all of the functionality in a code module and a separate interface in a Windows Form. The code module will be used here, but the add-in will provide a better looking interface through a PropertyManager page.

21. Select Project, Add Existing Item and browse to *AddComponentsMod.vb* in the AddComponents macro folder.

22. Open *AddComponentsMod.vb* to review the code.

The AddComponentAndMate procedure requires an argument of the full path to the part to be added. This module also needs an active reference to SOLIDWORKS through its m_swApp variable.

23. Switch back to *SwAddin.vb* and add a new procedure at the top of the UI Callbacks region named AddComponent as shown. Be sure to enter a valid part path in the partPath variable.

```
#Region "UI Callbacks"

  Sub AddComponent()
    AddComponentsMod.m_swApp = iSwApp
    Dim partPath As String
    partPath = "C:\Parts\washer.SLDPRT"
    AddComponentsMod.AddCompAndMate(partPath)
  End Sub

  Sub CreateCube()
...
```

The first pass at the code will add a part by a file path that is hard coded. The command group and commands need to be modified to run the new procedure.

24. Edit the command name and callback function of cmdIndex0 in the AddCommandMgr procedure to give it a new name and run the new callback procedure.

```
...
Dim menuToolbarOption As Integer = swCommandItemType_e.swMenuItem Or
swCommandItemType_e.swToolbarItem

cmdIndex0 = cmdGroup.AddCommandItem2("Add Component", -1,
  "Add a part and mate", "Add Component", 0, NameOf(AddComponent), "",
  mainItemID1, menuToolbarOption)
cmdIndex1 = cmdGroup.AddCommandItem2("Show PMP", -1,
  "Display sample property manager", "Show PMP", 2, "ShowPMP",
  "PMPEnable", mainItemID2, menuToolbarOption)

cmdGroup.HasToolbar = True
cmdGroup.HasMenu = True
cmdGroup.Activate()
...
```

Notice the use of the NameOf operator. Rather than entering hard-coded strings as the callback function names, NameOf returns the name of the procedure, variable or property at runtime. If you rename the callback functions at any point, you will not have to also remember to edit the callback name in the command definition.

25. To limit visibility of the button to assemblies, add an If statement surrounding the lines that add cmdIndex0 to the command box.

```
...
For Each docType As Integer In docTypes
  Dim cmdTab As ICommandTab = iCmdMgr.GetCommandTab(docType, Title)
  Dim bResult As Boolean

  If Not cmdTab Is Nothing And Not getDataResult Or ignorePrevious Then
  'if tab exists, but we have ignored the registry info, re-create the tab.
  'Otherwise the ids won't matchup and the tab will be blank

    Dim res As Boolean = iCmdMgr.RemoveCommandTab(cmdTab)
    cmdTab = Nothing
  End If

  If cmdTab Is Nothing Then
    cmdTab = iCmdMgr.AddCommandTab(docType, Title)

    Dim cmdBox As CommandTabBox = cmdTab.AddCommandTabBox

    Dim cmdIDs(3) As Integer
    Dim TextType(3) As Integer

    If docType = swDocumentTypes_e.swDocASSEMBLY Then
      cmdIDs(0) = cmdGroup.CommandID(cmdIndex0)
      TextType(0) = _
      swCommandTabButtonTextDisplay_e.swCommandTabButton_TextBelow
    End If

    cmdIDs(1) = cmdGroup.CommandID(cmdIndex1)
    TextType(1) = _
    swCommandTabButtonTextDisplay_e.swCommandTabButton_TextHorizontal
...
```

The add-in is now ready to test with the new callback function.

26. Start the add-in by selecting Debug, Start Debugging or by clicking ▶ Start. Open an assembly to test the Add Component function. Try opening a part or drawing to verify the Add Component button is hidden.

Troubleshooting

We have discussed a few potential problems when building add-ins. First, if you have any errors in the registration procedures, the add-in may fail to load. If you don't see the menus or commands for your add-in, check the add-ins list by selecting Tools, Add-ins. If your add-in is missing from the Other Add-ins section, there is a problem registering the assembly (dll) with Windows. Be sure to run Visual Studio As administrator to reduce this possible problem.

In rare cases, you may need to manually register the add-in. Follow these steps to verify the .NET Framework version and register the add-in dll.

27. Select Project, MyMacroAddIn Properties, and select the Application tab. Verify the Target framework value. This example references .NET Framework 4.

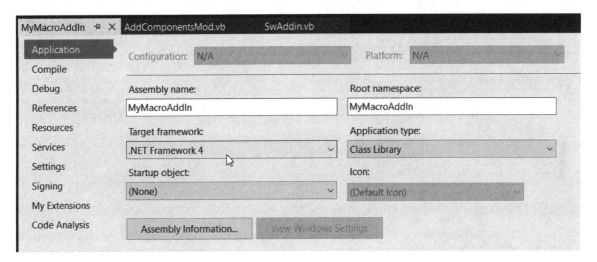

28. Start a CMD window and Run as Administrator.
29. Change the directory to the corresponding Microsoft.NET Framework folder "C:\Windows\Microsoft.NET\Framework64\v4.0.30319".

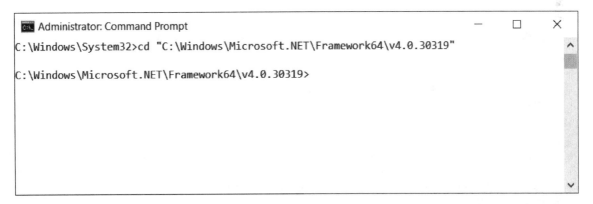

30. Type the command: `regasm /codebase` "`full path to your dll`" where you enter the full path to your add-in dll and Enter to register the add-in assembly.

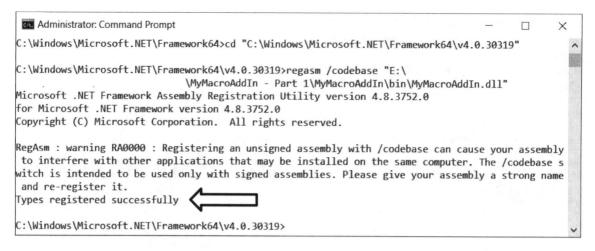

Ignore the warning about registering unsigned assemblies and verify the message, "Types registered successfully".

Hint: copy the full dll path from a File Explorer folder by right-clicking on the dll file and selecting Copy as path.

Your add-in should now run with the Visual Studio debugger.

Calling VBA Macros

Many of you may still have old VBA macros around that you would like to include in your add-in. Since VBA macros are not the same format, you cannot just add them to the project and reference the code. But you are not out of luck. A callback procedure can run a VBA macro, but the procedure to add them to the project and reference them in code is different.

You can still add a VBA macro to your Visual Studio project by selecting Project, Add Existing Item. Change the file type to All Files (*.*) and browse to your VBA macro(s). Click Add or Add As Link to bring them into the project.

The VBA macro must be copied into the project output (compiled file set) to enable deployment through an installer. Visual Studio does not know what to do with the macro on its own. You need to set this property manually.

Select your VBA macro in the Solution Explorer and change its Copy to Output Directory setting to Copy if newer. This will ensure the latest VBA macro version is deployed anytime you build your project. Also change its Build Action setting to Content. This trigger will be used later to make sure it gets into the installation package.

The following example shows how to call a VBA macro in a callback procedure named RunVBAMacro. You can name your callback procedure whatever you like, just make sure your command is set to use that callback.

```
Sub RunVBAMacro()
  Dim appDir As String = _
    IO.Path.GetDirectoryName(Assembly.GetExecutingAssembly().Location)
```

```
iSwApp.RunMacro2(appDir & "\3DPoints.swp", "Module3DPoints", "main", _
    swRunMacroOption_e.swRunMacroUnloadAfterRun, Nothing)
End Sub
```

This method will first determine the path to the add-in DLL. It uses that path to find a VBA macro named *3DPoints.swp* in the same directory. Explore the SOLIDWORKS API Help to review the structure of the RunMacro2 method of the ISldWorks interface for more detail.

PropertyManager Pages

When you are ready to get some real polish to your application and you want to follow the SOLIDWORKS interface as closely as possible, you will want to build PropertyManager Page controls for your user interface. This type of interface does take more code to create, but it will give your tool the familiarity of any other SOLIDWORKS feature.

Unlike building Visual Basic forms, building a PropertyManager Page interface is all code. There is no toolbar from which to drag and drop selection boxes, pull down lists or text boxes. This means it is best to have your user interface well planned before coding.

The Add Component tool needs an interface where the user can browse to a part. It would also be helpful to show them their selections for their selected edges. When finished, it should look like this.

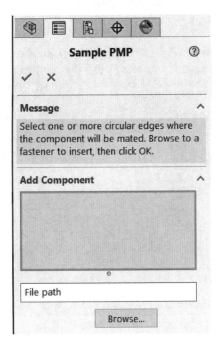

The template application has a ready-built PropertyManager page that is launched from the second CommandGroup command. We will use the pre-built structure for the PropertyManager page, so the command button should be customized to represent our new Add Command function.

31. Modify the AddCommandItem2 code for cmdIndex1 in AddCommandMgr as shown to give the command a new name, hint string and tooltip. Leave the callback and enable functions as is.

```
...
Dim menuToolbarOption As Integer = swCommandItemType_e.swMenuItem Or
swCommandItemType_e.swToolbarItem

cmdIndex0 = cmdGroup.AddCommandItem2("Add Component", -1,
  "Add a part and mate", "Add Component", 0, NameOf(AddComponent), "",
  mainItemID1, menuToolbarOption)
cmdIndex1 = cmdGroup.AddCommandItem2("Add Component PMP", -1,
  "Add a part and mate", "Add Component PMP", 2, "ShowPMP",
  "PMPEnable", mainItemID2, menuToolbarOption)

cmdGroup.HasToolbar = True
cmdGroup.HasMenu = True
cmdGroup.Activate()
...
```

PropertyManager Page Controls

Before starting the work, review the add-in template PropertyManager page that has already been created for us.

32. Start MyMacroAddIn. Open a part or assembly and select the Show PMP command from the MyMacroAddIn command tab.

The PropertyManager page includes several typical user interface elements, or controls. These are similar to those found in typical form controls, but all use the naming convention PropertyManagerPageGroup, PropertyManagerPageCheckbox, etc. The names below have been shortened for simplicity.

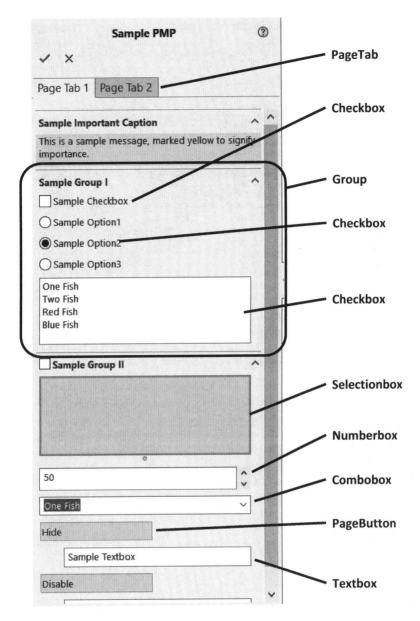

33. Stop the Visual Studio debugger and open *UserPMPage.vb* from the Solution Explorer. Review the code that builds the page.

Expand the Property Manager Page Controls code region and review the declared controls. Each is a variable that will be assigned a specific control in the AddControls procedure.

```
Public Class UserPMPage
    Dim iSwApp As SldWorks
    Dim userAddin As SwAddin
    Dim handler As PMPageHandler
    Friend PropMgrPage As PropertyManagerPage2
    Dim ppagetab1 As PropertyManagerPageTab
```

```
    Dim ppagetab2 As PropertyManagerPageTab

#Region "Property Manager Page Controls"
    'Groups
    Dim group1 As PropertyManagerPageGroup
    Dim group2 As PropertyManagerPageGroup

    'Controls
    Dim checkbox1 As PropertyManagerPageCheckbox
    Dim option1 As PropertyManagerPageOption
    Dim option2 As PropertyManagerPageOption
    Dim option3 As PropertyManagerPageOption
    Dim list1 As PropertyManagerPageListbox

    Dim selection1 As PropertyManagerPageSelectionbox
    Dim num1 As PropertyManagerPageNumberbox
    Dim combo1 As PropertyManagerPageCombobox

    Dim button1 As PropertyManagerPageButton
    Dim button2 As PropertyManagerPageButton
    Friend text1 As PropertyManagerPageTextbox
    Friend text2 As PropertyManagerPageTextbox

    'Control IDs
    Dim group1ID As Integer = 0
    Dim group2ID As Integer = 1
    Dim checkbox1ID As Integer = 2
    Dim option1ID As Integer = 3
    Dim option2ID As Integer = 4
    Dim option3ID As Integer = 5
    Dim list1ID As Integer = 6
    Dim selection1ID As Integer = 7
    Dim num1ID As Integer = 8
    Dim combo1ID As Integer = 9
    Dim TabID1 As Integer = 10
    Dim TabID2 As Integer = 11
    Friend buttonID1 As Integer = 12
    Friend buttonID2 As Integer = 13
    Dim textID1 As Integer = 14
    Dim textID2 As Integer = 15

#End Region
```

Each control also requires a unique ID. Those Integer values are declared and set in this same region of class-level variables.

The user interface for Add Component will only need a PageGroup, Selectionbox, PageButton and Textbox.

34. Simplify the project by deleting extra class-level variables. Your results should look like the following code.

```vb
Public Class UserPMPage
    Dim iSwApp As SldWorks
    Dim userAddin As SwAddin
    Dim handler As PMPageHandler
    Friend PropMgrPage As PropertyManagerPage2

#Region "Property Manager Page Controls"
    'Groups
    Dim group2 As PropertyManagerPageGroup

    'Controls
    Dim selection1 As PropertyManagerPageSelectionbox
    Dim button1 As PropertyManagerPageButton
    Friend text1 As PropertyManagerPageTextbox

    'Control IDs
    Dim group2ID As Integer = 1
    Dim selection1ID As Integer = 7
    Friend buttonID1 As Integer = 12
    Dim textID1 As Integer = 14

#End Region
```

The AddControls procedure creates all of the PropertyManager page controls before the page is shown to the user. Now that the extra variables for the controls have been deleted from the code, the AddControls procedure will show errors wherever these old controls were referenced.

```
AddComponentsMod.vb    UserPMPage.vb*  ⊟ ×  SwAddin.vb*
VB MyMacroAddin                    ▾  ⁴ˢ UserPMPage           ▾  ⊕ AddControls                  ▾  ╪
                      1 reference
    48    ⊟    Sub AddControls()
    49 ✎          Dim options As Integer
    50             Dim leftAlign As Integer
    51             Dim controlType As Integer
    52             Dim retval As Boolean
    53
    54             ' Add Message
    55             retval = PropMgrPage.SetMessage3("This is a sample message, marked y
    56                               swPropertyManagerPageMessageVisibility.sw
    57                               swPropertyManagerPageMessageExpanded.swMe
    58                               "Sample Important Caption")
    59
    60             'Add PropertyManager Page Tabs
    61             ppagetab1 = PropMgrPage.AddTab(TabID1, "Page Tab 1", "", 0)
    62             ppagetab2 = PropMgrPage.AddTab(TabID2, "Page Tab 2", "", 0)
    63
120%   ▾   ◎ 43   ⚠ 0   ← →   ◀                          ▶   Ln: 49   Ch: 9   SPC   CRLF
```

You will see the errors in underlined code as well as with red bars in the vertical scroll bar in the code window. A little clean-up will make the code easier to customize for our application.

35. Delete the AddTab and the first GroupBox code highlighted below.

```
' Add Message
retval = PropMgrPage.SetMessage3( _
    "This is a sample message, marked yellow to signify importance.", _
    swPropertyManagerPageMessageVisibility.swImportantMessageBox, _
    swPropertyManagerPageMessageExpanded.swMessageBoxExpand, _
    "Sample Important Caption")

'Add PropertyManager Page Tabs
ppagetab1 = PropMgrPage.AddTab(TabID1, "Page Tab 1", "", 0)
ppagetab2 = PropMgrPage.AddTab(TabID2, "Page Tab 2", "", 0)

'Add Groups
options = swAddGroupBoxOptions_e.swGroupBoxOptions_Expanded +
swAddGroupBoxOptions_e.swGroupBoxOptions_Visible
group1 = ppagetab1.AddGroupBox(group1ID, "Sample Group I", options)

options = swAddGroupBoxOptions_e.swGroupBoxOptions_Checkbox +
swAddGroupBoxOptions_e.swGroupBoxOptions_Visible
group2 = ppagetab1.AddGroupBox(group2ID, "Sample Group II", options)
```

36. Delete the entire "Add Controls to Group1" code section up to the comment "Add Controls to Group2".
37. Delete the "Num1", "Combo1", the second "Button" and "Textbox2" code groups under the "Add Controls to Group2" code section.

There should only be one remaining error under the "Add Groups" code section.

38. Edit the "Add Groups" code section as shown to replace the reference to ppagetab1 with PropMgrPage.

```
...
'Add Groups
options = swAddGroupBoxOptions_e.swGroupBoxOptions_Expanded +
swAddGroupBoxOptions_e.swGroupBoxOptions_Visible
group2 = PropMgrPage.AddGroupBox(group2ID, "Sample Group II", options)

'Add Controls to Group2
...
```

CreatePropertyManagerPage

The CreatePage procedure, just above the AddControls procedure, is responsible for creating the page. This would be the corollary to creating a Windows Form using code. The page has to be created before the controls can be added.

The CreatePropertyManagerPage method of the ISldWorks interface creates a PropertyManager page using the specified options. It must also have a PropertyManagerPageHandler class passed to it to understand how to handle the actions from the page. The page handler is another class in the project that we will discuss later in the chapter.

value = ISldWorks.CreatePropertyManagerPage(Title, Options, Handler, Errors)

- **Title** is a string that displays as a title on the page.

- **Options** are defined in swPropertyManagerPageOptions_e. These represent the buttons that display at the top of the page. These include whether the page will have multiple pages, OK and Cancel buttons or a Pushpin button.

- **Handler** must be an instance of a class that implements PropertyManagerPage2Handler9 as we have discussed.

- **Errors** is an integer used as an output in case there were errors when the page was created. The value of Errors can be compared to constants from swPropertyManagerPageStatus_e to evaluate the errors.

- **value** is an instance of the newly created PropertyManager page.

The PropertyManager page is set to the class-level variable PropMgrPage at the end of the CreatePage procedure.

39. Change the Title of PropMgrPage to "Add Component" in the CreatePage procedure as shown.

```
Sub CreatePage()
  handler = New PMPageHandler()
  handler.Init(iSwApp, userAddin, Me)
  Dim options As Integer
  Dim errors As Integer
  options = _
    swPropertyManagerPageOptions_e.swPropertyManagerOptions_OkayButton +
    swPropertyManagerPageOptions_e.swPropertyManagerOptions_CancelButton
  PropMgrPage = iSwApp.CreatePropertyManagerPage("Add Component",
    options, handler, errors)
End Sub
```

Let's continue reviewing the AddControls procedure to understand the different control types and their options.

SetMessage3

The SetMessage3 method of the IPropertyManagerPage2 interface allows you to create a standard message on the page. This is typically used to give basic instructions to the user. The following is an example from the SOLIDWORKS Insert Component tool.

value = IPropertyManagerPage2.SetMessage3(Message, Visibility, Expanded, Caption)

- **Message** is simply a text string with the desired message.

- **Visibility** is used to set how the message displays based on values from swPropertyManagerPageMessageVisibility.

- **Expanded** sets whether the message is expanded or collapsed based on values from swPropertyManagerPageMessageExpanded.

- **Caption** is simply the string at the top of the message.

40. Update the instructions for the user in the SetMessage3 line.

```
' Add Message
Dim message As String
message = _
    "Select one or more circular edges where the component will be mated. " _
    & "Browse to a fastener to insert, then click OK."
retval = PropMgrPage.SetMessage3(message, _
    swPropertyManagerPageMessageVisibility.swImportantMessageBox, _
    swPropertyManagerPageMessageExpanded.swMessageBoxExpand, _
    "Message")
...
```

AddGroupBox

A group allows you to assemble several controls together in a logical collection. Each group has an ID number, a Caption and Options that are set when the group is created.

IPropertyManagerPage2.AddGroupBox (Id, Caption, Options)

- **ID** is a long integer representing a unique identity.

- **Caption** is a string containing the text displayed at the top of the group.

- **Options** can be a combination of SOLIDWORKS constants defining group options. The following is a list of the different options available to the PropertyManagerGroup control defined by the SOLIDWORKS constant swAddGroupBoxOptions_e as shown below.

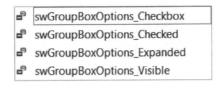

41. Update the Caption of the group2 GroupBox as shown.

```
...
'Add Groups
options = swAddGroupBoxOptions_e.swGroupBoxOptions_Expanded +
swAddGroupBoxOptions_e.swGroupBoxOptions_Visible
group2 = PropMgrPage.AddGroupBox(group2ID, "Add Component", options)
...
```

AddControl

The remaining controls are added to the group2 GroupBox using AddControl.

IPropertyManagerPageGroup.AddControl(ID, ControlType, Caption, LeftAlign, Options, Tip)

- **ID** is a unique long integer value that is used to identify the control. Think of this as the control's name. This value is used in the PropertyManagerPageHandler to distinguish each control of a particular type.

- **ControlType** is a long value that corresponds to the type of control from the SOLIDWORKS constants shown in the table above.

- **Caption** requires a string and is only applicable to the Button, Checkbox, Label and Option controls and sets the text shown on the resulting control.

- **LeftAlign** sets the position of the control by assigning it to a value from swPropertyManagerPageControlLeftAlign_e such as swControlAlign_LeftEdge or swControlAlign_Indent.

- **Options** sets any desired option or combination of options for the control from swAddControlOptions_e: swControlOptions_Visible, swControlOptions_Enabled, or swControlOptions_SmallGapAbove.

- **Tip** a string that sets the text for the popup tool tip.

42. Modify the selection1 SelectionBox caption and tip as shown.

```
'Add Controls to Group2
'Selection1
```

```
controlType = swPropertyManagerPageControlType_e.swControlType_Selectionbox
leftAlign = swPropertyManagerPageControlLeftAlign_e.swControlAlign_LeftEdge
options = swAddControlOptions_e.swControlOptions_Enabled _
  + swAddControlOptions_e.swControlOptions_Visible
selection1 = group2.AddControl(selection1ID, controlType, _
  "Select Circular Edges", leftAlign, options,
  "Select circular edges on flat faces where components will be added")
...
```

Control Properties and Methods

After creating a control you may want to set certain properties or use certain methods. For example, a Combobox control is only useful when it has a list to choose from. Numberbox controls frequently need units and precision to be defined. Selectionbox controls are easier to use if there is a predefined selection filter. The following table shows commonly used controls and some of their associated properties and methods.

Control Name
PropertyManagerPageControl (general)   • **Enabled** – gets or sets whether the control is enabled   • **SetPictureLabelByName** – adds an icon to the left of the control using the bitmap image at a specified location   • **SetStandardPictureLabel** – adds a standard icon to the left of the control using one of the following from swBitmapLabel_SelectVertex swBitmapLabel_SelectFace swBitmapLabel_SelectEdge or others found in swControlBitmapLabelType_e
PropertyManagerPageButton   • **Caption** – gets or sets the caption text
PropertyManagerPageCheckbox   • **Caption** – gets or sets the caption text   • **Checked** – True or False
PropertyManagerPageCombobox   • **AddItems** – adds an array of items   • **Clear** – clears the list   • **CurrentSelection** – gets or sets the item that is currently selected   • **Height** – the max height of the dropdown box   • **ItemText** – gets the text at a specified item (requires an integer input for the location or -1 for the currently selected item)   • **Style** – A value from swPropMgrPageComboBoxStyle_e such as swPropMgrPageComboBoxStyle_Sorted or swPropMgrPageComboBoxStyle_EditableText or both.
PropertyManagerPageLabel

- **Caption** – gets or sets the caption text
- **Syle** – A value from swPropMgrPageLabelStyle_e such as swPropMgrPageLabelStyle_LeftText, swPropMgrPageLabelStyle_CenterText, swPropMgrPageLableStyle_RightText or a combination.

PropertyManagerPageNumberbox

- **SetRange2** – applies several settings to a Numberbox. The first argument sets the units of the box using the a value from swNumberboxUnitType_e such as swNumberBox_UnitlessInteger, swNumberBox_UnitlessDouble, swNumberBox_Length, or swNumberBox_Angle. You may wish to limit the values the user can input by setting a minimum and maximum value. SetRange2 will also set the increment used by the spin box arrows to the right of the box. And finally, you must define whether your values are inclusive or exclusive of the maximum and minimum using a Boolean value.
- **Value** – get or set the value as a double.

PropertyManagerPageOption

- **Checked** – True or False
- **Style** – A value from swPropMgrPageOptionStyle_e. Only swPropMgrPageOptionStyle_FirstInGroup should be used to specify the first control in a group of options. The next group is defined by the next option control with this style.

PropertyManagerPageSelectionbox

- **GetSelectionFocus** – returns True if the specified selection box has the focus (the one with the focus is pink).
- **Height** – the height of the box.
- **Mark** – get or set the mark used for ordered selections such as a sweep path and profile.
- **SetSelectionFilters** – sets the selection filters. This method requires passing an array of integers from swSelectType. (*Hint: search for "swSelectType" in the Visual Studio Object Browser. Select **View, Object Browser**, click* .)
- **SetSelectionFocus** – set the focus on a specific selection box (this will turn it blue).
- **SingleEntityOnly** – set this to True if you only want to allow a single item to be selected. Otherwise, set it to False.

PropertyManagerPageTextbox

- **Text** –get or set the text string.

The Selectionbox control `selection1` should only allow the user to select edges for this application.

43. Modify the filter array for the `selection1` control to only allow selection of edges.

```
...
If Not selection1 Is Nothing Then
  Dim filter() As Integer = New Integer() {swSelectType_e.swSelEDGES}
  selection1.Height = 50
  selection1.SetSelectionFilters(filter)
```

```
End If
...
```

PropertyManagerPage Control Location

The location of each control on a PropertyManager page is based on the order it is added in the code. In the template code, the Button is added before the Textbox. If you prefer a control to be in a different position, simply move its code to a different position in the procedure.

44. Move the "Textbox1" code section just above the "Button" code section to create a Textbox for the file path above the button that will browse for the file. Additional modifications are made to left-align the Textbox and indent the button and change its caption.

```
...
'Textbox1
controlType = swPropertyManagerPageControlType_e.swControlType_Textbox
leftAlign = swPropertyManagerPageControlLeftAlign_e.swControlAlign_LeftEdge
options = swAddControlOptions_e.swControlOptions_Enabled +
  swAddControlOptions_e.swControlOptions_Visible
text1 = group2.AddControl2(textID1, controlType, "File path",
  leftAlign, options, "File path")

'Button
controlType = swPropertyManagerPageControlType_e.swControlType_Button
leftAlign = _
  swPropertyManagerPageControlLeftAlign_e.swControlAlign_DoubleIndent
options = swAddControlOptions_e.swControlOptions_Enabled +
  swAddControlOptions_e.swControlOptions_Visible
button1 = group2.AddControl2(buttonID1, controlType, "Browse...",
  leftAlign, options, "Browse to a part")

End Sub
```

Finally, to make the interface a little easier for the user to understand, we will add a label above the Textbox. As a shortcut, you can always add controls with all arguments in one line. This is especially useful with Labels. You will not typically need a reference to the Label control anywhere else in the project, so it does not need to be stored in a class-level variable. If you add more labels using this simplified technique, do not forget to give them each a unique ID.

45. Add a label above the "Textbox1" code section as shown. The options and alignment have been simplified as their numeric values.

```
'Label
group2.AddControl(100, 1, "File path", 1, 3, "")

'Textbox1
...
```

Debugging PropertyManager Pages

Since we cannot see the results of our PropertyManager interface in code, it must be run to verify the layout and test the functionality.

46. Start MyMacroAddIn 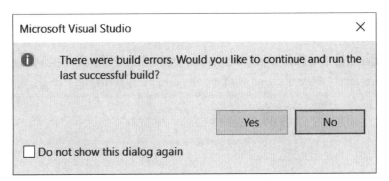 and review the build error message. Click No to return to code.

Microsoft Visual Studio ×

ⓘ There were build errors. Would you like to continue and run the last successful build?

Yes   No

☐ Do not show this dialog again

Because several class-level variables were deleted, there are now errors in the *PMPHandler.vb* code that must be corrected before seeing our PropertyManager page in action.

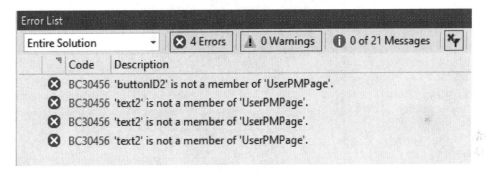

Error List

Entire Solution ▾ | ❌ 4 Errors | ⚠ 0 Warnings | ⓘ 0 of 21 Messages

	Code	Description
❌	BC30456	'buttonID2' is not a member of 'UserPMPage'.
❌	BC30456	'text2' is not a member of 'UserPMPage'.
❌	BC30456	'text2' is not a member of 'UserPMPage'.
❌	BC30456	'text2' is not a member of 'UserPMPage'.

47. Double-click on the first error to jump to the problem in code.

There used to be two buttons and two Textbox controls on the page. Since there is now only one of each, the code referencing the deleted controls must be either deleted or commented out.

48. Delete or comment out the ElseIf code block in OnButtonPress to remove the code error.

```
Sub OnButtonPress(ByVal id As Integer) Implements
PropertyManagerPage2Handler9.OnButtonPress
  If id = ppage.buttonID1 Then ' Toggle the textbox control visibility state

    If ppage.text1.Visible = True Then
      ppage.text1.Visible = False
    Else
      ppage.text1.Visible = True
    End If

    'ElseIf id = ppage.buttonID2 Then
```

```
'   If ppage.text2.Enabled = True Then
'      ppage.text2.Enabled = False
'   Else
'      ppage.text2.Enabled = True
'   End If

   End If
End Sub
```

49. Start MyMacroAddIn again. When the SOLIDWORKS session is live, open an assembly and click the Add Component PMP button to show the new PropertyManager page.

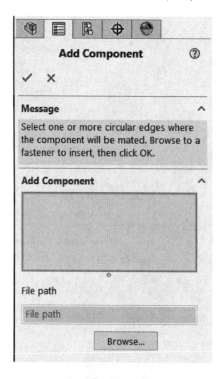

Test the Selectionbox. You should only be able to select edges in your assembly. The Browse button and OK buttons are not yet functional.

50. Stop debugging the project. The SOLIDWORKS session will automatically close.

PropertyManager Page Handler

Once the interface is built, PropertyManager page actions are handled through a class in your project that implements IPropertyManagerPage2Handler9. This interface will give you access to all of the properties, methods, event handlers and procedures related to a PropertyManager page from the SOLIDWORKS API. Using the Implements keyword tells your class that it must make use of the methods and procedures of that particular interface.

51. Open *PMPHandler.vb* from the Solution Explorer. Review the code at the top of the class, including the `Init` function and the declaration of the SOLIDWORKS interface, the SwAddin and UserPMPage classes.

Add two class-level declarations for `filePath` and `cancelled`. Add the bold line of code to pass the SOLIDWORKS application interface to the AddComponentsMod variable `m_swApp`.

```
Public Class PMPageHandler
    Implements PropertyManagerPage2Handler9

    Dim iSwApp As SldWorks
    Dim userAddin As SwAddin
    Dim ppage As UserPMPage

    Dim filePath As String
    Dim cancelled As Boolean

    Function Init(ByVal sw As SldWorks, ByVal addin As SwAddin, _
        page As UserPMPage) As Integer
        iSwApp = sw
        userAddin = addin
        ppage = page
        AddComponentsMod.m_swApp = sw
    End Function

    'Implement these methods from the interface
...
```

The `Implements` statement requires all public procedures involved with the referenced interface. In fact, if you leave any of these out of your Class, you will get an error like the one below when you try to run or debug.

❌   Class 'PMPageHandler' must implement 'Sub AfterActivation()' for interface 'SolidWorks.Interop.swpublished.IPropertyManagerPage2Handler9'.

Take some time to look through the list of actions that have associated handlers. Event handlers are procedures that are triggered by events. For example, you have already used the Button Click event handlers in forms that associate a specific procedure with the action of clicking on a specific form button. You can use any of these actions to get the desired functionality out of your PropertyManager interface. A few of the most common are listed below along with a brief description.

Method	Description
AfterClose	Code is run once the PropertyManager page has been completely closed. You may not be able to access elements of the PropertyManager page at this point.

OnButtonPress	Code is run when a button control in the PropertyManager page is clicked. The Id argument is the button's ID. Use this value to determine which button was pressed.
OnClose	Code is run when any action closes the PropertyManager page. The argument reason will contain information about why the page was closed from swPropertyManagerPageCloseReasons_e.

OnButtonPress

52. Navigate to the OnButtonPress procedure and modify the code to browse to a SOLIDWORKS part file, then display the selected file path in the Textbox control.

```
Sub OnButtonPress(ByVal id As Integer) _
  Implements PropertyManagerPage2Handler9.OnButtonPress
  If id = ppage.buttonID1 Then
    Dim ofd As New Windows.Forms.OpenFileDialog
    ofd.Filter = "SOLIDWORKS Parts (*.sldprt)|*.sldprt"
    Dim diaRes As Windows.Forms.DialogResult
    diaRes = ofd.ShowDialog()
    If diaRes = Windows.Forms.DialogResult.OK Then
      ppage.text1.Text = ofd.FileName
    End If
  End If
End Sub
```

All of the PropertyManager page controls are accessible through ppage, declared and initialized earlier in the PMPageHandler class. The OnButtonPress handler provides access to the control ID so you only need one procedure to handle all buttons on your PropertyManager page.

OnSubmitSelection

The OnSubmitSelection method is called every time a selection is made in SOLIDWORKS while a PropertyManagerPageSelectionbox control is active. It gives you the opportunity to validate a user's selection. This function is declared as a Boolean. If it returns True, the selection will be added to the selection box. If it returns False, the selection is blocked. For example, if you only wanted circular edges to be selected, this procedure could check the selected edge and reject straight edges. You have already limited selections to edges in the design of the PropertyManagerPageSelectionbox control so all that is needed in this example is a true value to be returned. The Add Component code will ignore anything that isn't circular.

53. Review the following code to the *OnSubmitSelection* to allow any selection by returning a True value. Setting a function name to a value is the same as the Return statement except any remaining code will still run.

```
Public Function OnSubmitSelection(ByVal Id As Integer, _
```

```
  ByVal Selection As Object, ByVal SelType As Integer, _
  ByRef ItemText As String) As Boolean _
  Implements SolidWorks.Interop.swpublished. _
  IPropertyManagerPage2Handler9.OnSubmitSelection
    OnSubmitSelection = True
End Function
```

OnClose

There are two similar procedures that control what happens when a PropertyManager page is closed. OnClose is triggered when the user clicks either OK or Cancel. It runs before the instance of the page is destroyed from memory. This is the place to find out whether the user clicked OK, or if they clicked Cancel. This is also the place to store user selections and control values from the PropertyManager page controls to pass to the processing code to do the work. AfterClose is where your processing should occur. It runs after OnClose completes successfully.

The OnClose procedure provides an argument named Reason. Its value is the cause of the page closure. If Reason is the constant swPropertyManagerPageClose_Okay then OK was clicked. If Reason is swPropertyManagerPageClose_Cancel, the user clicked Cancel. These values come from the enumeration swPropertyManagerPageCloseReasons_e.

The action to run AddCompAndMate should happen in AfterClose only if the user clicked OK. The cancelled variable will store the user's response while filePath will store the path to the selected part file.

54. Replace the code in the *OnClose* procedure with the following. If the selected part does not exist, tell the user with a message and do not close the PropertyManager page. Set the cancelled and filePath variables accordingly.

```
Sub OnClose(ByVal reason As Integer) _
  Implements PropertyManagerPage2Handler9.OnClose
  If reason =
    swPropertyManagerPageCloseReasons_e.swPropertyManagerPageClose_Okay Then
    cancelled = False
    filePath = ppage.text1.Text
    If Not System.IO.File.Exists(filePath) Then
      iSwApp.SendMsgToUser2("Browse to a valid part first.",
    swMessageBoxIcon_e.swMbInformation, swMessageBoxBtn_e.swMbOk)
      'prevent the Page from closing
      Err.Raise(0,, "cancel close")
    End If
  Else
    cancelled = True
  End If
End Sub
```

SendMsgToUser2

The SendMsgToUser2 method of the ISldWorks interface is much like the MsgBox function. Feel free to use either method for messages with responses.

Err.Raise

You have seen errors in your code many times, but until this point, you have not generated an intentional error. To prevent the OnClose command from closing the page, you have to raise an error in your code. A value of 0 will prevent the page from closing. The third argument provides a message back to the debugger so you can see the result of the error when debugging.

55. Start MyMacroAddIn. After SOLIDWORKS is active, open an assembly and select the Add Components PMP command button. Click OK without selecting edges or browsing to a file. You should see the message followed by the debugger stopping at the `Err.Raise` line.

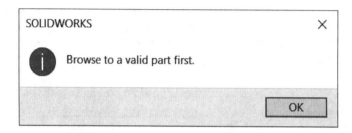

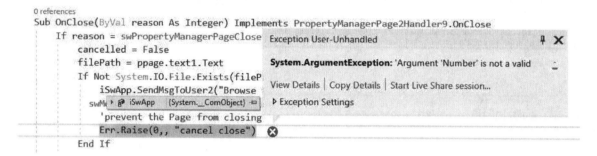

56. Select Debug, Continue or ▶ to ignore the error and continue running. Notice that the PropertyManager page stays open.

57. Stop the debugger ■.

AfterClose

Now that the selections can be made and the user can browse to a part, the `AddCompAndMate` method can be called from AddComponentsMod in the AfterClose procedure.

58. Edit the AfterClose procedure to call AddCompAndMate, passing `filePath` as an argument. Use an If block to only run if the user clicked OK, not Cancel.

```
Sub AfterClose() Implements PropertyManagerPage2Handler9.AfterClose
  If Not cancelled Then
    AddComponentsMod.AddCompAndMate(filePath)
  End If
End Sub
```

59. Start MyMacroAddIn. After SOLIDWORKS is active, open an assembly and select the Add Components PMP command button. Select one or more circular edges and browse to *washer.sldprt* and click OK to add the part.

The application is now a functional add-in with a PropertyManager page, all starting from the add-in template and your existing macro code.

Hint: When you need multiple PropertyManager pages in your application, you will need separate page and handler classes for each.

Add-In Development Simplified

Now that you know the hard way to develop PropertyManager pages, let me introduce you to a shortcut. Artem Taturevych of Xarial has been developing automation tools for SOLIDWORKS for over 10 years. He even worked as a member of the SOLIDWORKS API support team for a few of those.

Artem developed and offers XCAD.NET, an open-source simplified framework for SOLIDWORKS development. Make your development life a little easier by downloading the XCAD.NET toolkit today.

https://xcad.net

Conclusion

Implementing a PropertyManager style interface in an add-in takes more preparation and programming, but the results can be impressive. Try adding more macros to MyMacroAddIn and create more PropertyManager pages and handlers. The more you build the easier they become. The best way to simplify the process is to start with a good code template. After building PropertyManager classes, copy them to additional projects as needed.

C# Example

For the complete C# solution, open *MyMacroAddIn.sln* from the downloadable example files.

```csharp
public class PMPHandler : IPropertyManagerPage2Handler9
{
    ISldWorks iSwApp;
    SwAddin userAddin;
    UserPMPage ppage;

    string filePath;
    bool cancelled;

    public PMPHandler(SwAddin addin, UserPMPage page)
    {
        userAddin = addin;
        iSwApp = (ISldWorks)userAddin.SwApp;
        ppage = page;
        AddComps.m_swApp = (SldWorks)iSwApp;
    }

    //Implement these methods from the interface
```

```
    public void AfterClose()
    {
        if (!cancelled)
        {
            AddComps.AddCompAndMate(filePath);
        }
    }

    public void OnClose(int reason)
    {
        if (reason == (int)swPropertyManagerPageCloseReasons_e
            .swPropertyManagerPageClose_Okay)
        {
            cancelled = false;
            filePath = ppage.textbox1.Text;
            if (!System.IO.File.Exists(filePath))
            {
                iSwApp.SendMsgToUser2("Browse to a valid part first.",
                    (int)swMessageBoxIcon_e.swMbInformation,
                    (int)swMessageBoxBtn_e.swMbOk);
                //prevent the page from closing
                COMException ex = new COMException("cancel close", 1);
                throw ex;
            }
        }
        else
        {
            cancelled = true;
        }
    }

    public void OnButtonPress(int id)
    {
        if (id == UserPMPage.buttonID1)
        // Toggle the textbox control visibility state
        {
            OpenFileDialog ofd = new OpenFileDialog();
            ofd.Filter = "SOLIDWORKS Parts (*.sldprt)|*.sldprt";
            DialogResult diaRes = ofd.ShowDialog();
            if (diaRes == DialogResult.OK)
            {
                ppage.textbox1.Text = ofd.FileName;
            }
        }
    }
...
```

Add-In Events and Installation

- **Event Handling**

- **Installation**

Introduction

This chapter will build on the MyMacroAddIn developed in the last chapter. The new code will monitor SOLIDWORKS and run a checking procedure when a model is saved. If a "Description" custom property hasn't been added, the user will be reminded before saving.

Events

Thus far, the focus has been on forms and PropertyManager interfaces. Users interact with your macros and applications by clicking buttons. Events trigger code to run when a user performs standard operations in SOLIDWORKS like opening, saving or rebuilding. Events are the ideal technique when you want to run a procedure before or after an action in SOLIDWORKS. Unlike typical code, which runs after the user clicks a button or manually runs a macro, event handlers run continuously. Add-ins are the best package for event handling since they are always available during a SOLIDWORKS session.

1. Open your MyMacroAddIn project in Visual Studio.
2. Open *SwAddin.vb* from the Solution Explorer and expand the "Local Variables" code region.

```
Public Class SwAddin
    Implements SolidWorks.Interop.swpublished.SwAddin

#Region "Local Variables"
    Dim WithEvents iSwApp As SldWorks
    Dim iCmdMgr As ICommandManager
    Dim addinID As Integer
    Dim openDocs As Hashtable
. . .
```

WithEvents Declaration

Notice the unique declaration of `iSwApp`. Declaring a variable for an interface using the `WithEvents` keyword tells the code that this is an interface that has event triggers. Several interfaces in SOLIDWORKS support events, each of which has its own list that are specific to the interface. The interfaces that support events are listed here for reference.

IAssemblyDoc
IDrawingDoc
IFeatMgrView
IModelView
IMotionStudy
IMouse
IPartDoc
ISldWorks
ISWPropertySheet
ITaskpaneView

Within the SwAddin class, several procedures are involved in connecting event handling procedures to the SOLIDWORKS application, `iSwApp`, and all open documents.

First you will find a call to `AttachEventHandlers()` in `ConnectToSw` which runs when the add-in is turned on. Right-click on `AttachEventHandlers()` and select Go To Definition to find the procedure in its code window.

Within AttachEventHandlers there is a call to `AttachSWEvents()` which adds handler procedures for several ISldWorks application events, and `AttachEventsToAllDocuments()` that attaches event handlers, as you might guess from its name, to all open documents.

3. Review the AttachSWEvents procedure code shown below.

```
Sub AttachSWEvents()
  Try
    AddHandler iSwApp.ActiveDocChangeNotify, _
      AddressOf Me.SldWorks_ActiveDocChangeNotify
    AddHandler iSwApp.DocumentLoadNotify2, _
      AddressOf Me.SldWorks_DocumentLoadNotify2
    AddHandler iSwApp.FileNewNotify2, _
      AddressOf Me.SldWorks_FileNewNotify2
    AddHandler iSwApp.ActiveModelDocChangeNotify, _
      AddressOf Me.SldWorks_ActiveModelDocChangeNotify
    AddHandler iSwApp.FileOpenPostNotify, _
      AddressOf Me.SldWorks_FileOpenPostNotify
  Catch e As Exception
    Console.WriteLine(e.Message)
  End Try
End Sub
```

Specific event handling procedures have been created in the class with the following names. Their action should be clear based on their names. Some event handlers have Pre or Post in the names. These indicate events that are triggered before or after a specific action and can be used accordingly.

SldWorks_ActiveDocChangeNotify
SldWorks_DocumentLoadNotify2
SldWorks_FileNewNotify2
SldWorks_ActiveModelDocChangeNotify
SldWorks_FileOpenPostNotify

Review the AttachEventsToAllDocuments procedure. It loops through all open files in the SOLIDWORKS session starting with ISldWorks.GetFirstDocument, wich returns a ModelDoc2 interface. At the end of each loop, it uses ModelDoc2.GetNext to get the next ModelDoc2 interface until the returned value is `Nothing`.

For each open model, AttachModelDocEventHandler connects the associated event handling class to the model based on its type. The respective classes, `PartEventHandler`, `AssemblyEventHandler` and `DrawingEventHandler`, are defined in *EventHandling.vb*.

4. Open *EventHandling.vb* from the Solution Explorer. Select PartEventHandler from the class dropdown list.

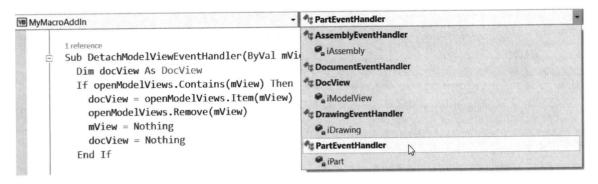

iPart has been declared using the WithEvents keyword as an IPartDoc interface to make its events available.

5. Select iPart from the class dropdown list and then click on the event list to view the possible IPartDoc events.

Parts, assemblies and drawings have many events that can trigger code in your add-in. For our example, we need to make use of the FileSaveAsNotify2 event.

6. Add the new event handler declaration in the AttachEventHandlers function in the PartEventHandler class.

```
Overrides Function AttachEventHandlers() As Boolean
  AddHandler iPart.DestroyNotify, AddressOf Me.PartDoc_DestroyNotify
  AddHandler iPart.NewSelectionNotify, _
    AddressOf Me.PartDoc_NewSelectionNotify
  AddHandler iPart.FileSaveAsNotify2, AddressOf Me.PartDoc_FileSaveAsNotify2
  ConnectModelViews()
End Function
```

7. Ensure the event handler will be disconnected when the add-in is turned off by updating DetachEventHandlers with the RemoveHandler line.

```
Overrides Function DetachEventHandlers() As Boolean
  RemoveHandler iPart.DestroyNotify, AddressOf Me.PartDoc_DestroyNotify
  RemoveHandler iPart.NewSelectionNotify, _
    AddressOf Me.PartDoc_NewSelectionNotify
  RemoveHandler iPart.FileSaveAsNotify2, _
    AddressOf Me.PartDoc_FileSaveAsNotify2
  DisconnectModelViews()

  userAddin.DetachModelEventHandler(iDocument)
End Function
```

The two functions now reference a procedure that doesn't yet exist, `PartDoc_FileSaveAsNotify2`. Visual Studio can also help you generate the expected function syntax.

8. Hover your cursor over the underlined code and select Show Potential Fixes 💡. Then select Generate method to create the function.

```
4 references
Overrides Function AttachEventHandlers() As Boolean
  AddHandler iPart.DestroyNotify, AddressOf Me.PartDoc_DestroyNotify
  AddHandler iPart.NewSelectionNotify, AddressOf Me.PartDoc_NewSelectionNotify
  AddHandler iPart.FileSaveAsNotify2, AddressOf Me.PartDoc_FileSaveAsNotify2
  ConnectModelViews()
End Function
```

💡 ▾

Generate method 'PartEventHandler.PartDoc_FileSaveAsNotify2' ▸

The following new function is added in the class.

```
Private Function PartDoc_FileSaveAsNotify2(FileName As String) As Integer
  Throw New NotImplementedException()
End Function
```

Use this technique when you need event handlers for other IPartDoc, IAssemblyDoc or IDrawingDoc events.

Now that we have an event handler in the add-in, it is time to build the custom property checking class. The following class checks any ModelDoc2 for a custom property. If its value is empty, or it does not exist, the function will return False. If it does exist with a value, the function will return True.

9. Add a new class to MyMacroAddIn named *CustomProps.vb*. Modify the class as shown, adding a PropExists function to check for the custom property.

```
Imports SolidWorks.Interop.sldworks
Public Class CustomProps
  Public Function PropExists(swDoc As ModelDoc2,
    propName As String, Optional config As String = "") As Boolean
```

```
      Dim cpm As CustomPropertyManager
      cpm = swDoc.Extension.CustomPropertyManager(config)
      Dim val As String
      val = cpm.Get(propName)
      If val = "" Then
        Return False
      End If
      Return True
    End Function
End Class
```

The custom property checking code should look familiar. Even though there is a newer version of the CustomPropertyManager.Get method, this older and simpler method does all we need for this application.

Use this new class and function in the event handler.

10. Switch back to *EventHandling.vb* and update the code in the PartDoc_FileSaveAsNotify2 function, removing the NotImplementedException line.

```
Private Function PartDoc_FileSaveAsNotify2(FileName As String) As Integer
  Dim cp As New CustomProps
  Dim propName As String = "Description"
  If cp.PropExists(iPart, propName) Then
    Return 0 'continue the save as action
  Else
    iSwApp.SendMsgToUser2("Please add a " & propName _
      & " property before saving.", _
      swMessageBoxIcon_e.swMbStop, swMessageBoxBtn_e.swMbOk)
    Return 1 'stop the save as action
  End If
End Function
```

The new function uses the newly created CustomProps class and its PropExists function to check for a Description property. By design, if the function returns 0, the save operation continues. If it returns 1, the save operation is cancelled, preventing the user from saving if the Description is blank or does not exist.

11. Start MyMacroAddIn to test the new event handler. Open a new part and try to save. The initial save of a new part is a save as operation. You should see the message shown below.

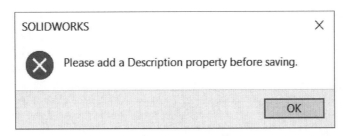

No save dialog is shown after the message is closed because the event handler effectively bypassed the operation. Add a Description property to the part and save again. The save should be successful without the warning. You can turn off the add-in to bypass the save check, confirming that the DetachEventHandlers function was also successful.

12. Stop the Visual Studio debugger ■.

The goal of checking a part for a Description property is successful. Use the same technique to add event handlers to the AssemblyEventHandler and DrawingEventHandler classes to check them as well. You will be able to re-use the code in PartDoc_FileSaveAsNotify2 function in your new AssemblyDoc_FileSaveAsNotify2 and DrawingDoc_FileSaveAsNotify2 functions, just replace the reference to `iPart` with `iAssembly` and `iDrawing` like the example shown below.

```
Private Function AssemblyDoc_FileSaveAsNotify2(FileName As String) As Integer
    Dim cp As New CustomProps
    Dim propName As String = "Description"
    If cp.PropExists(iAssembly, propName) Then
        Return 0 'continue the save as action
    Else
        iSwApp.SendMsgToUser2("Please add a " _
                & propName & " property before saving.",
                swMessageBoxIcon_e.swMbStop, swMessageBoxBtn_e.swMbOk)
        Return 1 'stop the save as action
    End If
End Function
```

Event handlers are invaluable for SOLIDWORKS automation. Whenever you think of checking selections or validating user operations, think of event handlers. There are so many different event handlers, the possibilities are nearly endless.

Installation

There are several options to get your newly built add-in deployed and installed on other computers. The SOLIDWORKS forums, accessible from My.SolidWorks.com, have a few great posts you can review to learn the techniques. This book will focus on a traditional Visual Studio Installer project.

If you don't have the Visual Studio Installer project available in Visual Studio, download and install it. The Visual Studio Installer Project download can be accessed through the following shortened url, or search the web for Visual Studio Installer Projects.

https://bit.ly/3UUUL2V

Adding an Installer to your Solution

Once you have the Visual Studio Installer Project available, use the following steps to create an installation package for your add-in.

1. Add a Setup Project to your add-in solution by selecting File, Add, New Project. Search for "setup" from the search bar. Select Setup Project and tap Next.

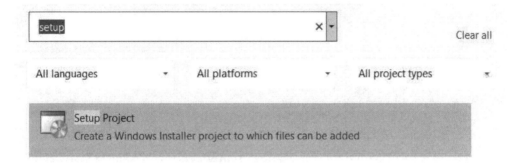

2. Name it *MyMacroAddin Setup*, verify the Location and tap Create.

The Solution Explorer will now show both the MyMacroAddin and MyMacroAddin Setup projects.

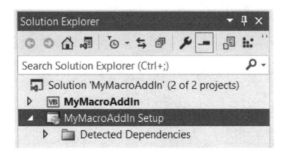

3. From the Properties of *MyMacroAddin Setup*, change the TargetPlatform to x64 since SOLIDWORKS add-ins must be 64bit. Set the Manufacturer to your name or company name. This will set the installation folder path. Change the Product Name and Title to MyMacroAddin.

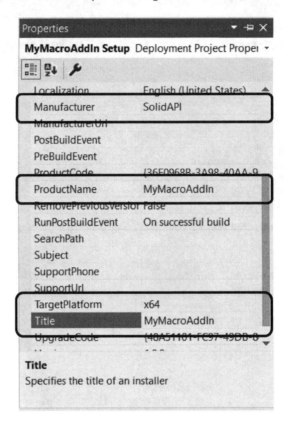

4. With MyMacroAddIn Setup still selected in the Solution Explorer, add the output from MyMacroAddIn to the setup project by selecting Project, Add, Project Output. Select Primary output and Content Files (Ctrl select multiples) from MyMacroAddin and click OK.

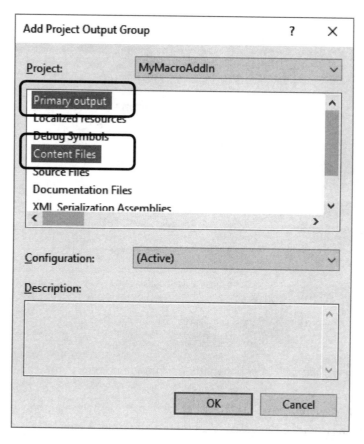

Register the Add-In

Since the add-in dll is a .NET assembly, it must be registered on each client computer just as it was when you were debugging on your own system. .NET assemblies use Regasm.exe to do the registration. This takes an extra step in both the MyMacroAddIn application and the setup project.

The first step is to add code to MyMacroAddIn that will be triggered by the installer and will register the application.

5. Add an Installer Class to MyMacroAddIn by selecting the project in the Solution Explorer, then Project, Add New Item. Select Installer Class from the General category. Leave the name of Installer1.vb and click Add.

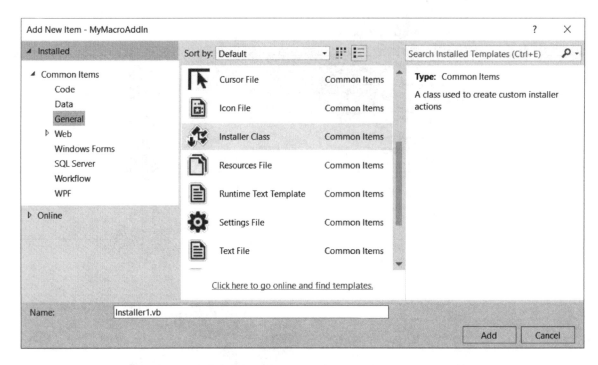

6. Right-click on *Installer1.vb* from the Solution Explorer and select View code to open its code window.

7. Select Installer1 Events from the class list at the top of the page and then AfterInstall and BeforeUninstall from the events list to add these two event handlers.

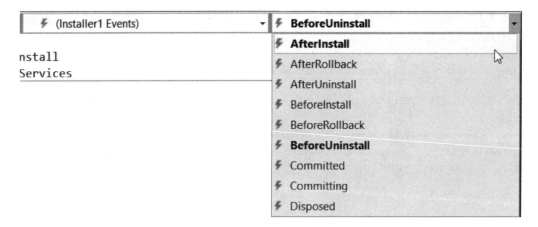

8. Add the following code to the new procedures to register and unregister your add-in whenever these events are triggered.

```
Private Sub Installer1_AfterInstall(sender As Object, _
  e As InstallEventArgs) Handles Me.AfterInstall
  Dim regsrv As New RegistrationServices
  regsrv.RegisterAssembly(MyBase.GetType().Assembly, _
    AssemblyRegistrationFlags.SetCodeBase)
End Sub
```

```
Private Sub Installer1_BeforeUninstall(sender As Object, _
  e As InstallEventArgs) Handles Me.BeforeUninstall
  Dim regsrv As New RegistrationServices
  regsrv.UnregisterAssembly(MyBase.GetType().Assembly)
End Sub
```

The RegistrationServices class from System.Runtime.InteropServices includes the RegisterAssembly and UnregisterAssembly functions. Both functions are passed an instance of the MyMacroAddIn assembly.

9. Add the Imports statement for System.Runtime.InteropServices either manually, or by hovering over the underlined text and using "Show potential fixes" (Ctrl+.).

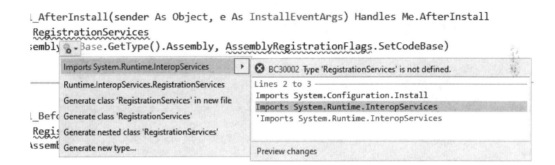

The last step is to trigger these new functions through custom events in the installer project.

10. Right-click on MyMacroAddIn Setup in the Solution Explorer and select View, Custom Actions.

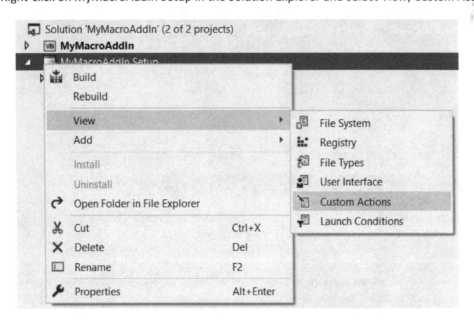

11. In the Custom Actions list, right-click on Install and select Add Custom Action. Open the Application Folder and select Primary output from MyMacroAddIn (Active) and click OK.

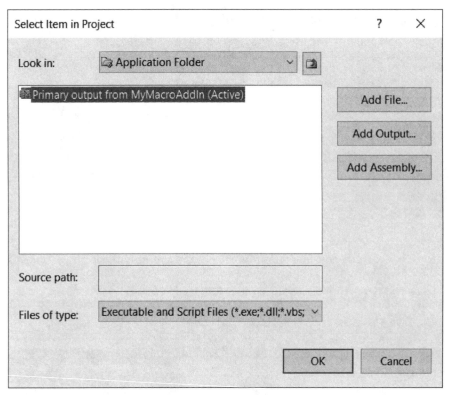

12. Repeat the same process for Uninstall.

After adding the two custom actions, you should see the following.

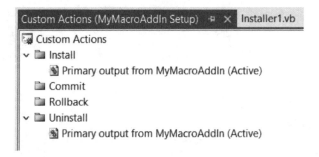

Compile and Build the Installer

The project is now ready to be compiled for distribution.

13. Build (compile) MyMacroAddin by right-clicking on it in the Solution Explorer and selecting Build. Also build MyMacroAddin Setup by right-clicking on it in the Solution Explorer and selecting Build.

If you run into errors while compiling, make sure SOLIDWORKS is not running, either visibly or in the background by checking the Windows Task Manager.

14. Browse to the newly created installation files by right-clicking on MyMacroAddin Setup in the Solution Explorer and selecting Open Folder in File Explorer. Open the *Debug* folder to view the files.

 MyMacroAddin Setup.msi
Windows Installer Package
712 KB

 setup.exe
Setup
14.0.23107.0

15. Copy the installation files to another computer and test the installation by running *setup.exe*.

Conclusion

You now have a functional add-in that references previously written macro code. Use the Visual Studio add-in template to explore the possibilities for deployment and installation.

C# Example

For a full C# example, refer to *MyMacroAddin.sln* in the Part2 folder.

```
//Event Handlers

public int OnFileSaveAs(string filename)
{
    CustomProps cp = new CustomProps();
    string propName = "Description";
    if (cp.PropExists((ModelDoc2)doc, propName))
    {
        return 0;  //continue the save action
    }
    else
    {
        iSwApp.SendMsgToUser2("Please add " + propName
            + " property before saving.",
            (int)swMessageBoxIcon_e.swMbStop, (int)swMessageBoxBtn_e.swMbOk);
        return 1; //stop the save action
    }
}

class CustomProps
```

```
{
    public bool PropExists(ModelDoc2 swDoc,
        string propName, string config = "")
    {
        CustomPropertyManager cpm;
        cpm = swDoc.Extension.CustomPropertyManager[config];
        string val = cpm.Get(propName);
        if (val == "")
        {
            return false;
        }
        return true;
    }
}
```

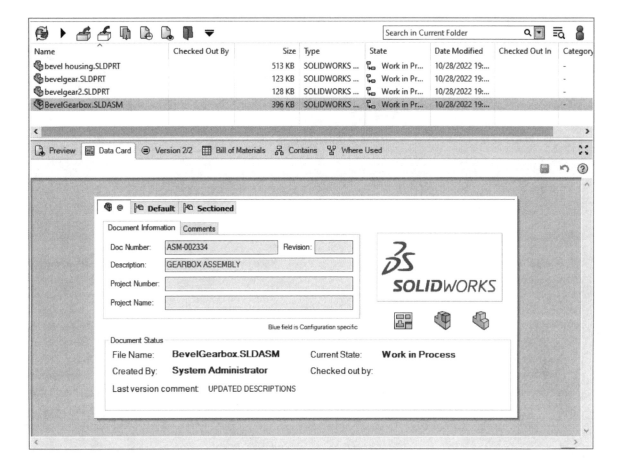

- **PDM Professional API**

- **PDM Professional Interfaces**

- **Basic File Methods and Properties**

Introduction

This chapter will focus on the SOLIDWORKS PDM Professional API fundamentals. The next chapters will focus on variables and bills of materials.

SOLIDWORKS PDM Professional Type Library

Since we will be using a SOLIDWORKS macro to communicate with another application, we will need to add a reference to its API library. Its namespace is EPDM.Interop.epdm. EPDM is an old acronym from the name Enterprise PDM. There is no macro recording for PDM Professional so we will start with a new macro.

1. Start a new macro in SOLIDWORKS and save it as *PDMConnection.vbproj*.

2. Add a reference to the PDM Professional Type Library by selecting Project, Add Reference. Select the Browse tab. Browse to the PDM installation directory, typically *C:\Program Files \SOLIDWORKS PDM* and select *EPDM.Interop.epdm.dll*.

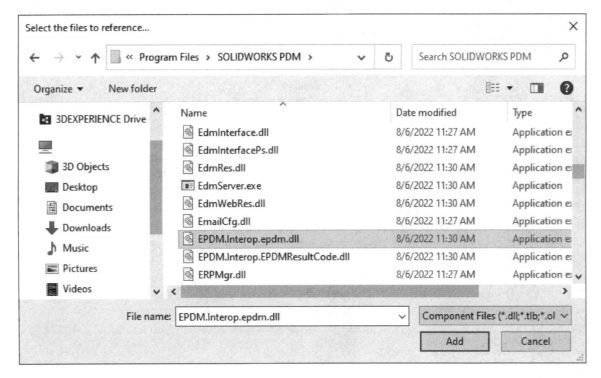

PDM Professional API Help

The PDM Professional API help documentation is included in the SOLIDWORKS API Help like many of the other applications from SOLIDWORKS.

The Getting Started page under the SOLIDWORKS PDM Professional API Help chapter provides an overview of common tasks and explains that you can create standalone applications or add-ins. SOLIDWORKS macros are treated as stand-alone applications in relation to the API help. Add-ins require additional code that will not be covered in this chapter.

3. Open the API Help from the SOLIDWORKS Help menu and browse to the SOLIDWORKS PDM Professional API Help topic under the Contents tab. Click on the Getting Started page. View the basic instructions for starting a stand-alone application by clicking on the hyperlink in the Getting Started page.

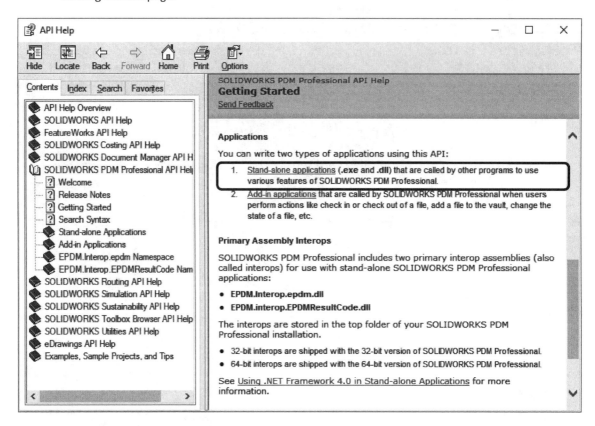

You can skip over the first paragraph referencing creating a Windows Forms Application. We have essentially done the same startup by creating a new macro in SOLIDWORKS. We have also already added the reference to the type library.

The PDM Professional API help is full of sample code in common languages. The sample code on the help page is intended to be added behind a button on a form, but we will copy it into the main procedure in the new macro. For now, the error handling section will be ignored and some errors will need to be corrected.

Logging In

4. Copy the following code from the help Getting Started page sample code into the main procedure. Copy all code from the first commented line to just before Exit Sub. The ErrHandler code section can be ignored. The following shows the main procedure code after pasting it into your macro.

```
Public Sub main()
    'Create a file vault interface and log in to a vault
    Dim vault As IEdmVault5 = New EdmVault5
    vault.LoginAuto("MyVaultName", Me.Handle.ToInt32)
```

```
'Get the vault's root folder interface
Dim message As String
Dim file As IEdmFile5
Dim folder As IEdmFolder5
folder = vault.RootFolder

'Get position of first file in the root folder
Dim pos As IEdmPos5
pos = folder.GetFirstFilePosition
If pos.IsNull Then
    message = "The root folder of your vault does not contain any files."
Else
    message = "The root folder of your vault contains these files:" + vbLf
End If

'For all files in the root folder, append the name to the message
While Not pos.IsNull
    file = folder.GetNextFile(pos)
    message = message + file.Name + vbLf
End While

'Show the names of all files in the root folder
MsgBox(message)
End Sub
```

Fixing Errors

After pasting in the code, you will see several errors that need to be corrected. This code was supposed to be pasted into a Windows application form rather than a macro. We also skipped one more critical step.

5. Point your cursor to IEdmVault5 and click on the recommendations icon to view the error details.

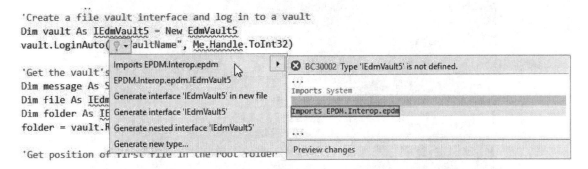

The error "Type 'IEdmVault5' is not defined" means that Visual Studio cannot find that interface in any of the imported namespaces. The error recommends importing the EPDM.Interop.epdm namespace where it found the definition of IEdmVault5.

6. Add the Imports statement for EPDM.Interop.epdm namespace by clicking the recommendation or typing in the statement.

```
Imports SolidWorks.Interop.sldworks
Imports SolidWorks.Interop.swconst
Imports System
Imports EPDM.Interop.epdm
```

7. Fix the final error by changing the LoginAuto code as shown below. This change will be explained later.

```
Public Sub main()
  'Create a file vault interface and log in to a vault
  Dim vault As IEdmVault5 = New EdmVault5
  vault.LoginAuto("MyVaultName", 0)
  ...
```

Now that the errors are corrected, we will start to examine the first few calls that make the connection to PDM.

EdmVault5

The EdmVault5 interface is much like the ISldWorks interface. It is the root interface to all other interfaces in the PDM Professional API. However, unlike the SOLIDWORKS application, you must connect to the EdmVault5 interface by declaring it explicitly and then calling its New method.

EdmVault5.LoginAuto

The LoginAuto method of EdmVault5 is essentially the same as a user opening the local vault view folder. The default login operation occurs. If automatic login is enabled, the user does not see any login screen and the macro will have a valid vault connection. Otherwise, the user will see the standard PDM Professional login screen.

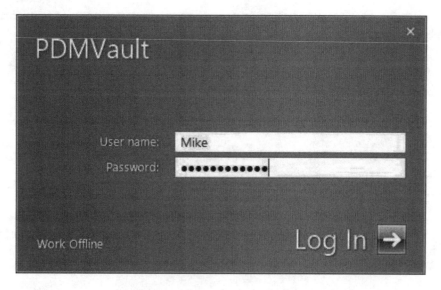

There are two arguments required for this method. The first is the name of the local vault view as a string, not including the path to the view. The second is an integer that references the calling applications window handle. Rather than discussing window handles in this book, using a value of 0 indicates the macro will not have a tie to a calling application. There are minor disadvantages to making this assumption. The PDM Professional API may send back messages or dialogs that may not appear on top of the calling application.

8. Edit the string value in the LoginAuto line and enter a valid local vault view name. The example below will connect to a vault named *PDMPro*.

```
Public Sub main()
    'Create a file vault interface
    'and log in to a vault.
    Dim vault As IEdmVault5 = New EdmVault5
    vault.LoginAuto("PDMPro", 0)

    ...
```

Debug

At this point the macro should be ready for a test run. There are a couple common problems that might occur at this point.

You will get the following error message on the LoginAuto line if either a) there is no local view on your computer or b) you have entered the wrong name string from step 8 above. This could also indicate a network problem as the error message indicates.

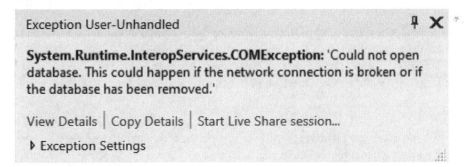

If the macro runs successfully, it will report back any root level files in a message. If there are no root level files, it will tell you.

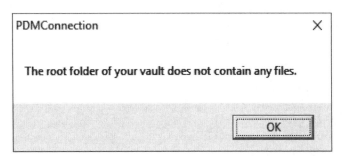

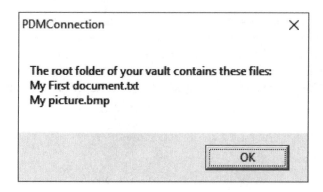

The list of files shown in the second message will depend on the files you have in your vault at the root level. The first message is a little more common since many users keep files in sub folders in their vault.

9. Dismiss the message box from your test run and stop the macro.

PDM Professional Interfaces

There are several other PDM Professional API methods and interfaces introduced by this macro that deserve some discussion.

IEdmObject5

Although we do not see a direct reference to this PDM Professional interface, the IEdmObject5 interface is the basis for most of its interfaces. This is an example of the programming concept of inheritance. Any class or interface can be created by first inheriting another base class or interface. It is essentially a programming shortcut to create a base class that contains all of the typical requirements of its children.

For example, in PDM Professional, both folders and files have a name, a database ID, and are related to a specific vault. These are all properties of their base class IEdmObject5. When you are reviewing properties of interfaces in the PDM Professional API help, do not forget that they have a name and ID property available if they inherit from IEdmObject5. The API help will mention if they do.

IEdmFolder5

Beyond the vault interface itself, the first interface used is the IEdmFolder5 interface. As you might guess, this is the interface to a folder in the vault. There are a few ways to access an instance of the IEdmFolder5 interface depending on what you are trying to get. The code example gets the root folder from the IEdmVault5 property RootFolder.

10. Review some of the other ways to get a folder in the vault by typing IEdmFolder5 into the Index in the API help. Select IEdmFolder5 Interface in the list and review the Accessors. The list gives you all of the different ways to access this interface. The IEdmVault5 methods and properties are listed here.

   IEdmVault5.BrowseForFolder
   IEdmVault5.GetFileFromPath
   IEdmVault5.GetFolderFromPath
   IEdmVault5.GetObject
   IEdmVault5.RootFolder

As you search through the API help, you might also notice many variations of IEdmFolder. These variations from IEdmFolder5 to IEdmFolder11 each inherit from one another. As new functionality was added through the API, new versions of IEdmFolder were created that inherit from the previous.

IEDMVault5.GetFolderFromPath

One of the most helpful methods for getting a folder is IEdmVault5.GetFolderFromPath, though it is not used in this example. You simply pass one argument, a string representing the full path to the folder relative to the local view. For example, if the local view of the vault in this example were *C:\PDMPro* and we wanted to get a top level folder named *Projects*, you could use the following code to get the Projects folder.

```
Dim folder As IEdmFolder5
Dim folderName As String = "C:\PDMPro\Projects"
folder = vault.GetFolderFromPath(folderName)
```

Traversing Files in a Folder

The PDM Professional API uses a unique technique for traversing groups of objects such as files in a folder or results from a search. Its mechanism is similar to the SOLIDWORKS API structure for traversing features in the FeatureManager. You use a kind of GetFirst, then GetNext method.

IEdmFile5

Before getting into the process, the IEdmFile5 interface should be introduced. You can move, copy and rename as well as access card variables, where-used and contains lists from IEdmFile5. This interface has many inherited interfaces of which IEdmFile5 is the basis.

11. Type IEdmFile into the API Help Index. Double-click IEdmFile5, then select the Members link at the top of the page to view its methods and properties.

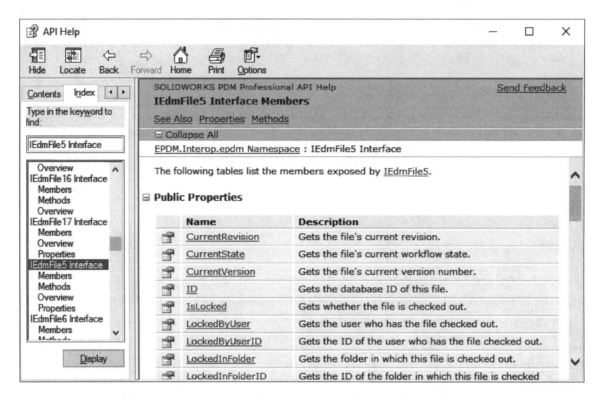

In addition to IEdmFile5, there are several other IEdmFile interfaces with IEdmFile17 as the latest. These newer versions of the IEdmFile5 interface all inherit from the one before. Like the way IEdmFile5 inherits from IEdmObject5, they share the underlying structure of their parent. The API help only documents the added methods and properties for the child versions of the interface. Unless you need access to a newer method or property, you can stick to using a parent interface (earlier version) as we have in this example.

Take some time to review the methods and properties. Some of these will be used later in this chapter to expand the example. For now, we will return to traversing the files in the root folder.

IEdmPos5

The first step in traversing files in a folder is to get the IEdmPos5 interface. Think of it as the index of an array. It is a counter of sorts. It is a required argument to get the actual file interface when traversing, and the next index is returned.

The only property of IEdmPos5 is IsNull. IsNull is a Boolean value that verifies there is a file at that position. The following section of code gets the IEdmPos5 interface from the IEdmFolder5 interface of the root folder. If the IEdmPos5 interface, represented in the code by the variable name pos, has an IsNull value of True, then there are no files in the root folder of the vault. If the IsNull property of pos is False, then there are files in the folder. However, we do not know how many.

```
'Get position of first file in the root folder
Dim pos As IEdmPos5
pos = folder.GetFirstFilePosition
If pos.IsNull Then
    message = "The root folder of your vault does not contain any files."
```

```
Else
    message = "The root folder of your vault contains these files:" + vbLf
End If
```

The String variable `message` is used to build a message for later use in a Message Box dialog. Notice how the Visual Basic constant `vbLf` is used at the end of the message if there are files in the folder. As a reminder, this is a line feed character. `vbCr` or `vbCrLf` could have been substituted to represent a return character in the message box to start a new line in the string as well.

IEdmFolder5.GetFirstFilePosition

From an IEdmFolder5 interface, call GetFirstFilePosition to return the IEdmPos5 interface for possible files in the folder.

IEdmFolder5.GetNextFile

The While loop checks if there are any files at the current position. If there are, the IEdmFile5 interface is returned for that file by using the IEdmFolder5.GetNextFile method. It is important to point out here that the GetNextFile method also changes the `pos` variable. It essentially indexes the IEdmPos5 interface to the next file position, which may be no file at all if we are at the end of the group or list.

```
'For all files in the root folder, append the name to the message
While Not pos.IsNull
    file = folder.GetNextFile(pos)
    message = message + file.Name + vbLf
End While
```

IEdmFile5.Name

If there is a file in the next position, thinking of the group of files as a list, the Name property of IEdmFile5, actually the underlying IEdmObject5 interface, named `file` in this code, is appended to the message string. I hope you followed that! By the way, the Name property of a file does not include its path. It does include its extension.

More File Properties

You now have a successful connection to a PDM Professional vault view and have reported back some file names. The next section will expand the macro, adding the ability to browse for a file and return common properties such as whether the file is checked out, its current revision and workflow state.

Automatic Vault Login Method

In some cases your code may need to run using a different user's permissions. The original code uses the LoginAuto method which relies on the local user to enter an appropriate user name and password. All transactions then run based on their vault permissions.

12. Edit the login section of the macro to use an alternative login method. Make sure to change the `username`, `password` and `vaultname` variable values for your PDM Professional system.

```
Public Sub main()
    'Create a file vault interface and log in to a vault
    Dim vault As IEdmVault5 = New EdmVault5
```

```
'vault.LoginAuto("PDMPro", 0)
'set username, password and vault name and login
Dim username As String = "Admin"
Dim password As String = "password"
Dim vaultname As String = "PDMPro"
vault.Login(username, password, vaultname)
```
...

IEdmVault5.Login

The Login method of IEdmVault5 allows you to send a user name and password in your code. Use this method if you need to run under rights that are different from the users expected to run the macro. Don't forget to modify the username and password to an actual user and password from your vault along with the vault name.

The next step will be to add the ability for the user to browse to one or more files in the vault.

13. Add the following code at the end of the procedure just after the MsgBox(message) line.

...
```
'Show the names of all files in the root folder
MsgBox(message)

'Let the user select one or more files that are
'in the file vault to which we are logged in.
Dim PathList As EdmStrLst5
PathList = vault.BrowseForFile(0, , , , , , _
"Select a file to show its information")
```

End Sub

IEdmStrLst5

The BrowseForFile method returns the IEdmStrLst5 interface. You can think of this interface essentially like an array of string values. However, similar to traversing files in a folder, values in the list can only be accessed using the IEdmPos5 interface. The example will show how it is done.

IEdmVault5.BrowseForFile

The BrowseForFile method of the IEdmVault5 interface is very similar to a standard Windows open file dialog. However, it is limited to only selecting files in the vault. Its structure is described below.

There are seven arguments passed into the method described below.

Many methods in the PDM Professional API have optional values for some arguments that allow you to simplify your code. In cases where you don't care about that specific argument, you can leave it empty as shown in the example code by entering a sequence of commas. For example, if you were not concerned about any of the optional values for the BrowseForFile dialog in the code above, you could simplify it with the following line.

```
PathList = vault.BrowseForFile(0)
```

Value = IEdmVault5.BrowseForFile(hParentWnd, IEdmBrowseFlags, bsFilter, bsDefaultExtension, bsDefaultFileName, bsDefaultFolder, bsCaption)

- **hParentWnd** is a Long value representing the calling application's window handle. For macros we are simply passing 0 to represent no application window handle. This may cause dialog boxes to fall behind the calling application as a result.

- **IEdmBrowseFlags** is an enumeration and can be a combination of the following values. You can add them together with "+". Start typing EdmBrowseFlags followed by a period and you will see the following list.

⊞ EdmBws_ForOpen
⊞ EdmBws_ForSave
⊞ EdmBws_Help
⊞ EdmBws_PermitExternalFiles
⊞ EdmBws_PermitLocalFiles
⊞ EdmBws_PermitMultipleSel
⊞ EdmBws_PermitVaultFiles

- **bsFilter** is a string to allow selection of different file extensions as you would typically see in an open file dialog. For example, if you wanted to allow selection of SOLIDWORKS files or All Files, you would pass the following string. Unlike a Windows File Open dialog filter, the string for this dialog must be closed with a double pipe (| |).

```
"SOLIDWORKS Files (*.SLD*)|*.SLD*|All Files (*.*)|*.*||"
```

- **bsDefaultExtension** would only be used if this were a Save dialog. The default empty string is typically used.

- **bsDefaultFileName** is a string and would be used to pre-select a specific file by default.

- **bsDefaultFolder** is a string value representing the default folder that is active when the dialog is shown.

- **bsCaption** is a string used to change the caption at the top of the dialog.

14. Run the macro with the changes that were applied in previous steps. After dismissing the initial message, a dialog similar to the following should be displayed.

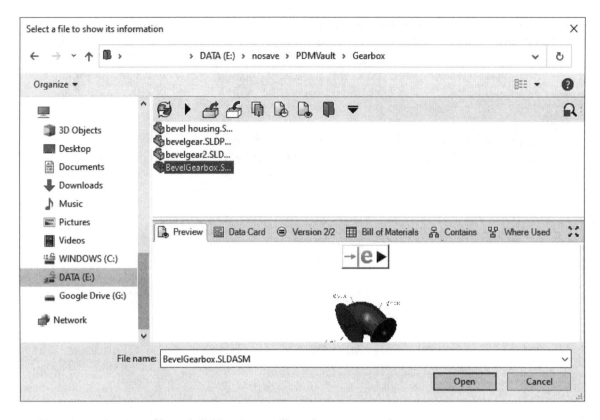

At this point, selecting a file and clicking Open will not have any result.

15. Add the following code before the End Sub statement to handle the list of selected files.

```
...

'Let the user select one or more files that must be
'part of the file vault to which we are logged in.
Dim PathList As EdmStrLst5
PathList = eVault.BrowseForFile(0, , , , , , _
"Select a file to show its information")

'Check if the user pressed Cancel
If PathList Is Nothing Then
    'do nothing - user pressed cancel
Else
    'Display a message box with
    'the paths of all selected files.
    message = "You selected the following files:" + vbLf
    pos = PathList.GetHeadPosition
    While Not pos.IsNull
        Dim filePath As String
        filePath = PathList.GetNext(pos)

        'connect to the file object
```

```
        Dim eFile As IEdmFile5
        eFile = vault.GetFileFromPath(filePath)
        message = filePath
        message = message & vbLf & "State: " & eFile.CurrentState.Name
        message = message & vbLf & "Is Checked Out: " _
            & eFile.IsLocked.ToString
        If eFile.IsLocked Then
            message = message & vbLf & "Is Checked Out By: " _
                & eFile.LockedByUser.Name
        End If
        'show information about the selected file
        MsgBox(message)
    End While
End If

End Sub
```

If the user cancels the dialog or does not select a file before clicking Open, the value returned by BrowseForFile will be Nothing. An If, Then statement is used to check the results of the PathList variable. If the user selected files, the code uses a While loop to get the IEdmFile5 interface for each and evaluates some properties.

Note: if you want to allow the user to select more than one file in the BrowseForFile dialog, you must change the first argument in BrowseForFile to 2 or IEdmBrowseFlags.EdmBws_PermitMultipleSel.

IEdmStrLst5.GetHeadPosition

The first step in looping through an IEdmStrLst5 interface is to use its GetHeadPosition property. Just like traversing files in a folder, IEdmStrLst5 uses the IEdmPos5 interface to define the position in the list. GetHeadPosition returns the position of the first selected file. Though it may seem awkward at first, mastering the use of IEdmPos5 for groups in the PDM Professional API is necessary.

If you need another explanation, try this one. GetHeadPosition gets the location of the first string in an IEdmStrLst5 interface, not the string itself. If you had a string array of five values and used this method to get the position of the first, it would return 1. When you call its GetNext method (described below), the position would increment to 2 while the method returns the string at that position. Remember that this is an analogy. IEdmPos5 is an interface rather than an integer, so you cannot actually get a number out of it.

IEdmStrLst5.GetNext

The GetNext method will return the actual string value from the list. In this example, the first use of GetNext will return the first filename string from the list. Though this may seem counter-intuitive, it gets the string value at the current IEdmPos5 location and also increments the IEdmPos5 counter so that the next use of GetNext returns the next string value. There is a subtle detail in the description of the GetNext method in the PDM Professional API help. If you look up IEdmStrLst5.GetNext, you will notice that in the description of the IEdmPos5 argument, it states that it will be forwarded one position on each call to GetNext.

This process of getting the next position and the current string continues until there are no values left and IEdmPos5 is set to Nothing. The While loop ends when the IEdmPos5 interface is Nothing.

IEdmVault5.GetFileFromPath

Once we have the first file path string from the IEdmStrLst5 interface returned by the BrowseForFile method, we can get to an IEdmFile5 interface. The GetFileFromPath method makes it easy.

value = IEdmVault5.GetFileFromPath(bsFilePath, ppoRetParentFolder)

- **bsFilePath** is a string representing the full file path in the local vault view. For example, "*C:\PDMPro\Projects\MyPart.sldprt*" where "*C:\PDMPro*" is the path of the local view.

- **ppoRetParentFolder** is an optional variable that will return the IEdmFolder5 interface of the folder the file is in. Since it is optional, you can pass only the file path argument if you do not need the folder. The IEdmFolder5 interface of the parent folder is helpful when you need both the file and the folder interface for additional steps such as checking out a file (the LockFile method of the IEdmFile5 interface).

- **Value** returns an IEdmFile5 interface.

The message string is finally populated with information pulled directly from the IEdmFile5 interface. The values used here are the file's current workflow state name, whether it is checked out and its owner if it is checked out. The first relates to its workflow state.

IEdmFile5.CurrentState

This property of a file returns an IEdmState5 interface. IEdmState5 inherits from IEdmObject5. If you recall from our earlier discussion, all IEdmObject5 interfaces have a Name property. This literally returns the name of the current workflow state of the file. The code sample does not store the IEdmState5 interface in a variable, but goes directly for its Name property.

If you need to get to the workflow the state belongs to, you will need to first get to the IEdmState6 interface. IEdmState6 inherits from IEdmSate5, so you need to declare another variable as IEdmState6 and equate it to your IEdmState5 variable as shown below. This code assumes you have a variable named eFile that is already connected to an IEdmFile5 interface.

```
Dim State5 As IEdmState5
Dim State6 As IEdmState6
State5 = eFile.CurrentState
State6 = State5
```

IEdmVault5.GetObject

The code in the example does not use the GetObject method, but since I brought up the idea of getting a workflow interface, I should explain it. You can only get the workflow's ID from an IEdmState6 interface; you cannot get the workflow interface itself. The ID is an integer returned by the IEdmState6.WorkflowID property. Any time you have an object's ID, you can get its interface using the IEdmVault5.GetObject method outlined below.

Value = IEdmVault5.GetObject(eType, lObjectID)

- **eType** is from a PDM Professional constant enumeration named EdmObjectType and defines what kind of an object is to be returned. The following image shows the IntelliSense list that appears when calling this method. The object names are generally self-explanatory and should be selected from the list.

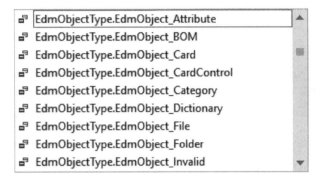

- **lObjectID** is an integer representing the object's database ID. Every object in a vault has a unique ID number.

- **Value** is the returned interface to the desired object type.

IEdmFile5.IsLocked

The IsLocked property is the way to determine if the file is checked out. The "locked" term came from the early days of PDM Professional when they referred to files as being locked and unlocked rather than checked out and checked in. The API still uses that older terminology. IsLocked returns a Boolean True or False value, so it is being converted to a string to add to the message.

IEdmUser5

The last part of the message is added only if the selected file is checked out. A call to the LockedByUser property of IEdmFile5 returns an IEdmUser5 interface. Again, this interface inherits from IEdmObject5, so it also has a Name property. The example adds this Name string to the message.

If needed, IEdmUser5 has a method to send an email message to the user as well as a property to check if that user is currently logged in.

Debug

16. Run the edited macro, browse to a file and click Open. Correct any errors that occur.

It may be worth using breakpoints and Step Into (F11) to walk through your macro line-by-line. Review the values of each step and watch how the variable message changes by adding it to the Watch list.

After any required debugging, you will see a message reporting the details of the selected file.

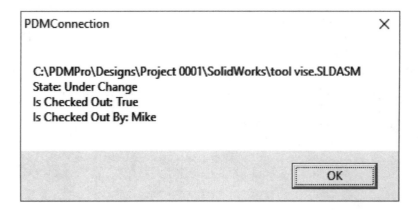

Conclusion

The key to using the PDM Professional API effectively is to first establish a connection to the application and then connect to its major interfaces. Now you know how to connect. It's time to read and write data. More to come in the following chapters.

C# Example

```
//console application code
static void Main()
{
  // Create a file vault interface and log in to a vault
  IEdmVault5 vault = new EdmVault5();
  //vault.LoginAuto("PDMPro", 0);
  string username = "Admin";
  string password = "admin";
  string vaultname = "PDMPro";
  vault.Login(username, password, vaultname);

  // Get the vault's root folder interface
  string message;
  IEdmFile5 file;
  IEdmFolder5 folder;
  folder = vault.RootFolder;

  // Get position of first file in the root folder
  IEdmPos5 pos;
  pos = folder.GetFirstFilePosition();
  if (pos.IsNull)
  {
    message = "The root folder of your vault does not contain any files.";
  }
  else
  {
    message = "The root folder of your vault contains these files:\n";
  }

  // For all files in the root folder, append the name to the message
```

```
while (!pos.IsNull)
{
  file = folder.GetNextFile(pos);
  message = message + file.Name + "\n";
}

// Show the names of all files in the root folder
Console.WriteLine(message);

//Let the user select one or more files that are
//n the file vault to which we are logged in.
EdmStrLst5 PathList;
PathList = vault.BrowseForFile(hParentWnd:0,
  bsCaption:"Select a file to show its information");
//Check if the user pressed Cancel
if (PathList == null)
{
  //do nothing - user pressed cancel

}
else
{
  //Display a message with the paths of all selected files
  message = "You selected the following files:\n";
  pos = PathList.GetHeadPosition();
  while (!pos.IsNull)
  {
    string filePath;
    filePath = PathList.GetNext(pos);
    //connect to the file
    IEdmFile5 eFile;
    IEdmFolder5 eFolder;
    eFile = vault.GetFileFromPath(filePath, out eFolder);
    message += filePath;
    message += "\nState: " + eFile.CurrentState.Name;
    message += "\nIs Checked Out: " + eFile.IsLocked.ToString();
    if (eFile.IsLocked)
    {
      message += "\nIs Checked Out By: " + eFile.LockedByUser.Name;
    }
    //show information about the selected file
    Console.WriteLine(message);
  }
}

Console.WriteLine("Press any key to close.");
Console.Read();
}
```

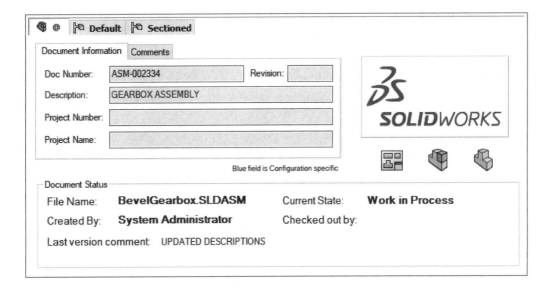

- **Reading Card Variables**

- **File Check Out**

- **Changing Card Variables**

- **File Check In**

- **Add Files to the Vault**

Introduction

Whether you call them custom properties, metadata, or card variables, reading and writing information about your files is common. This chapter is dedicated to the PDM Professional API methods involved in reading and writing card variables. Check out and check in operations fit nicely into the discussion. As you might already know, if you want to change a card variable, you must have the file checked out. Preserving those changes requires a check in. This example will only be an outline of the methods involved. I will leave a full application implementation up to you and your creativity.

Reading Variables

The first step will be to create a macro that reads a variable named Description from a SOLIDWORKS part file card. Since our last chapter built a nice macro for browsing for a file, we will start by reusing its code.

1. Create a new macro in SOLIDWORKS and save it as *PDMVariables.vbproj*.

2. Add a reference to the PDM Professional library. Select Project, Add Reference... Select the Browse tab and browse to EPDM.Interop.epdm.dll from the SOLIDWORKS PDM installation directory.

3. Add the following code to login and browse for a file. Include the Imports statement for EPDM.Interop.epdm. Make sure to enter a valid vault name. Most of this code can be copied from the last macro example.

   Note: if you prefer, you can edit your PDMConnection macro and continue rather than starting a new macro. Remove or comment out the section that reports on root folder files.

```
Imports SolidWorks.Interop.sldworks
Imports SolidWorks.Interop.swconst
Imports System
Imports EPDM.Interop.epdm

Partial Class SolidWorksMacro

  Public Sub main()
    'Create a file vault interface and log in to a vault
    Dim vault As EdmVault5 = New EdmVault5
    Dim vaultName As String = "PDMPro"
    vault.LoginAuto(vaultName, 0)

    'Let the user select one or more files that are
    'in the file vault to which we are logged in
    Dim PathList As EdmStrLst5
    PathList = vault.BrowseForFile(0, , , , , , _
    "Select a file to show its information")

    'Check if the user pressed Cancel
    If PathList Is Nothing Then
      'do nothing - user pressed cancel
```

```
      Else
        Dim message As String
        Dim pos As IEdmPos5
        'Display a message box with
        'the paths of all selected files.
        message = "You selected the following files:" + vbLf
        pos = PathList.GetHeadPosition
        While Not pos.IsNull
          Dim filePath As String
          filePath = PathList.GetNext(pos)

          'connect to the file object
          Dim eFile As IEdmFile5
          eFile = vault.GetFileFromPath(filePath)
          message = filePath
          message = message & vbLf & "State: " & eFile.CurrentState.Name
          message = message & vbLf & "Is Checked Out: " _
            & eFile.IsLocked.ToString
          If eFile.IsLocked Then
            message = message & vbLf & "Is Checked Out By: " _
            & eFile.LockedByUser.Name
          End If

          'show information about the selected file
          MsgBox (message)
        End While

      End If
End Sub
```

4. Edit the code in the main procedure as follows, just before the MsgBox is called.

...

```
      If eFile.IsLocked Then
          message = message & vbLf & "Is Checked Out By: " _
          & eFile.LockedByUser.Name
      End If

      'get the file's EnumeratorVariable interface
      'for working with its card variables
      Dim eVar As IEdmEnumeratorVariable10
      eVar = eFile.GetEnumeratorVariable

      'show information about the selected file
      MsgBox(message)
    End While

  End If
End Sub
```

IEdmEnumeratorVariable10

This interface is the source to getting and setting file and folder card variable values. It can be compared loosely to the SOLIDWORKS ICustomPropertyManager interface. As an additional benefit, this interface can also return a thumbnail image of the file.

From an IEdmFile5 interface you call its GetEnumeratorVariable method to get to the IEdmEnumeratorVariable10 interface. This version of the interface inherits from IEdmEnumeratorVariable5, so it shares all of its parent properties and methods. GetEnumeratorVariable has an optional argument for a string representing the full path to the local file but it is not typically needed. It would only be helpful if you allow users to share files to another folder which creates two files of the same name that are tied to the same database records.

5. Add the following code to get the value of the Description variable, adding it to the message string. The code should be added between the call to GetEnumeratorVariable and the MsgBox call. Make sure to choose a variable name that exists on your local file cards.

```
...
    'get the file's EnumeratorVariable interface
    'for working with its card variables
    Dim eVar As IEdmEnumeratorVariable8
    eVar = eFile.GetEnumeratorVariable

    'get the description
    Dim varValue As String
    eVar.GetVar("Description", "@", varValue)
    message = message & vbLf & "Description: " & varValue

    'show information about the selected file
    MsgBox(message)
End While
```

IEdmEnumeratorVariable5.GetVar

The GetVar method will return the value of a named card variable from a specific configuration tab on the data card. Its structure is described here.

Value = IEdmEnumeratorVariable5.GetVar(bsVarName, bsCfgName, poRetValue)

- **bsVarName** is a string representing the name of the card variable. If you are not sure what variable name is referenced by the card field, click ⑦ and then click the card field. You will be presented with a callout similar to the following.

> **Editbox**
> This control is connected to the variable "Number".
> The variable is of type Text.
>
> The variable is connected to the following attributes:
> CustomProperty : Number (*.slddrw, *.sldasm, *.sldprt)
>
> Changing the value of this field in one configuration will update the value in all configurations.

- **bsCfgName** should be passed a string representing the configuration name. This should match the file card tab name for the desired variable. In this example, "@" is used to represent the file variable value. If you need to access variables from a non-SOLIDWORKS file, pass the empty string "". From the image below, if you wanted to get the variable value from the *Default* configuration, you would pass "Default" as the string.

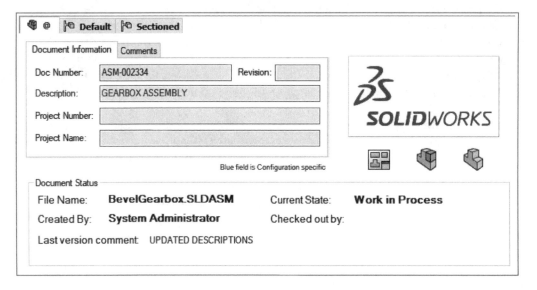

- **poRetValue** should be passed an empty variable that is then set to the return value of the card variable. This approach to providing a returned value in a method argument is not the traditional approach, but is still valid. This argument is defined in the API help as an Object type. The reason is that it may return a string, integer, double or Boolean depending on the type of the card variable. In the example, the variable is expected to be a string, so the empty variable passed to the argument has been declared as a string.

- **Value** will return a Boolean. It will be True only if the method was successful.

Run the macro at this point and review the results. The resulting message should include the value of the "@" tab's Description as long as you have selected a SOLIDWORKS file that has a Description.

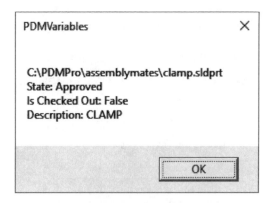

Check Out and Editing Card Variables

Now that you can read card variables, we can move on to editing. When using the PDM Professional API, you have to follow the same rules that apply to a user. For example, to edit a card variable, you must first check out the file. To check out the file, you must be logged in as a user who has those rights. If you lack permissions, or if you perform steps out of order, expect the code to throw errors. Any good application needs good error trapping to avoid errors if any methods or properties might fail. Do not forget to make regular use of Try, Catch blocks to handle possible errors.

For the next step, assume that the file is not checked out. The earlier chapter reviewed how to determine if the file is checked out and which user has it checked out. For a full application, check if the current user has the file checked out. If they do, carry on with the editing operations. But for this example, assume that a check out is required.

IEdmVault5.GetFileFromPath

Before we attempt to check out the file we need to get its parent folder's ID. This is a required argument to the LockFile method (check out) that we will discuss a little later. As mentioned in the previous chapter, GetFileFromPath can return the file's parent folder as an IEdmFolder5 interface. Since IEdmFolder5 inherits from IEdmObject5 (there it is again), we can get the folder ID.

6. Add the following code change to the GetFileFromPath line to also retrieve the parent folder as an IEdmFolder5 interface.

```
...
'connect to the file object
Dim eFile As IEdmFile5
Dim parentFolder As IEdmFolder5
eFile = vault.GetFileFromPath(filePath, parentFolder)
message = filePath
...
```

7. Add the following code in Try, Catch blocks to check out and then check in the file. If an error is caught, send a message to the user describing the error and exit the application.

```
...
    'get the description
    Dim varValue As String
    eVar.GetVar("Description", "@", varValue)
    message = message & vbLf & "Description: " & varValue

    'try to check out the file
    Try
        eFile.LockFile(parentFolder.ID, 0)
    Catch ex As Exception
        MsgBox(ex.Message & vbLf & eFile.Name)
        Exit Sub
    End Try

    'edit the file here
```

```
    'try to check in the file
    Try
        eFile.UnlockFile(0, "API check in")
    Catch ex As Exception
        MsgBox(ex.Message & vbLf & eFile.Name)
        Exit Sub
    End Try

    'show information about the selected file
    MsgBox(message)
End While
```

IEdmFile5.LockFile

A PDM Professional user knows this operation as a check out. But to the API it is known as lock. As mentioned earlier in the book, PDM Professional used to use "lock" in the user interface rather than check out. The LockFile method requires the parent folder's ID as well as the calling application's window handle.

IEdmFile5.LockFile(lParentFolderID, lParentWnd, lEdmLockFlags)

- **lParentFolderID** is an integer representing the file's parent folder's ID. If you do not pass the right ID integer here, the LockFile method will fail.

- **lParentWnd** is an integer representing the calling application's window handle. Since we are using a SOLIDWORKS macro, we are simply passing a 0, meaning the method is not aware of its calling application. This will not be critical for the operation.

- **lEdmLockFlags** is an optional integer. The funny thing is that there is only one option for this method, EdmLock_Simple, and that is the default. In other words, there is no reason to ever worry about this optional value.

The LockFile method may fail for several reasons. The user may lack permission, an add-in may prevent the operation, the file may already be checked out or the file may not be found. This makes it very important to use good error handling to avoid crashing or code failures. The following messages might be caught by the Try block.

Return Codes	Description
S_OK	The method was successfully executed.
E_EDM_PERMISSION_DENIED	The user lacks permission to check out this file.

E_EDM_OPERATION_REFUSED_BY_PLUGIN	One of the installed EdmCmd_PreLock hooks didn't permit the operation.
E_EDM_FILE_IS_LOCKED	The file is already checked out.
E_EDM_FILE_NOT_FOUND	The file wasn't found in the vault (someone probably just deleted it).

For example, if a file is checked out by another user, you would get the following MsgBox based on the current code.

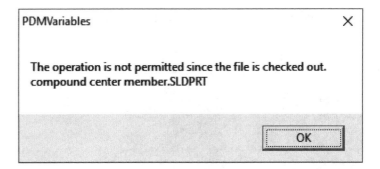

IEdmFile5.UnlockFile

The UnlockFile method checks the file in. If changes have been made to the card or the file, a new version will be created. If nothing has changed, the owner's name will simply be removed.

IEdmFile5.UnlockFile (lParentWnd, bsComment, lEdmUnlockFlags, poIEdmRefCallback)

- **lParentWnd** is the calling application's window handle, represented by an integer. We have seen this in several places and will again use a 0 to represent no window handle.

- **bsComment** is the comment string that displays in the file's history.

- **lEdmUnlockFlags** is an optional integer created from a combination of values from the EdmUnlockFlag enumeration. The default is EdmUnlock_Simple corresponding to the integer 0.

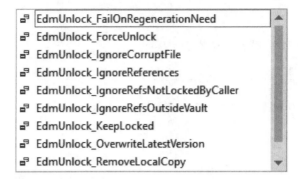

- **polEdmRefCallback** is optional, not typically used and is beyond the scope of this example.

Notice that the code you added passes a simple string for the check in comment and leaves all other arguments as defaults by ignoring them.

8. Add the following code to change the Description card variable for the "@" configuration (file tab).

```
'try to check out the file
Try
    eFile.LockFile(parentFolder.ID, 0)
Catch ex As Exception
    MsgBox(ex.Message & vbLf & eFile.Name)
    Exit Sub
End Try


'reconnect to the file's EnumeratorVariable
'after check out
eVar = eFile.GetEnumeratorVariable

'edit the file here
Try
    eVar.SetVar("Description", "@", "NEW DESCRIPTION")
    'close the file and save (flush) any updates
    eVar.CloseFile(True)
Catch ex As Exception
    MsgBox(ex.Message & vbLf & eFile.Name)
    Exit Sub
End Try


'try to check in the file
Try
    eFile.UnlockFile(0, "API check in")
Catch ex As Exception
    MsgBox(ex.Message & vbLf & eFile.Name)
    Exit Sub
End Try
```

The first step before attempting to edit the card variable is to reconnect to the IEdmEnumeratorVariable8 interface. The first connection was made while the file was checked in. If you do not reconnect to the interface after checking the file out, you may get errors about the file not being checked out by you. The process of reconnecting to the interface is identical to connecting in the first place.

IEdmEnumeratorVariable5.SetVar

The SetVar method is straight forward. You must pass the card variable name, the configuration name and the desired value. Each is a string. If you pass an invalid card variable name in the first argument, the method returns an error. The Try block will catch the error and display the following message.

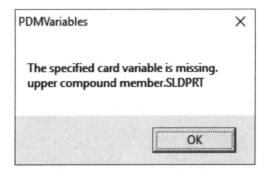

Be aware that you will not get any errors from an invalid configuration name. The method will run, but you will not see any change to the card variables.

IEdmEnumeratorVariable8.CloseFile

As a user, if you edit a field on a card, you must click the save button to store the values in the database. In the same respect, after editing any card variables, you must close the file and flush, or save, the values you have changed. If you skip this step, check in will fail. The argument passed is a Boolean. If it is True, the modified values are saved to the database. False would essentially undo your changes.

Debug

It is time to run the macro and test your new code. Use stepping methods and breakpoints as needed to view the results as the macro runs. The resulting file card will have a new Description variable value and a new version, complete with a new line in the history with the comment "API check in."

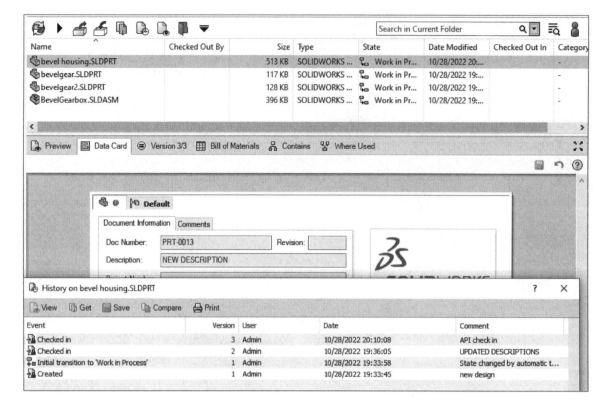

Adding Files to the Vault

My favorite feature of PDM Professional is that it utilizes Windows Explorer as its interface. One of the benefits is simple automation of many standard file operations – with one caveat. If the file is not cached locally, standard file methods will fail because there would be no local file.

Adding files to the vault is unique. You cannot simply copy files using Windows file system API methods. The PDM Professional database has to know about the file being added. The following example explains how to add new files to the vault and check them in.

9. Start a new macro and name it *PDMAddFiles.vbproj*.

10. Add a reference to the SOLIDWORKS PDM Professional Type Library. See the earlier instructions for the procedure if needed. It should now be accessible in the Recent list.

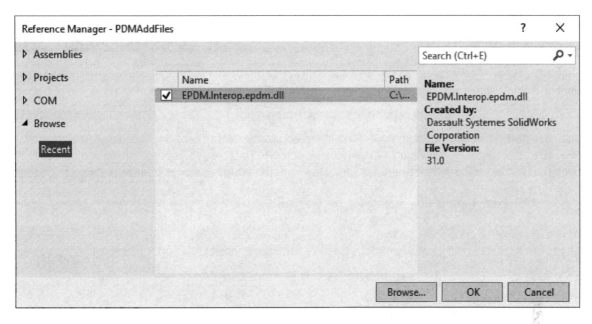

11. Add an `Imports EPDM.Interop.epdm` statement at the top of the code file.

12. Add the following code to make a vault connection and login. Replace *PDMPro* with your local vault view name.

```
Imports SolidWorks.Interop.sldworks
Imports SolidWorks.Interop.swconst
Imports System
Imports EPDM.Interop.epdm

Partial Class SolidWorksMacro

    Public Sub main()
        Dim vault As New EdmVault5
        vault.LoginAuto("PDMPro", 0)
    End Sub
```

13. Add code to allow the user to browse to a destination folder for the file to be added.

```
Public Sub main()
    Dim vault As New EdmVault5
    vault.LoginAuto("PDMPro", 0)

    Dim eFolder As IEdmFolder5
    eFolder = vault.BrowseForFolder(0, "Select the destination folder")
```

EdmVault5.BrowseForFolder

The vault's BrowseForFolder method is similar to the BrowseForFile method. By name, as you might guess, it allows the user to browse for a vault folder. It is very similar to the FolderBrowserDialog method of the Windows.Forms namespace. Like many other PDM Professional methods, the first argument is the calling application's window handle as an integer. Again, we use a 0 in our example. The second argument is a string that displays a caption in the dialog to provide instructions for the user.

Unlike the BrowseForFile method, the user can only select one folder in the dialog. The method returns an IEdmFolder5 interface. There is no need for the same GetFirst, GetNext type of looping needed when using BrowseForFile.

The next step will be to prompt the user for a file to add to the vault. Since the file should be outside the vault, we cannot use the vault's BrowseForFile method. That method is restricted to vault files. The code that will be added will use the OpenFileDialog from the Windows.Forms namespace instead.

   14. Add code to let the user browse for any file type and add it to the vault.

```
Dim eFolder As IEdmFolder5
eFolder = vault.BrowseForFolder(0, "Select the destination folder")

Dim filePath As String
Dim fileID As Integer
Dim eFile As IEdmFile5
Dim OpenDia As New Windows.Forms.OpenFileDialog
Dim diaRes As Windows.Forms.DialogResult
diaRes = OpenDia.ShowDialog
If diaRes = Windows.Forms.DialogResult.OK Then
    filePath = OpenDia.FileName
End If
```

In preparation to add the file, two additional variables have been declared: `fileID` and `eFile`. The first is an integer while the second, as you might guess, is an IEdmFile5 interface. Until now, there has not been a need for the actual file ID which will be used in the process of adding the file to the vault.

   15. Insert the following code to add the desired file to the vault.

```
Dim filePath As String
Dim fileID As Integer
Dim eFile As IEdmFile5
Dim OpenDia As New Windows.Forms.OpenFileDialog
Dim diaRes As Windows.Forms.DialogResult
diaRes = OpenDia.ShowDialog
If diaRes = Windows.Forms.DialogResult.OK Then
    filePath = OpenDia.FileName
    'try to add the file to the selected folder
    Try
        fileID = eFolder.AddFile(0, filePath)
        eFile = vault.GetObject(EdmObjectType.EdmObject_File, fileID)
    Catch ex As Exception
        'failed to add the file
        MsgBox(ex.Message)
        Exit Sub
    End Try
```

```
End If
```

IEdmFolder5.AddFile

The process of adding a file to the vault runs through the destination IEdmFolder5 interface. The AddFile method can add a file from an outside location, as if it were copied into the vault, or a file from within the vault folder structure as long as it is a local file and not yet part of the vault. The second condition is rare, but is worth noting.

Value = IEdmFolder5.AddFile (lParentWnd, bsSrcPath, bsNewFileName, lEdmAddFlags)

- **lParentWnd** is again an integer representing the parent application's window handle.

- **bsSrcPath** is a string that must be the full path of the source file to be copied and added to the vault. If the file is in the vault, but a local file, its full path is still required.

- **bsNewFileName** is an optional string to be used if you need to rename the file that is being copied in. If left empty or not used, the file retains its original name.

- **lEdmAddFlags** is again an optional integer entry. It can be a combination of several options that are worth reviewing from the EdmAddFlag enumeration. The default value is EdmAdd_Simple which corresponds to the integer 0. You may need to use a combination of these values to filter what you add as well as generate new serial numbers.

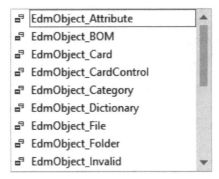

- **Value** is an integer returned by the method. This is the database ID of the file.

In order to get the IEdmFile5 interface, the code uses the file's ID and the GetObject method from the EdmVault5 interface.

The code can now be run and tested. If you have appropriate permissions, the file you select from the OpenFileDialog will be added to the vault. If you lack permissions, an error will be displayed.

The last step will be to have the file automatically checked into the vault and a message returned of its success or failure.

16. Add a Try block with the following code to check in the newly added file.

```
'try to add the file to the selected folder
Try
```

```
                    fileID = eFolder.AddFile(0, filePath)
                    eFile = vault.GetObject(EdmObjectType.EdmObject_File, fileID)
            Catch ex As Exception
                'failed to add the file
                MsgBox(ex.Message)
                Exit Sub
            End Try

            'try to check in the file
            Try
                eFile.UnlockFile(0, "Automatic check in")
                MsgBox(eFile.Name & " added to the vault.")
            Catch ex As Exception
                MsgBox(ex.Message & vbLf & eFile.Name)
                Exit Sub
            End Try
        End If

End Sub
```

The check in operation is identical to the previous example. You simply use the UnlockFile method of the IEdmFile5 interface. The Try block will return any errors in a message to the user.

If you need to load variable values into new files, it should be done between adding the file and checking it in. This is extremely helpful if you need to load many files and merge in searchable values from another source. We will load the file name into the Description variable as a simple example.

   17. Add the following code to write the file name without its extension to the Description variable of the file (@ tab) before the check in code.

```
'try to add the file to the selected folder
Try
    fileID = eFolder.AddFile(0, filePath)
    eFile = vault.GetObject(EdmObjectType.EdmObject_File, fileID)
Catch ex As Exception
    'failed to add the file
    MsgBox(ex.Message)
    Exit Sub
End Try

'update the Description with the file name
Try
  Dim eVar As IEdmEnumeratorVariable8
  eVar = eFile.GetEnumeratorVariable
  Dim newDescription As String
  newDescription = IO.Path.GetFileNameWithoutExtension(eFile.Name)
  eVar.SetVar("Description", "@", newDescription)
  eVar.CloseFile(True)
Catch ex As Exception
  MsgBox(ex.Message & vbLf & eFile.Name)
```

```
  Exit Sub
End Try
...
```

Advanced Check In

This file modification process works great, unless you need to copy in a completely new file version. When you need to check in a new file version, beyond just updating variables, use the following strategy.

The first and last steps are the same. Instead of editing the file after checking it out, you can use the Windows File.Copy method from System.IO namespace. The method lets you copy a file over the top of an existing file, as long as you have the vaulted file checked out. Check in the file to finish the procedure. There is no need to use the AddFile method since you are working with an existing vaulted file rather than adding a new file. This is a great option for loading in file versions from an outside source.

Conclusion

Now that you have the ability to read and write properties and check files in and out of the vault, you can start automating. I have used these same methods to help companies load metadata into file cards from ERP databases and other PDM systems. And hopefully, your new familiarity with the PDM Professional API basics can jumpstart your exploration of how to move files through workflow states and creating add-ins.

C# Example

```csharp
//console application
static void Main()
{
  // Create a file vault interface and log in to a vault
  IEdmVault5 vault = new EdmVault5();
  //vault.LoginAuto("PDMPro", 0);
  string username = "Admin";
  string password = "admin";
  string vaultname = "PDMPro";
  vault.Login(username, password, vaultname);

  //Let the user select one or more files that are
  //in the file vault to which we are logged in
  EdmStrLst5 PathList;
  PathList = vault.BrowseForFile(0,
    bsCaption: "Select a file to show its information");

  //Check if the user pressed Cancel
  if (PathList == null)
  {
    //do nothing - user pressed cancel
  }
  else
  {
    string message;
```

```
IEdmPos5 pos;
//Display a message box with
//the paths of all selected files.
message = "You selected the following files:\n";
pos = PathList.GetHeadPosition();
while (!pos.IsNull)
{
  string filePath;
  filePath = PathList.GetNext(pos);
  //connect to the file object
  IEdmFile5 eFile;
  IEdmFolder5 eFolder;
  eFile = vault.GetFileFromPath(filePath, out eFolder);
  message += filePath;
  message += "\nState: " + eFile.CurrentState.Name;
  message += "\nIs Checked Out: " + eFile.IsLocked.ToString();

  //get the file's EnumeratorVariable interface
  //for working with its card variables
  IEdmEnumeratorVariable10 eVar;
  eVar = (IEdmEnumeratorVariable10)eFile.GetEnumeratorVariable();

  //get the description
  object varValue;
  eVar.GetVar("Description", "@", out varValue);
  message += "\nDescription: " + varValue.ToString();

  //try to check out the file
  try
  {
    eFile.LockFile(eFolder.ID, 0);
  }
  catch (Exception ex)
  {
    Console.WriteLine(ex.Message);
    Console.WriteLine(eFile.Name);
    return;
  }

  //edit the file here
  //reconnec to the file's EnumeratorVariable after check out
  eVar = (IEdmEnumeratorVariable10)eFile.GetEnumeratorVariable();
  try
  {
    eVar.SetVar("Description", "@", "NEW DESCRIPTION");
    //close the file and save (flush) any updates
    eVar.CloseFile(true);
  }
  catch (Exception ex)
  {
```

```
          Console.WriteLine(ex.Message);
          Console.WriteLine(eFile.Name);
          return;
        }

        try
        {
          eFile.UnlockFile(0, "API check in");
        }
        catch (Exception ex)
        {
          Console.WriteLine(ex.Message);
          Console.WriteLine(eFile.Name);
          return;
        }

        //show information about the file
        Console.WriteLine(message);
      }
      Console.WriteLine("Press any key to close.");
      Console.Read();
    }
}

//Add Files example
[STAThreadAttribute]
static void Main()
{
  EdmVault5 vault = new EdmVault5();
  vault.LoginAuto("PDMPro", 0);

  IEdmFolder5 eFolder;
  eFolder = vault.BrowseForFolder(0, "Select the destination folder");

  string filePath;
  int fileID;
  IEdmFile5 eFile;
  OpenFileDialog ofd = new OpenFileDialog();
  DialogResult diaRes;
  diaRes = ofd.ShowDialog();
  if (diaRes == DialogResult.OK)
  {
    filePath = ofd.FileName;
    //try to add the file to the selected folder
    try
    {
      fileID = eFolder.AddFile(0, filePath);
      eFile = (IEdmFile5)vault.GetObject(
        EdmObjectType.EdmObject_File, fileID);
    }
```

```
    catch (Exception ex)
    {
      //failed to add the file
      MessageBox.Show(ex.Message);
      return;
    }

    //update the Description with the file name
    try
    {
      IEdmEnumeratorVariable8 eVar;
      eVar = (IEdmEnumeratorVariable8)eFile.GetEnumeratorVariable();
      string newDescription;
      newDescription =
        System.IO.Path.GetFileNameWithoutExtension(eFile.Name);
      eVar.SetVar("Description", "@", newDescription);
      eVar.CloseFile(true);
    }
    catch (Exception ex)
    {
      MessageBox.Show(ex.Message + "\n" + eFile.Name);
      return;
    }

    //check in the file
    try
    {
      eFile.UnlockFile(0, "API check in");
      MessageBox.Show(eFile.Name + " added to the vault.");
    }
    catch (Exception ex)
    {
      MessageBox.Show(ex.Message);
      return;
    }

  }
}
```

PDM Professional Bills of Materials

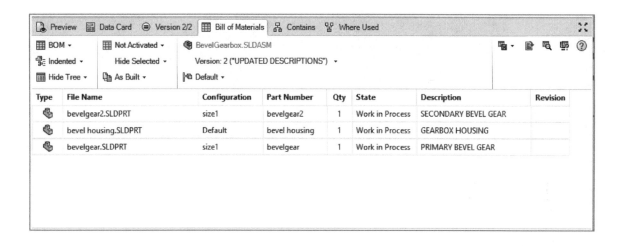

- **Named and Computed BOMs**

- **BOM Views**

- **Rows and Columns**

- **Cell Values**

Introduction

SOLIDWORKS PDM Professional is an outstanding tool for reporting bills of materials (BOMs) for drawings and assemblies without having to open them in SOLIDWORKS. Multiple column arrangements, or BOM Views, can be created for different visualization needs. They can be exported by the user in CSV format or automated by workflows in XML. However, you may still have a need to query BOMs for other uses. This chapter focuses on how to access bill of materials table information using the SOLIDWORKS PDM Professional API.

This exercise will explore reading both Computed and Named BOMs from a file selected by the user. The table text will be written to the Output window, but could easily be modified to write to a file.

Named and Computed BOMs

SOLIDWORKS PDM Professional can generate two different kinds of bills of materials. The Computed BOM is the most common. Computed BOMs are the actively reported, or computed, bill of materials reported directly from the database. It will report the current structure of a selected configuration and version of any file that has child references. Common fields such as file name, description and quantities are based on the current variable values and the reference counts from the parent file. Cells from a Computed BOM cannot be directly edited unless the associated child component is checked out. Rows cannot be added or removed.

A Named BOM is a separate table created by saving a Computed BOM. It is independent of the data structure of its parent drawing or assembly and can be edited beyond its origin. Named BOMs allow for rows to be added and removed. Named BOMs can have their own independent workflow process separate from their parent file. They can also be re-computed based on changes to a parent document.

Accessing Computed BOMs

We will start the exercise with Computed BOMs.

18. Create a new macro in SOLIDWORKS and save it as *PDMBOM.vbproj*.

19. Add a reference to the PDM Professional library.

20. Add the following code to login and browse for a file. Include the Imports statement for EPDM.Interop.epdm. Replace *PDMPro* with your vault name. *Note: most of this code can be copied from the last macro example.*

```
Imports SolidWorks.Interop.sldworks
Imports SolidWorks.Interop.swconst
Imports System
Imports EPDM.Interop.epdm
Imports System.Diagnostics

Partial Class SolidWorksMacro

  Public Sub main()
    'Create a file vault interface
    'and log in to a vault.
```

```
        Dim vault As EdmVault5 = New EdmVault5
        Dim vaultName As String = "PDMPro"
        vault.LoginAuto(vaultName, 0)

        'Let the user select a SOLIDWORKS file
        Dim PathList As EdmStrLst5
        PathList = vault.BrowseForFile(0, , _
          "SOLIDWORKS Files (*.sld*)|*.sld*||", , , , _
          "Select a file to show its information")

        'Check if the user pressed Cancel
        If PathList Is Nothing Then
          'do nothing - user pressed cancel
        Else
          Dim pos As IEdmPos5
          pos = PathList.GetHeadPosition
          While Not pos.IsNull
            Dim filePath As String
            filePath = PathList.GetNext(pos)

            'connect to the file object
            Dim eFile As IEdmFile7
            Dim parentFolder As IEdmFolder5
            eFile = vault.GetFileFromPath(filePath, parentFolder)

          End While
        End If
    End Sub
```

BrowseForFile Filter

The code is only slightly different from the previous macro. There is no need for the message variable and a filter string is applied to the BrowseForFile method. The filter string is the third argument in BrowseForFile and allows you to restrict the types of files a user can select. This example filters only SOLIDWORKS files. You can add as many filters as needed to give the user the options they need. The filter string format is *nearly* identical to the OpenFileDialog in the Windows.Forms namespace, with emphasis on nearly. The entire filter string must end with a double vertical bar ||.

BOM Views

A PDM administrator can create several BOM styles available to the user. BOM table styles can be organized using different fields in varying order. The table styles, or views, are visible through the Bill of Materials dropdown in the top-left corner of the Bill of Materials tab as shown below.

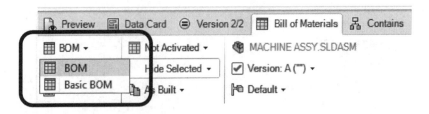

This example shows two different BOM Views. Each view can also be accessed through the API by its name.

21. Add the following code to access a computed bill of materials with a specific name. Be sure to enter the name of a bill of materials view from your vault.

```
...
    While Not pos.IsNull
      Dim filePath As String
      filePath = PathList.GetNext(pos)
      'connect to the file object
      Dim eFile As IEdmFile7
      eFile = vault.GetFileFromPath(filePath)

      'get a specific BOM view
      Dim bom As IEdmBomView3
      bom = eFile.GetComputedBOM("Basic BOM", 0, _
        "@", EdmBomFlag.EdmBf_AsBuilt)

    End While
  End If
End Sub
```

IEdmFile7.GetComputedBOM

The GetComputedBOM method of IEdmFile7 will return an EdmBomView interface. EdmBomView has several methods for traversing the content of the BOM view by columns and rows. Using the child EdmBomView3 interface provides the ability to save the BOM to a csv file. GetComputedBOM has the following structure.

Value = IEdmFile7.GetComputedBOM (oBomLayoutNameOrID, lVersionNo, bsConfiguration, lEdmBomFlags)

- **oBomLayoutNameOrID** can be either the database ID of the BOM view or its name as a string. A name string is more suitable for known BOM styles. A BOM view ID is best suited to a user selecting a BOM view style from a selection list.

- **lVersionNo** indicates which file version to use to retrieve the BOM. Use 0 or -1 to use the latest version.

- **bsConfiguration** is a string representing the configuration. The string "@" will always indicate the file-level reference for SOLIDWORKS files. Use an empty string, "", for BOMs from files that do not have configuration tabs on their data cards.

- **lEdmBomFlags** can be a combination of Long values from the EdmBomFlag enumeration. It currently includes the following. These correspond to the AsBuilt and ShowSelected options in the user interface. This example uses the EdmBf_AsBuilt flag to retrieve references and values based on the last checked-in structure.

⊟ EdmBf_AsBuilt	
⊟ EdmBf_ShowSelected	

There is currently no support for indicating visibility of all components in an indented list, top level only, or parts only. Your code will need to filter the resulting data to achieve the same results.

Traversing BOM Content

There are two primary methods for traversing the content of an IEdmBomView. GetColumns is used to retrieve the column layout of the BOM view, including column headers and what data and type is visible in each. Use GetRows to traverse all of the references in the BOM, row by row, and retrieve the string values that are visible per column. GetRows will also return the ID of the PDM object reference if you need to retrieve the file itself.

22. Add the following code to traverse the columns and rows in the bill of materials view and write it to the Output window using Debug.Print.

...

```
'get a specific BOM view
Dim bom As IEdmBomView3
bom = eFile.GetComputedBOM("Basic BOM", 0, _
  "@", EdmBomFlag.EdmBf_AsBuilt)

'get column headers
Dim columns() As EdmBomColumn
bom.GetColumns(columns)
Dim header As String = "LEVEL" & vbTab
For Each column As EdmBomColumn In columns
  header = header & column.mbsCaption & vbTab
Next
Debug.Print(header)

'read each BOM row
Dim rows() As Object
bom.GetRows(rows)
Dim row As IEdmBomCell
For Each row In rows
  If IsNothing(row) Then Exit For
  Dim rowString As String = _
    row.GetTreeLevel.ToString & vbTab
  Dim varVal As String = ""
  For Each column As EdmBomColumn In columns
    row.GetVar(column.mlVariableID, _
      column.meType, varVal, Nothing, _
      Nothing, Nothing)
    If IsNothing(varVal) Then varVal = ""
    rowString = rowString & varVal & vbTab
  Next
  'print the row
  Debug.Print(rowString)
```

```
        Next
    End While
  End If
End Sub
```

IEdmBomView.GetColumns

The GetColumns method returns an array of the type EdmBomColumn. Since an array is returned, a simple For, Each, Next loop can be used to iterate through the columns. They exist in the array in the same order as they do in the bill of materials visible to the user.

To build the string to be printed to the Output window, the variable header is created with an initial value of "LEVEL." Since the results will be Tab delimited, a Tab character is appended to the end of the string. The LEVEL column in the results will represent the indenture level visible in the bill of materials through the user interface.

EdmBomColumn, the returned data type from GetColumns, is formally a Structure. This means that it contains values only. There are no methods that can operate on a Structure. The first value used in the sample is mbsCaption. This is simply the string value of the column header. As the code traverses each column, its caption is appended to the existing header string, followed by a Tab character. When the loop is finished, the resulting header string is printed to the Output window using Debug.Print.

IEdmBomView.GetRows

The next, larger section of code will traverse each row in the bill of materials. Column data will again be needed to extract row values based on column. Again, a string variable will be populated with Tab delimited values for the entire row and printed successively to the Output window.

GetRows will return an array of type Object. Each Object represents an IEdmBomCells interface. This does not mean that you have individual cell interfaces for each cell or field in the bill of materials. Each IEdmBomCells interface represents a single row. So a BOM with 10 rows will return an array of at least 10 elements. However, GetRows may also return an empty, or null Object as the last element of the array. The addition of the IsNothing statement takes care of the inclusion of a null array element and exits the loop.

The IEdmBomCells interface has several methods for evaluating BOM data. The example code uses both GetTreeLevel and GetVar. GetTreeLevel simply returns an integer representing the bill of materials indenture level. Top level components will have a value of 1. The rowString variable that will be output using Debug.Print is initialized by converting the tree level integer to a string value. A Tab character is again added for Tab delimited output.

IEdmBomCells.GetVar

The GetVar method can be used to retrieve the text value of any BOM field. GetVar requires input arguments that can only be retrieved from the EdmBomColumn structure. Another For, Each, Next loop is used to traverse each column of the row to get the cell values.

Value = IEdmBomCell.GetVar(lVariableID, eColumn, poValue, poComputedValue, pbsConfiguration, pbReadOnly)

- **lVariableID** is an integer value best retrieved from the column using EdmBomColumn.mlVariableID.

- **eColumn** is long value representing the type of column. The column type can include file name, path, version, variables and many others. This argument is again best retrieved from the column using EdmBomColumn.meType.

- **poValue** is populated with the returned string value displayed in the specific column for this row. A corresponding variable must be declared prior to calling GetVar. poValue may return an empty value. In the example code, empty values are replaced by an empty string to avoid runtime errors.

- **poComputedValue** is similar to poValue. There is not typically a difference between these two values, but you may wish to retrieve this value to compare. In this example, `Nothing` is passed to GetVar so no return value will be available.

- **pbsConfiguration** is also a returned string value representing the configuration of the row being referenced. It is not used in this example.

- **pbReadOnly** is a Boolean value that indicates whether the cell is writable. This is typically only the case when the row's file is checked out by the current user. It is also a return value and is not used in this example.

Debug

Your macro is ready for testing and debugging.

23. Add a breakpoint at End Sub. This will allow you to read the Output window results.

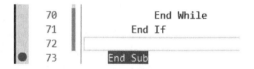

```
70                        End While
71                End If
72
73      End Sub
```

24. Run the macro and browse to any drawing or assembly. When the macro reaches the breakpoint, view the Output window to see the bill of materials result.

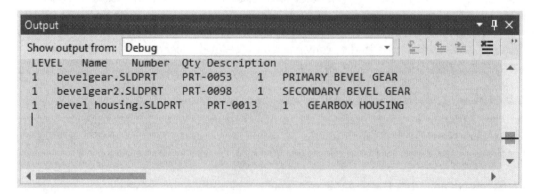

25. Stop and save the macro.

Saving BOMs as CSV

EdmBomView3 includes the ability to save the BOM directly to a CSV file. If you simply need to save the BOM outside of PDM, this is the simplest process.

26. Add the following code to save the Computed BOM to a CSV file.

```
...
'get a specific BOM view
Dim bom As IEdmBomView3
bom = eFile.GetComputedBOM("Basic BOM", 0, _
  "@", EdmBomFlag.EdmBf_AsBuilt)

If Not bom Is Nothing Then
  bom.SaveToCSV("C:\temp\" & eFile.Name & ".csv", True)
End If

'get column headers
Dim columns() As EdmBomColumn
```

IEdmBomView3.SaveToCSV

The SaveToCSV method of IEdmBomView3 is uncomplicated. The first argument must be a fully qualified path to the resulting CSV file. In this example, the path is hard-coded and the file name is the name of the source file with its extension and a new CSV extension added. For example, if the file selected were MyDrawing.SLDDRW, the resulting BOM would be saved to *C:\temp\MyDrawing.SLDDRW.CSV*. The second argument is a Boolean. Pass True to have an Item column automatically added to the CSV output. This is similar to saving the BOM from the PDM interface.

Accessing Named BOMs

Named bills of materials include the standard Named BOMs saved from a Computed BOM as well as bills of materials that display on the drawing or assembly in SOLIDWORKS. The following procedure focuses on extracting a saved, Named BOM, but will also report the other if it exists.

Named BOMs can be checked out and run through workflows like files. This means they also need an additional interface that can be used to make these transactions as well as report checked-out and workflow status. Named BOMs can have any name imaginable, so they are not accessed through a named view. Rather, an array of named BOMs can be retrieved from a file. Then each resulting BOM can be evaluated.

This section will use the same macro and modify it for Named BOMs. If you would like two separate macros, create a new macro and set it up exactly as has been done to this point.

27. Modify the macro as shown to get an array of Named BOMs from the user selected file.

```
...
    'get a specific BOM view
    Dim bom As IEdmBomView3
```

```
        bom = eFile.GetComputedBOM("Basic BOM", 0, _
          "@", EdmBomFlag.EdmBf_AsBuilt)

        If Not bom Is Nothing Then
          bom.SaveToCSV("C:\temp\" & eFile.Name & ".csv", True)
        End If

        Dim namedBOMs() As EdmBomInfo
        eFile.GetDerivedBOMs(namedBOMs)

        If namedBOMs.Length > 0 Then
          'found at least one Named BOM

          'get column headers
          Dim columns() As EdmBomColumn
          bom.GetColumns(columns)
          Dim header As String = "LEVEL" & vbTab
...
            'print the row
            Debug.Print(rowString)
          Next
        End If
      End While
    End If
End Sub
```

IEdmFile7.GetDerivedBOMs

GetDerivedBOMs returns an array of EdmBomInfo structures. However, it is returned through the argument passed in rather than as a result of the method. Recall this as a ByRef argument.

The EdmBomInfo structure has two values. The first, mlBomID, is the database ID of the specific Named BOM represented as an integer. The next, mbsBomName, is a string value representing the actual BOM name. These values will be used in additional code to get the BOM and subsequently its view so the existing code can report the contents.

The If statement checks for a resulting array with values. If no Named BOM is found related to the selected file, the routine ends.

28. Add a For, Each, Next loop to evaluate any named bills of materials associated with the selected file.

...

```
      'get a specific BOM view
      Dim bom As IEdmBomView3
      bom = eFile.GetComputedBOM("Basic BOM", 0, _
        "@", EdmBomFlag.EdmBf_AsBuilt)

      If Not bom Is Nothing Then
        bom.SaveToCSV("C:\temp\" & eFile.Name & ".csv", True)
```

```
        End If

        Dim namedBOMs() As EdmBomInfo
        eFile.GetDerivedBOMs(namedBOMs)

        If namedBOMs.Length > 0 Then
          'found at least one Named BOM
          For Each bomInf As EdmBomInfo In namedBOMs
            Dim nbom As EdmBom = vault.GetObject( _
              EdmObjectType.EdmObject_BOM, bomInf.mlBomID)
            If nbom.CurrentState.Name <> "" Then
              'found a current Named BOM
              bom = nbom.GetView
              Debug.Print("Named BOM: " & nbom.Name)

              'get column headers
              Dim columns() As EdmBomColumn
              bom.GetColumns(columns)
              Dim header As String = "LEVEL" & vbTab
...

                'print the row
                Debug.Print(rowString)
            Next
          End If
        Next
      End If
    End While
  End If
End Sub
```

IEdmBom Interface

Now that we know the BOM ID of each Named BOM, EdmVault5.GetObject can be used to its interface. IEdmBom is comparable to an PDM file or folder interface. It has various methods and properties that can be reported or queried to get its workflow state, current checked out user, or to check it in and out as well as change its workflow state.

The IEdmBom interface's CurrentState property along with the state's Name property are used to verify we have a current BOM. There may be additional BOMs associated with the file from previous versions that will not have a workflow state. These are typically BOMs from a drawing or assembly in older versions of the parent file.

IEdmBom.GetView

The GetView method returns an IEdmBomView interface. This is the same interface used with Computed BOMs and is the source for the actual contents of the Named BOM. The BOM name is printed to the Output window to indicate which BOM is reported when multiple BOMs are found.

Debug

The macro is ready to test. You will first need a Named BOM or an assembly or drawing with an included SOLIDWORKS BOM table.

29. Save a Named BOM from an assembly or drawing in your vault. Use Save As from the Bill of Materials tab controls.

30. Check in the resulting Named BOM using the Check In button in the same area of the Bill of Materials tab.

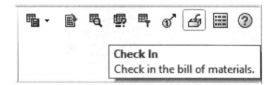

31. Set a breakpoint at End Sub and run the macro. Browse to the parent file of the saved Named BOM and review the Output window results.

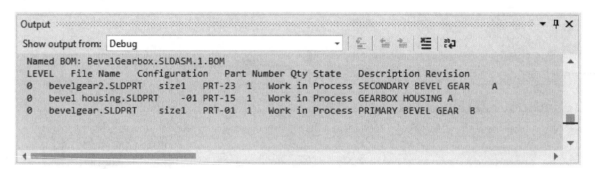

32. Debug and edit as needed.

Conclusion

There are many ways and reasons to report and share bill of materials information. This chapter should help you get started building tools to extract the data so you can manipulate and present it as needed.

If you need further BOM manipulation, review the additional IEdmBomView2 interface for the InsertRow method as well as the SetVar method of IEdmBomCell. If you are working with a Named BOM, remember to include the CheckOut and CheckIn methods of IEdmBom to enable and preserve changes. These all apply primarily to Named BOMs.

If you need to manipulate properties in a Computed BOM, you will either need to save it as a Named BOM, or check out the referenced files and modify their variables as discussed in the previous chapter.

C# Example

```csharp
//read computed BOM console application
static void Main()
{
  EdmVault5 vault = new EdmVault5();
  vault.LoginAuto("PDMPro", 0);

  //Let the user select a SOLIDWORKS file
  EdmStrLst5 PathList;
  PathList = vault.BrowseForFile(0,
    bsCaption: "Select a file to show its information");

  //check if the user pressed cancel
  if (PathList == null)
  {
    //do nothing, user cancelled
  }
  else
  {
    IEdmPos5 pos;
    pos = PathList.GetHeadPosition();
    while (!pos.IsNull)
    {
      string filePath;
      filePath = PathList.GetNext(pos);
      //connect to the file object
      IEdmFile7 eFile;
      IEdmFolder5 eFolder;
      eFile = (IEdmFile7)vault.GetFileFromPath(filePath, out eFolder);

      //get a specific BOM view
      IEdmBomView3 bom;
      bom = (IEdmBomView3)eFile.GetComputedBOM("Basic BOM",
        0, "@", (int)EdmBomFlag.EdmBf_AsBuilt);

      //get column headers
      EdmBomColumn[] columns;
      bom.GetColumns(out columns);
      string header = "LEVEL\t";
      foreach (EdmBomColumn column in columns)
      {
        header += column.mbsCaption + "\t";
      }
      Debug.Print(header);

      //read each BOM row
      object[] rows;
      bom.GetRows(out rows);
      foreach (IEdmBomCell row in rows)
```

```
    {
      if (row  == null){break;}
      string rowString = row.GetTreeLevel().ToString() + "\t";
      object varVal = "";
      foreach (EdmBomColumn column in columns)
      {
        object compVal;
        string config;
        bool readOnly;
        row.GetVar(column.mlVariableID, column.meType,
          out varVal, out compVal, out config, out readOnly);
        if (varVal == null) { varVal = ""; }
        rowString += varVal + "\t";
      }
      //print the row
      Debug.Print(rowString);
    }
  }
  }
  }
}
```

Favorite Code Examples

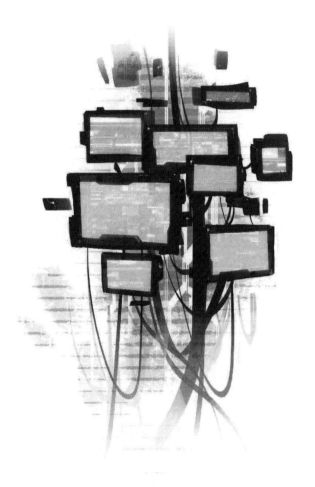

- **Batch Process Files**

- **Feature Traversal and Selection**

- **Assembly Traversal**

Introduction

Over the years I have learned several tips and shortcuts for typical tasks in Visual Basic.NET. I have used several of them throughout this book. But there are still more. In fact, I seem to learn another trick or shortcut with every programming project. Some are specific to the SolidWorks API and some are Visual Studio tools or namespaces. This chapter will repeat some that I have shown in other chapters and introduce others. I could not figure out a way to fit these all into the other chapters so you get the shotgun approach here.

Batch Process Files

It is not uncommon to need to batch process files for a variety of reasons. You might want to add or update custom properties. You may wish to publish PDFs, IGES, DWG or other output types. Or you may simply want to edit a bunch of similar models. This example is a repeat of common batch processing.

The drawing automation chapter introduced batch processing by creating drawings from every part in a specified folder using the System.IO namespace and the DirectoryInfo and FileInfo members.

1. Edit the existing macro project *BatchProcess.vbproj* from the downloaded example files.

2. Review the `main` procedure of *SolidWorksMacro.vb*. Notice that it creates an instance of the form named *Form1*. The form is shown as a dialog. If the user clicks OK, the procedure named Process of the Module1 module runs.

3. Right-click on the word `Process` in the line `Module1.Process` and select Go To Definition to switch to that procedure. Review the code as shown below.

Process Procedure

There are three procedures that take care of the batch processing. The procedure named Process is the one that initiates processing by verifying that the folder paths passed to the method really exist and creates the output directory if necessary. It initiates a log file named *Process.log.* It then calls the procedure named SaveAs.

```
Sub Process(ByVal FromFolder As String, _
ByVal ToFolder As String)

    'get the folders as DirectoryInfo interfaces
    Dim FromDir As New DirectoryInfo(FromFolder)
    Dim ToDir As New DirectoryInfo(ToFolder)

    'if the processing directory doesn't exist,
    'tell the user and exit
    If Not FromDir.Exists Then
        MsgBox(FromDir.FullName & " does not exist.", _
          MsgBoxStyle.Exclamation)
        Exit Sub
    End If
```

```
    'if the output folder doesn't exist, create it
    If Not ToDir.Exists Then
        ToDir.Create()
    End If

    'initialize the log file
    LogFile = New StreamWriter(ToDir.FullName _
      & "\Process.log")

    'save as another file format
    'change the extension value to
    'whatever you want to export
    Dim Extension As String = ".PDF"
    Dim FileTypeToProcess As Long = _
      swDocumentTypes_e.swDocDRAWING
    SaveAs(FromDir, ToDir, Extension, FileTypeToProcess)

    ''save as DXF
    'Extension = ".DXF"
    'FileTypeToProcess = swDocumentTypes_e.swDocDRAWING
    'SaveAs(FromDir, ToDir, Extension, FileTypeToProcess)

    ''save parts as IGS
    'Extension = ".IGS"
    'FileTypeToProcess = swDocumentTypes_e.swDocPART
    'SaveAs(FromDir, ToDir, Extension, FileTypeToProcess)

    'add other processing as needed

    'finish up by closing out the log file
    LogFile.Flush()
    LogFile.Close()
End Sub
```

Batch Process Error Handling

Most batch processes are run without someone watching for error messages. You can imagine the last thing you would want to do is babysit a batch process that takes hours or even days to click OK any time a warning or error occurs. Instead, it is common practice to create a log file of the errors and warnings.

It really becomes critical that you put effective error trapping in place when doing any type of batch processing. It can be frustrating to run a batch process and return to find that it only partially completed because of an error that was not handled well.

The Process procedure initiates a log file in the output directory. At the end of the Process procedure, the log file is flushed and closed. Flushing a text file ensures that anything in memory is written to the file.

SaveAs Procedure

The SaveAs procedure is where the action happens. As you might guess by the name, this procedure will be used to save files as a different type. In this case, the uncommented code will create a PDF of every drawing in the selected folder. The code is also in place, but commented out, to save DXF files of each drawing and IGES models of every part.

4. Review the SaveAs procedure code as shown below.

```
Sub SaveAs(ByVal dinf As DirectoryInfo, _
ByVal outDir As DirectoryInfo, _
ByVal Ext As String, ByVal TypeToProcess As Long)
    Dim longErrors As Long
    Dim longWarnings As Long

    Dim SWXWasRunning As Boolean

    'get the file extension based on file type
    Dim files() As FileInfo
    Dim filter As String = ""
    Select Case TypeToProcess
        Case swDocumentTypes_e.swDocASSEMBLY
            filter = ".SLDASM"
        Case swDocumentTypes_e.swDocPART
            filter = ".SLDPRT"
        Case swDocumentTypes_e.swDocDRAWING
            filter = ".SLDDRW"
    End Select

    'get all files in the directory
    'that match the filter
    files = dinf.GetFiles(filter)
    For Each f As FileInfo In files

      Dim swApp As SldWorks
      Dim Part As ModelDoc2

      'if solidworks is running, use the active session
      'if not, start a new hidden session
      Try
          swApp = GetObject(, "sldworks.application")
          SWXWasRunning = True
      Catch ex As Exception
          swApp = CreateObject("sldworks.application")
          SWXWasRunning = False
      End Try

      Try
        Part = swApp.OpenDoc6(f.FullName, TypeToProcess, _
        swOpenDocOptions_e.swOpenDocOptions_Silent _
```

```
                + swOpenDocOptions_e.swOpenDocOptions_ReadOnly, _
                "", longErrors, longWarnings)
            Dim newFileName As String
            newFileName = Path.ChangeExtension(f.FullName, Ext)
            Part.SaveAs2(newFileName, _
                swSaveAsVersion_e.swSaveAsCurrentVersion, True, True)
            swApp.CloseDoc(Part.GetTitle)
            Part = Nothing
            If Not SWXWasRunning Then
                'exit SolidWorks if it was
                'not previously running
                swApp.ExitApp()
                swApp = Nothing
            End If
        Catch ex As Exception
            WriteLogLine(Date.Now & vbTab & "ERROR" _
                & vbTab & f.Name & vbTab & ex.Message)
        End Try
    Next
End Sub
```

The SaveAs procedure loops through all files of a specified type in the selected folder. DirectoryInfo and FileInfo classes are used to get the files and their names. The Path interface is used to quickly change the extension of the file so that the new name is the desired extension.

After each file is saved, it is closed. If a new instance of SolidWorks was created, it exits. Otherwise, the SolidWorks session stays active. Even though this macro uses a single session of SolidWorks, large processes may require you to do some careful memory management. Closing and opening multiple files can overrun the memory of a computer very quickly. For better memory management, you might need to consider creating a stand-alone application rather than a macro so you can exit the SolidWorks application after each file is processed to avoid memory overloads.

Writing to the Log File

Writing to the already created log file is very simple. We have discussed writing to text files in several chapters. So hopefully this looks familiar.

5. Review the WriteLogLine procedure.

```
Sub WriteLogLine(ByVal LogString As String)
    LogFile.WriteLine(Now() & vbTab & LogString)
End Sub
```

Simply pass the string you would like to log to this procedure. It writes to the log file with the date appended before your string in a tab separated format. I use this procedure in most applications I create to get effective error logging.

6. Right-click on the text LogFile in the LogFile.WriteLine code line and select Go To Definition. Notice that LogFile is declared globally to this module so you do not have to pass an instance to each procedure where it is used.

Expanding BatchProcess

If you need a batch processing tool for other operations simply create new procedures in the Module1 code module and add them to the Process procedure. The formatting will likely be similar to the SaveAs procedure so you can copy and paste to get the looping through files. Then simply change what action occurs on each file.

Traverse Features of a Part

I have found that I regularly need to access a certain feature type from a part. It is also very common to need to access a specific named feature. As an example, SolidWorks Routing uses features that have a specific name to help automate parts. I have built model library insertion applications that rely heavily on features with specific names. If you need to insert parts into assemblies, it is easy when you can get planes by their names for mates.

This example focuses on traversing the FeatureManager to find certain types of features. It also includes selecting a feature by its name as well as an important concept of persistent IDs that can be invaluable for repeat selection of a specific feature.

1. Edit the existing macro project named *FeatureTraversal.vbproj*. Review the main procedure from *SolidWorksMacro.vb. Note: this code is based off "Traverse Sub Features" example code from the API Help.*

```
Public Sub main()
  Dim Model As ModelDoc2 = swApp.ActiveDoc
  'make sure the active document is a part
  If Model Is Nothing Then
    MsgBox("Please open a part first.", MsgBoxStyle.Exclamation)
    Exit Sub
  ElseIf Model.GetType <> swDocumentTypes_e.swDocPART Then
    MsgBox("For parts only.", MsgBoxStyle.Exclamation)
    Exit Sub
  End If
  'get the PartDoc interface from the model
  Dim Part As PartDoc = Model

  'get the first feature in the FeatureManager
  Dim feat As Feature = Part.FirstFeature()
  Dim featureName As String
  Dim featureTypeName As String
  Dim subFeat As Feature = Nothing
  Dim subFeatureName As String
  Dim subFeatureTypeName As String
  Dim message As String

  ' While we have a valid feature
  While Not feat Is Nothing
    ' Get the name of the feature
    featureName = feat.Name
    featureTypeName = feat.GetTypeName2
    message = "Feature: " & featureName & vbCrLf & _
```

```
        "FeatureType: " & featureTypeName & vbCrLf & _
        " SubFeatures:"

        'get the feature's sub features
        subFeat = feat.GetFirstSubFeature

        ' While we have a valid sub-feature
        While Not subFeat Is Nothing

            ' Get the name of the sub-feature
            ' and its type name
            subFeatureName = subFeat.Name
            subFeatureTypeName = subFeat.GetTypeName2
            message += vbCrLf & " " & _
            subFeatureName & vbCrLf & " " & _
            "Type: " & subFeatureTypeName

            subFeat = subFeat.GetNextSubFeature
            ' Continue until the last sub-feature is done
        End While

        ' Display the sub-features in the
        System.Diagnostics.Debug.Print(message)
        ' Get the next feature
        feat = feat.GetNextFeature()
        ' Continue until the last feature is done
    End While
End Sub
```

This code gets most of its functionality by using methods related to the IFeature interface. SolidWorks does not provide a collection of features in the FeatureManager for immediate access. Rather, you call IPartDoc.GetFirstFeature to get the first IFeature interface. Then from that IFeature interface, you call GetFirstSubFeature to get its first child. For example, a sketch is a sub feature of a cut. Once you have run out of sub features, you call the GetNextFeature method of the existing IFeature interface to get the next feature in the FeatureManager. You continue this process until GetNextFeature returns Nothing.

The IFeature interface has a couple properties that are very useful for finding what you need. I use Name and GetTypeName2 all the time. If your application requires a specific feature naming convention, use the Name property. If you are first looking for a specific type of feature such as a sketch, cut, loft, plane or sheet metal bend, the GetTypeName2 property will help you. The list of feature type names is quite long, so I would recommend either running this macro on your parts to find out what the type names are or look to the API Help. Find GetTypeName2 in the Index. The list of type names are displayed under that topic. If you have a hard time trying to determine which type name a certain feature returns, just run this macro on a part that contains the feature.

The macro pushes the information to the Immediate Window.

2. Close the macro project, saving any changes.

3. Edit the existing macro *FeatureSelection.vbproj*. Review the code for the main procedure as shown below.

```
Public Sub main()
    Dim Model As ModelDoc2 = swApp.ActiveDoc
    'make sure the active document is a part
    If Model Is Nothing Then
      MsgBox("Please open a part first.", MsgBoxStyle.Exclamation)
      Exit Sub
    ElseIf Model.GetType <> swDocumentTypes_e.swDocPART Then
      MsgBox("For parts only.", MsgBoxStyle.Exclamation)
      Exit Sub
    End If
    Dim Part As PartDoc = Model
    'get the Extension interface from the model
    Dim Ext As ModelDocExtension = Model.Extension

    'get the first feature in the FeatureManager
    Dim feat As Feature = Part.FeatureByName("Extrude1")
    If Not feat Is Nothing Then
      'select it
      feat.Select2(False, -1)
      'get persistent ID
      Dim IDCount As Integer = Ext.GetPersistReferenceCount3(feat)
      Dim ID(IDCount) As Byte
      ID = Ext.GetPersistReference3(feat)
      Stop 'check selected feature and ID

      'get rid of the feature reference
      feat = Nothing
      'clear the selection
      Model.ClearSelection2(True)
      'select the feature by its ID
      Dim errors As Integer
      'get the feature by its persistent ID
      feat = Ext.GetObjectByPersistReference3(ID, errors)
      feat.Select2(False, -1)
      Stop 'check selected feature
    End If
  End Sub
```

This example illustrates a couple different ways to select features. Once you have a reference to the feature, face, edge or other topology, you can typically select it using a similar method to what is shown in the code.

IPartDoc.FeatureByName

The FeatureByName method of the IAssemblyDoc, IDrawingDoc and IPartDoc interfaces allows you to get to a feature in the respective FeatureManager tree simply by passing its name as an argument. The main limitation of this method is that it assumes the features are named consistently. This is not always the case. For example, the three default planes may be named *Plane1, Plane2* and *Plane3*. Or they could be named *Front, Top* and *Right*. In fact, they could really be named anything. If you use this method to get features, make sure you put good error trapping in place in case the feature was not found.

IFeature.Select2

Once you have the IFeature interface, a simple call to the Select2 method will cause the feature to be selected. The two arguments for Select2 are first a Boolean value that determines whether the new selection is added to the current selection, or whether a new selection is created. Pass False to create a new selection. The second argument is an integer that marks the selection. A value of -1, used in this example, indicates no mark. Some SolidWorks methods require selections with specific marks to understand complex selection requirements.

Persistent IDs

Sometimes it may be helpful to keep a reference to a specific feature, face, or other selectable entity even after your application closes and a new session of SolidWorks is used. The Persistent Reference ID is an array of Byte values that is unique to an entity.

IModelDocExtension.GetPersistentReferenceCount3

The trick to using Persistent IDs is that the size of the array can be different for different entities. So to effectively use them, you must size your array based on the IModelDocExtension interface's GetPersistentReferenceCount3 function. The argument for this method is simply a reference to the interface for which you are getting the ID. In this example it is the feature itself.

IModelDocExtension.GetPersistentReference3

Once you have the size of the required array, call the GetPersistentReference3 method of the IModelDocExtension interface to return the array. Again, the argument for this method is just the desired entity.

IModelDocExtension.GetObjectByPersistReference3

Once you have the ID, you can get the interface to the entity again at any time by calling GetObjectByPersistReference3. The first argument is the ID array and the second is an integer variable that will be populated with any errors that occur when trying to get the interface to the entity. If it is successful, the method returns the desired interface. If not, the errors variable will tell you what the problem was and the method will return a value of Nothing to the entity.

If you are going to use this method of entity retrieval, you would need to store the ID array to either the applications settings or to a file for later use.

Traverse Assembly Components

Another common need is to get specific components from an assembly. You may need to put parts together, edit components, suppress components or perform any other kind of assembly function. The process of traversing an assembly is nearly identical to the method used to traverse features in a part. You again have the concept of parents and children. In fact, the child levels can go quite deep with an assembly when compared to the child levels of features in parts which do not typically go more than one or two levels deep.

IComponent2 versus IModelDoc2

There are a couple things to keep in mind when working with assemblies. First, if you want to change something about a part or sub assembly, you must deal with two different interfaces that each have different functionality – IComponent2 and IModelDoc2. You can relate the differences to how you work with assemblies and parts. The IComponent2 interface is used for things you would modify through Component Properties in SolidWorks. IModelDoc2 is the interface you need to access the model's features, custom properties and dimensions. This concept was introduced through the custom properties chapter of this book.

The following example is again based on sample code from the API Help.

4. Edit the existing macro named *AssemblyTraversal.vbproj*. Review the two procedures – main and TraverseComponent as shown below.

```
Sub main()
  Dim swModel As ModelDoc2
  Dim swConf As Configuration
  Dim swRootComp As Component2

  swModel = swApp.ActiveDoc
  swConf = swModel.GetActiveConfiguration
  swRootComp = swConf.GetRootComponent

  Diagnostics.Debug.Print("File = " & swModel.GetPathName)

  TraverseComponent(swRootComp, 1)
End Sub

Sub TraverseComponent(ByVal swComp As Component2, ByVal nLevel As Long)

  Dim ChildComps() As Object
  Dim swChildComp As Component2
  Dim sPadStr As String = ""
  Dim i As Long

  For i = 0 To nLevel - 1
    sPadStr = sPadStr + "   "
  Next i

  ChildComps = swComp.GetChildren
  For i = 0 To ChildComps.Length - 1
    swChildComp = ChildComps(i)

    TraverseComponent(swChildComp, nLevel + 1)

    Diagnostics.Debug.Print(sPadStr & swChildComp.Name2 _
      & " <" & swChildComp.ReferencedConfiguration & ">")
  Next i
End Sub
```

IConfiguration.GetRootComponent

After a few declarations, the root component of the assembly is retrieved through the GetRootComponent method of the IConfiguration interface. Each configuration of an assembly can have a different component structure, so traversal must start from the IConfiguration interface. The root component is not actually the first component in the assembly. It is more like the IComponent2 interface of the top assembly itself. However, a root component cannot be manipulated in the same way as other components in the assembly. It is a bit of an odd case, but it must be used as the starting point for assembly component traversal.

Recursion

Sometimes there is a need in a macro or program to have a function that needs to essentially call itself, or be repeated. This concept is referred to as recursion. For example, the top assembly has components which may be assemblies themselves. We need the macro to traverse the children of the top assembly as well as the children of the sub assemblies. Those sub assemblies may also contain sub assemblies, and sub assemblies, and so on until we reach the last part of the last sub assembly. There is no way to know ahead of time how many sub assemblies a user might create, so the code simply calls itself if there are any children of the current component instance.

TraverseComponent Function

The TraverseComponent function is built to get all children of the component that is passed to the function as an argument. The top assembly's root component is passed as the first component to traverse.

IComponent2.GetChildren

After a few declarations and some string preparation used for displaying the list of components with a two space indent structure, an array of generic objects is retrieved using the GetChildren method of IComponent2. The array that is returned is an array of components. However, because of the definition of the return value of this function, you cannot declare your array explicitly as an array of IComponent2 type. This is likely due to the way that this method was developed in the SolidWorks API.

Each element of the `ChildComps` array is an IComponent2 interface that also must be traversed for more children. So the TraverseComponent function is called again inside of itself and it is passed each child component to traverse.

That is it! This code works on even the most complex assembly structure because of the use of recursion. The code prints the component name as well as its configuration to the Immediate window.

Additions

Once you have the code to traverse the components of an assembly, you can manipulate it in several ways. I have listed several common methods and properties below. Look up IComponent2 in the API Help for more.

- **IComponent2.Visible** gets whether or not the component is hidden.

- **IComponent2.Transform2** gets or sets the component's transform matrix. This is used to set the position and orientation of the component.

- **IComponent2.Select4** selects the component.

- **IComponent2.SetTexture** sets the texture to display on this component.

- **IComponent2.RemoveTexture** removes the texture from this component.

- **IComponent2.GetBodies3** gets the bodies in this component (surfaces or solids).

- **IComponent2.GetSuppression** gets the suppression state of the component.

If you wish to manipulate the underlying IModelDoc2 interface of the component, you must get the component's ModelDoc. We used this method in the custom properties chapter, but the method for getting the ModelDoc is referenced here again.

A quick modification of the TraverseComponent function will give you access to each component's ModelDoc.

```
Sub TraverseComponent(ByVal swComp As Component2, ByVal nLevel As Long)

    Dim ChildComps() As Object
    Dim swChildComp As Component2
    Dim sPadStr As String = ""
    Dim i As Long
    Dim Model As ModelDoc2

    For i = 0 To nLevel - 1
      sPadStr = sPadStr + "  "
    Next i

    ChildComps = swComp.GetChildren
    For i = 0 To ChildComps.Length - 1
      swChildComp = ChildComps(i)

      TraverseComponent(swChildComp, nLevel + 1)

      Model = swChildComp.GetModelDoc2
      'process the modeldoc here if needed
      'use Model.GetType to determine part or assembly

      Diagnostics.Debug.Print(sPadStr & swChildComp.Name2 _
        & " <" & swChildComp.ReferencedConfiguration & ">")
    Next i
End Sub
```

If you need to initiate topology traversal to get edges and faces of the model, use the GetBodies3 method to get a list of IBody2 interfaces to begin.

C# Examples

```
//Batch Processing
```

425

```
internal static class Batch
{
    internal static SldWorks swApp;
    private static StreamWriter LogFile;

    public static void Process(string FromFolder, string ToFolder)
    {

        // get the folders as DirectoryInfo interfaces
        DirectoryInfo FromDir = new DirectoryInfo(FromFolder);
        DirectoryInfo ToDir = new DirectoryInfo(ToFolder);

        // if the processing directory doesn't exist,
        // tell the user and exit
        if (!FromDir.Exists)
        {
            MessageBox.Show(FromDir.FullName + " does not exist.",
                "Batch Process", MessageBoxButtons.OK,
                MessageBoxIcon.Exclamation);
            return;
        }

        // if the output folder doesn't exist, create it
        if (!ToDir.Exists)
        {
            ToDir.Create();
        }

        // initialize the log file
        LogFile = new StreamWriter(ToDir.FullName + @"\Process.log");

        // save as another file format
        // change the extension value to whatever you want to export
        string Extension = ".PDF";
        int FileTypeToProcess = (int)swDocumentTypes_e.swDocDRAWING;
        SaveAs(FromDir, ToDir, Extension, FileTypeToProcess);

        // //save as DXF
        // Extension = ".DXF"
        // FileTypeToProcess = (int)swDocumentTypes_e.swDocDRAWING;
        // SaveAs(FromDir, ToDir, Extension, FileTypeToProcess);

        // //save parts as IGS
        // Extension = ".IGS";
        // FileTypeToProcess = (int)swDocumentTypes_e.swDocPART;
        // SaveAs(FromDir, ToDir, Extension, FileTypeToProcess);

        // add other processing as needed

        // finish up by closing out the log file
```

```
        LogFile.Flush();
        LogFile.Close();
}

public static void SaveAs(DirectoryInfo dinf,
    DirectoryInfo outDir, string Ext, int TypeToProcess)
{
    int longErrors = 0;
    int longWarnings = 0;

    // get the file extension based on file type
    FileInfo[] files;
    string filter = "";
    switch (TypeToProcess)
    {
        case (int)swDocumentTypes_e.swDocASSEMBLY:
        {
                filter = ".SLDASM";
                break;
            }

        case (int)swDocumentTypes_e.swDocPART:
        {
                filter = ".SLDPRT";
                break;
            }

        case (int)swDocumentTypes_e.swDocDRAWING:
        {
                filter = ".SLDDRW";
                break;
            }
    }

    // get all files in the directory
    // that match the filter
    files = dinf.GetFiles(filter);
    foreach (FileInfo f in files)
    {
        ModelDoc2 Part;
        try
        {
            Part = swApp.OpenDoc6(f.FullName, TypeToProcess,
                (int)swOpenDocOptions_e.swOpenDocOptions_Silent
                + (int)swOpenDocOptions_e.swOpenDocOptions_ReadOnly,
                "", ref longErrors, ref longWarnings);

            string newFileName;
            newFileName = Path.ChangeExtension(f.FullName, Ext);
            Part.SaveAs2(newFileName,
```

```
                            (int)swSaveAsVersion_e.swSaveAsCurrentVersion,
                            true, true);

                    swApp.CloseDoc(Part.GetTitle());
                    Part = null;
                    }
                catch (Exception ex)
                {
                    WriteLogLine(DateTime.Now + Constants.vbTab
                        + "ERROR" + Constants.vbTab
                        + f.Name + Constants.vbTab + ex.Message);
                }
            }
        }
    }

    public static void WriteLogLine(string LogString)
    {
        LogFile.WriteLine(DateAndTime.Now + Constants.vbTab + LogString);
    }
}


//Feature Traversal
public void Main()
{
    ModelDoc2 Model = (ModelDoc2)swApp.ActiveDoc;
    // make sure the active document is a part
    if (Model == null)
    {
        swApp.SendMsgToUser2("Please open a part first.",
            (int)swMessageBoxIcon_e.swMbWarning,
            (int)swMessageBoxBtn_e.swMbOk);
        return;
    }
    else if (Model.GetType() != (int)swDocumentTypes_e.swDocPART)
    {
        swApp.SendMsgToUser2("For parts only.",
            (int)swMessageBoxIcon_e.swMbWarning,
            (int)swMessageBoxBtn_e.swMbOk);
        return;
    }
    // get the PartDoc interface from the model
    PartDoc Part = (PartDoc)Model;

    // get the first feature in the FeatureManager
    Feature feat = (Feature)Part.FirstFeature();
    string featureName;
    string featureTypeName;
    Feature subFeat = null;
    string subFeatureName;
```

```
    string subFeatureTypeName;
    string message;

    // While we have a valid feature
    while (feat != null)
    {
        // Get the name of the feature
        featureName = feat.Name;
        featureTypeName = feat.GetTypeName2();
        message = "Feature: " + featureName + "\nFeatureType: "
            + featureTypeName + "\n SubFeatures:";


        // get the feature's sub features
        subFeat = (Feature)feat.GetFirstSubFeature();

        // While we have a valid sub-feature
        while (subFeat is object)
        {

            // Get the name of the sub-feature
            // and its type name
            subFeatureName = subFeat.Name;
            subFeatureTypeName = subFeat.GetTypeName2();
            message += "\n " + subFeatureName + "\n "
                + "Type: " + subFeatureTypeName;

            subFeat = (Feature)subFeat.GetNextSubFeature();
            // Continue until the last sub-feature is done
        }

        // Display the sub-features in the
        Debug.Print(message);
        // Get the next feature
        feat = (Feature)feat.GetNextFeature();
        // Continue until the last feature is done
    }
}


//Feature Selection
public void Main()
{
    ModelDoc2 Model = (ModelDoc2)swApp.ActiveDoc;
    // make sure the active document is a part
    if (Model == null)
    {
        swApp.SendMsgToUser2("Please open a part first.",
            (int)swMessageBoxIcon_e.swMbWarning,
            (int)swMessageBoxBtn_e.swMbOk);
```

```
                return;
        }
        else if (Model.GetType() != (int)swDocumentTypes_e.swDocPART)
        {
            swApp.SendMsgToUser2("For parts only.",
                (int)swMessageBoxIcon_e.swMbWarning,
                (int)swMessageBoxBtn_e.swMbOk);
            return;
        }

        PartDoc Part = (PartDoc)Model;
        // get the Extension interface from the model
        ModelDocExtension Ext = Model.Extension;

        // get a feature in the FeatureManager
        // assuming a feature named "Extrude1"
        Feature feat = (Feature)Part.FeatureByName("Extrude1");
        if (feat != null)
        {
            // select it
            feat.Select2(false, -1);
            // get persistent ID
            int IDCount = Ext.GetPersistReferenceCount3(feat);
            object ID = null;
            ID = Ext.GetPersistReference3(feat);
            Debugger.Break(); // check selected feature and ID

            // get rid of the feature reference
            feat = null;
            // clear the selection
            Model.ClearSelection2(true);
            // select the feature by its ID
            int errors = 0;
            // get the feature by its persistent ID
            feat = (Feature)Ext.GetObjectByPersistReference3(ID, out errors);
            feat.Select2(false, -1);
            Debugger.Break(); // check selected feature
        }
}


//Assembly Traversal
public void Main()
{
    ModelDoc2 swModel;
    Configuration swConf;
    Component2 swRootComp;
    swModel = (ModelDoc2)swApp.ActiveDoc;
    swConf = (Configuration)swModel.GetActiveConfiguration();
    swRootComp = (Component2)swConf.GetRootComponent();
```

```
    Debug.Print("File = " + swModel.GetPathName());
    TraverseComponent(swRootComp, 1);
}

public void TraverseComponent(Component2 swComp, int nLevel)
{
    object[] ChildComps;
    Component2 swChildComp;
    string sPadStr = "";
    for (int i = 0; i < nLevel; i++)
        sPadStr = sPadStr + "   ";
    ChildComps = (object[])swComp.GetChildren();

    for (int i = 0; i < ChildComps.Length; i++)
    {
        swChildComp = (Component2)ChildComps[i];
        TraverseComponent(swChildComp, nLevel + 1);
        Debug.Print(sPadStr + swChildComp.Name2 + " <"
            + swChildComp.ReferencedConfiguration + ">");
        ModelDoc2 Model = (ModelDoc2)swChildComp.GetModelDoc2();
        //process the model here if needed

    }
}
```

Index

L

M